W9-CCE-195

genetics

The National Medical Series for Independent Study

genetics

J. M. Friedman, M.D., Ph.D.

Professor and Head
Department of Medical Genetics
University of British Columbia
Vancouver, British Columbia

Fred J. Dill, Ph.D.

Associate Professor
Department of Medical Genetics
University of British Columbia
Vancouver, British Columbia

Michael R. Hayden, M.B., Ch.B., Ph.D.

Professor
Department of Medical Genetics
University of British Columbia
Vancouver, British Columbia

Barbara McGillivray, M.D.

Associate Professor
Department of Medical Genetics
University of British Columbia
Vancouver, British Columbia

National Medical Series from Williams & Wilkins
Baltimore, Hong Kong, London, Sydney

Harwal Publishing Company, Malvern, Pennsylvania

Library of Congress Cataloging-in-Publication Data

Genetics / J.M. Friedman . . . [et al.].
p. cm.—(The National medical series for independent study)
Includes index.
ISBN 0-683-06218-2 (pbk. : alk. paper)
1. Medical genetics. I. Friedman, Jan Marshall, 1947-
II. Series.
[DNLM: 1. Genetics, Medical—examination questions. QZ 18 G328]
RB155.G3874 1992
616'.042'076—dc20
DNLM/DLC
for Library of Congress 91-35379
CIP

ISBN 0-683-06218-2

© 1992 by Williams & Wilkins

Printed in the United States of America. All rights reserved. Except as permitted under the Copyright Act of 1976, no part of this publication may be reproduced or distributed in any form or by any means or stored in a data base or retrieval system, without the prior written permission of the publisher.

10 9 8 7 6 5 4 3 2

Dedication

To our spouses and children

Peg, Laura, and David

Barbara

Sandy, Sarah, Anna, Jessica, and Gideon

Bob, Karen, and Andrea

Contents

Preface

NMS *Genetics* has been written for medical students and house officers by four experienced medical geneticists, who have grown increasingly excited as modern genetic research has transformed our understanding of human biology and pathophysiology. Our goal in writing this book has been to help tomorrow's physicians gain a basic understanding of medical genetics so that as family physicians and specialists of all types they will be able to incorporate genetic medicine into their practices. Moreover, we hope to encourage all aspiring physicians to share in the excitement that comes from "thinking genetically."

J.M. Friedman

Acknowledgments

This book grew out of our combined experience of more than 40 years in teaching genetics to medical students. Our approach has been greatly influenced by our students, by our colleagues in the University of British Columbia Department of Medical Genetics, and by Drs. James Miller and Patricia Baird, who developed and taught the UBC Medical Genetics 440 course before us.

We are grateful to Donna Siegfried for her excellent editorial work and to Matt Harris for his encouragement and support. This book never would have been completed had not Deb Furlong-Raher kept kicking us in the seat of the pants to get it done. We are grateful for her kindness and good humor but also for her persistence.

To the Reader

Since 1984, the *National Medical Series for Independent Study* has been helping medical students meet the challenge of education and clinical training. In today's climate of burgeoning information and complex clinical issues, a medical career is more demanding than ever. Increasingly, medical training must prepare physicians to seek and synthesize necessary information and to apply that information successfully.

The *National Medical Series* is designed to provide a logical framework for organizing, learning, reviewing, and applying the conceptual and factual information covered in basic and clinical sciences. Each book includes a comprehensive outline of the essential content of a discipline, with up to 500 study questions. The combination of an outlined text and tools for self-evaluation allows easy retrieval of salient information.

All study questions are accompanied by the correct answer, a paragraph-length explanation, and specific reference to the text where the topic is discussed. Study questions that follow each chapter use the current National Board format to reinforce the chapter content. Study questions appearing at the end of the text in the Comprehensive Exam vary in format depending on the book. Wherever possible, Comprehensive Exam questions are presented as clinical cases or scenarios intended to simulate real-life application of medical knowledge. The goal of this exam is to challenge the student to draw from information presented throughout the book.

All of the books in the *National Medical Series* are constantly being updated and revised. The authors and editors devote considerable time and effort to ensure that the information required by all medical school curricula is included. Strict editorial attention is given to accuracy, organization, and consistency. Further shaping of the series occurs in response to biannual discussions held with a panel of medical student advisors drawn from schools throughout the United States. At these meetings, the editorial staff considers the needs of medical students to learn how the *National Medical Series* can better serve them. In this regard, the Harwal staff welcomes all comments and suggestions.

1
Nature of the Genetic Material

J. M. Friedman, Fred J. Dill, and Michael R. Hayden

I. GENETICS IN MEDICINE

A. Introduction. Since our understanding of the role that genetic factors play in pathologic processes has improved, the importance of genetics in medicine has increased. Scientific advances have made it possible to identify genetic diseases more precisely, to provide more accurate genetic counseling and prenatal diagnosis for genetic diseases, and to improve the health of many people affected with such conditions. Some areas of genetic medicine are listed below.

1. **Medical genetics** is a branch of medicine dealing with the inheritance, diagnosis, and treatment of diseases due to single gene mutations, chromosome abnormalities, and multifactorial predispositions. Genetic counseling and screening are also part of the medical genetics field.
2. **Clinical genetics** is the application of genetics to clinical problems in individual families and patients.
3. **Dysmorphology** is the study of abnormalities of morphologic development.

B. Major types of genetic disease

1. **Single gene (mendelian or monogenic) disorders** (see Ch 3) are those traits that are produced by the effects of one gene or gene pair. Such traits are transmitted in simple patterns as originally described by Mendel.
   a. **Autosomal dominant** traits are transmitted on the nonsex chromosomes (i.e., chromosomes other than the X or Y) and are expressed when only a single copy of the mutant gene is present. Huntington's disease is an example of an autosomal dominant disorder.
   b. **Autosomal recessive** traits are transmitted on the nonsex chromosomes and are only expressed when both copies of a gene are mutant. Cystic fibrosis is an autosomal recessive disorder.
   c. **X-linked** traits are transmitted on the X chromosome. An example of an X-linked disorder is Duchenne muscular dystrophy.
2. **Chromosome abnormalities** (see Ch 2) are deviations from the normal chromosome number or structure. Examples include Down syndrome and Turner syndrome.
3. **Multifactorial traits** (see Ch 7 III) result from the combined effects of multiple genetic and nongenetic influences. Examples include common kinds of cancer and atherosclerotic heart disease.

C. The incidence of genetic disease and other congenital anomalies apparent by age 25 is about 79 per 1000 livebirths. This excludes most of the common diseases of adulthood such as hypertension, noninsulin-dependent diabetes mellitus, coronary artery disease, and cancer, which have a multifactorial etiology but usually onset after 25 years of age.

1. **Single gene (mendelian) disorders** occur with an incidence of 3.6/1000 livebirths.
   a. **Autosomal dominant** diseases have an incidence of 1.4/1000 livebirths.
   b. **Autosomal recessive** conditions have an incidence of 1.7/1000 livebirths.
   c. **X-linked** disorders have an incidence of 0.5/1000 livebirths.
2. **Chromosome abnormalities** occur with an incidence of 1.8/1000 livebirths.
3. **Multifactorial conditions** with onset before age 25 have an incidence of 46.4/1000 livebirths.

4. **Conditions that appear to be genetic but for which no precise mechanism has been defined** have an incidence of 1.2/1000 livebirths.
5. The incidence of **other congenital anomalies** is about 26/1000 livebirths.

D. Effect of genetic disease on society

1. The **societal burden** of genetic disorders and congenital anomalies is particularly great because they often produce disease beginning in childhood and continuing throughout life.
2. **Hospital admissions.** Genetic diseases are responsible for **30% to 50% of pediatric hospital admissions** and **10% of adult hospital admissions**.
3. **Infant mortality.** Congenital anomalies, many of which have a genetic cause at least in part, are now the leading cause of infant mortality.

E. Thinking genetically

1. **The etiologic approach to disease.** In medical genetics, determination of the etiology of a patient's illness is a primary goal. However, consideration of etiology is a principle that is applicable throughout all branches of medicine. Physicians often treat the physical or functional manifestations of disease in a patient symptomatically, but disease prevention or cure requires an understanding of etiology and pathogenesis.
2. **The family as the unit of concern.** All physicians need to be aware that a disease in one person can have important repercussions for the entire family. This is especially true in medical genetics.
   - **a.** There may be substantial risk for similar disease in the relatives of an affected individual.
   - **b.** In addition to these risks for recurrence, diagnosis of a genetic disease or congenital anomaly in one family member often has major psychological and social implications for the patient's parents, siblings, and children.

II. STRUCTURE AND FUNCTION OF GENES

A. Gene structure. Genes are composed of deoxyribonucleic acid (DNA), which codes for the production of specific polypeptide chains. Polypeptide chains are then joined in specific orders to produce proteins that make up living organisms.

1. The **human genome** contains 23 chromosomes comprised of DNA, which consists of approximately 3.6×10^9 base pairs. In contrast, the chromosome of *Escherichia coli* contains only 4×10^6 DNA base pairs.
2. **DNA molecules** consist of two complementary chains twisted about each other in the form of a double helix—the "twisted ladder" model.
   - **a. Composition of DNA** (Figure 1-1). Each chain is composed of four nucleotides that contain a deoxyribose residue, a phosphate, and a pyrimidine or a purine base. The pyrimidine bases are thymine (T) and cytosine (C); the purine bases are adenine (A) and guanine (G).
     - **(1)** The **"sides" of the ladder** consist of the deoxyribose residues linked by phosphates.
     - **(2)** The **"rungs" of the ladder** are made up of an irregular order of the pyrimidine and purine bases. The two strands of DNA are joined together by hydrogen bonds existing between the pyrimidine and purine bases: Adenine is always paired with thymine (AT), and guanine is always paired with cytosine (GC).
     - **(3)** The **ends of the DNA strands** are designated 5′ and 3′. By convention, the 5′ end of a sequence is written to the left and indicates the sequence closer to the beginning of the gene; the 3′ end is written to the right and indicates the sequence closer to the end of the gene. New DNA is synthesized in the 5′ to 3′ direction (Figure 1-2).
     - **(4) Two chemical differences exist between DNA and ribonucleic acid (RNA).**
       - **(a)** Ribose is the sugar in the structure of RNA, rather than deoxyribose in DNA.
       - **(b)** RNA has the pyrimidine base uracil (U) instead of thymine as in DNA. So, in RNA, the correlating purine and pyrimidine bases are GC and AU.
   - **b. Functions of DNA**
     - **(1)** During **cell division,** the DNA content of the parent cell is replicated by separation of the two strands of the double helix and synthesis of two new complementary strands according to the stated rules of base pairing. The DNA acts as a template for messenger RNA (mRNA).

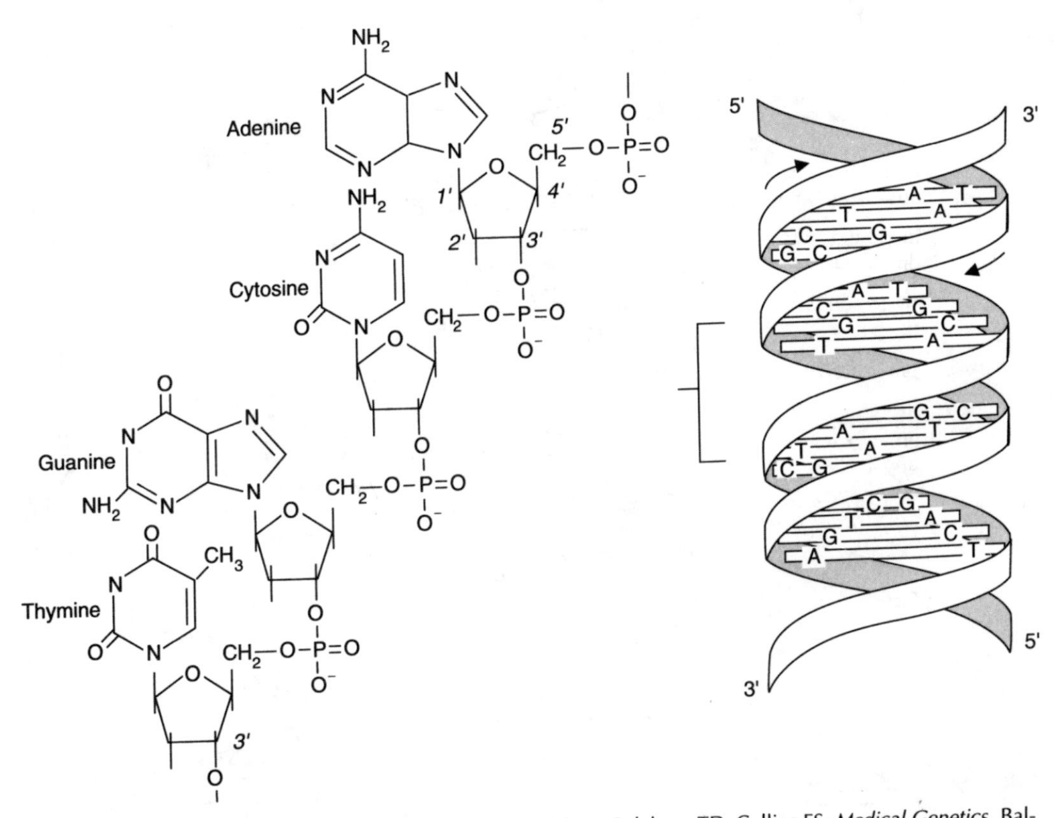

Figure 1-1. DNA molecule and nucleotide bases. (Reprinted from Gelehrter TD, Collins FS: *Medical Genetics*. Baltimore, Williams & Wilkins, 1990, p 10.)

(2) **Transcription.** The information contained in human DNA is transported from the nucleus to the cytoplasm by mRNA. Less than 10% of human DNA is transcribed into mRNA, which is translated into amino acids—the "building blocks" of protein.

(3) **Translation** is a process that occurs on the ribosomes, which translate the information coded by the mRNA into a chain of particular amino acids. The amino acids constitute a polypeptide.

(a) The first and second bases within a mRNA triplet are fixed with regard to the specified amino acid. However, the base in the third position is less crucial. For example, the mRNA triplet specifying arginine can be CGU, CGC, CGA, or CGG.

(b) All amino acids are coded for by three base pairs. Therefore, the number of base pairs in a gene is three times the number of amino acids in the polypeptide chain.

c. Classes of DNA

(1) **Repetitive DNA.** The vast majority of DNA does not encode genes but is present as short or long interspersed, **repeated DNA sequences**.

(a) SINES (short interspersed repeated sequences). The major human SINE is the *Alu* DNA sequence family, which is repeated between 300,000 and 900,000 times in the human genome. The function of *Alu* sequences is unknown, and the reason for their very high frequency in the human genome remains a mystery.

(b) LINES (long interspersed repeated sequences). The major human LINES family has a consensus sequence of 6400 base pairs. The 5′ ends of many LINE sequences differ, but most share homology on the 3′ end. The 5′ end is present approximately 4000 to 20,000 times in the genome, while the 3′ end is present at a greater frequency of 50,000 to 100,000. As with the *Alu* sequence, the function of these sequences are unknown.

(2) **Unique sequence (nonrepetitive) DNA** contains sequences that code for mRNA. In general, genes are comprised of unique sequence DNA that encodes information for RNA and protein synthesis.

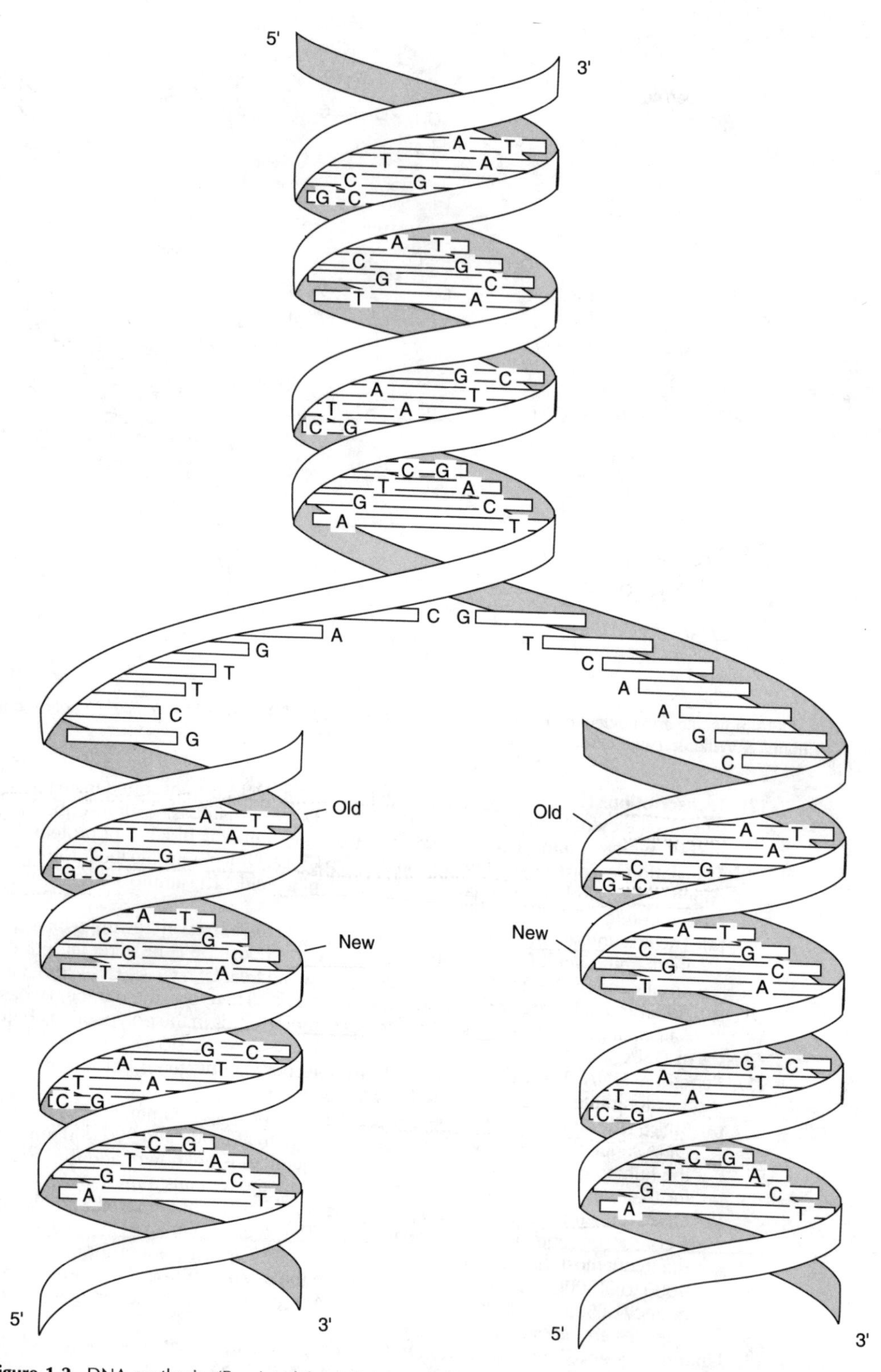

Figure 1-2. DNA synthesis. (Reprinted from Gelehrter TD, Collins FS: *Medical Genetics*. Baltimore, Williams & Wilkins, 1990, p 13.)

- **(a)** Most unique sequence DNA occurs once in the human genome.
- **(b)** Diploid cells, which have two sets of 23 chromosomes, have two copies of each gene, and therefore two copies of each unique DNA sequence.
- **(c)** **Pseudogenes** are stretches of DNA that are homologous to normal genes. However, there are minor changes which prevent either transcription or translation of these genes.

3. Organization of genes (Figure 1-3)

- **a.** **Exons** are the functional portions of gene sequences that code for proteins.
- **b.** **Introns** are the noncoding DNA sequences of unknown function that interrupt most mammalian genes. The **number and size of introns vary** in different genes.
  - **(1)** In certain instances, the **evolutionary relationship** of specific genes can be discerned by their similar organization with equivalent numbers of introns at appropriate locations within the gene.
  - **(2)** Examples include the **apolipoprotein genes,** which have a similar number and size of introns and share a common ancestry. All, however, have different functions that affect lipoprotein metabolism.
- **c.** The **boundaries between exons and introns** are not random base sequences. In most instances, the first two bases at the 5′ end of each intron are GT and the last two bases of each intron are AG.
- **d.** The **open reading frame** is the sequence beginning with the triplet **ATG—the universal translation initiation codon,** which specifies the initiation of protein synthesis. ATG resides at the 5′ end of genes.
- **e.** **TATA boxes. AT-rich regions** are found in most genes. These regions are about 20–30 bases to the 5′ end (left) of the open reading frame (ATG). TATA boxes are thought to help direct important enzymes to the correct initiation site for transcription.
- **f.** **CCAAT boxes** are sequences that occur 70–90 base pairs upstream (prior to the ATG starting point) and are also thought to be important in regulating transcription.
- **g.** **Termination codon.** The end of transcription is signified by a termination triplet at the 3′ end of genes. The triplet could be **TAA, TAG,** or **TGA.**

B. From gene to protein (Figure 1-4)

1. Transcription is the process that synthesizes a strand of mRNA with the same sequences as the original strand (the coding strand) of DNA.

- **a.** The two strands that compose DNA each serve as separate templates for transcription.
- **b.** The TATA box and the CCAAT sequence that are 5′ to the initiation site are usually required for transcription.
- **c.** The **primary transcript** is the large strand of RNA synthesized during transcription, which extends from the original 5′ to 3′ ends and comprises the exons and introns of the gene. This large strand is very unstable and is quickly modified at its 5′ and 3′ ends into mRNA.

2. RNA processing. At the 5′ end of mRNA, a cap structure is added, while a string of adenylic acid (poly A) residues is added to the 3′ end. These modifications help to stabilize the mRNA.

- **a.** The **poly A** sequence acts as a stop signal to terminate transcription.
- **b.** The introns are then removed, and the exons are spliced together to form the **mature mRNA,** which can then be transported to the cytoplasm for translation.
- **c.** The precise mechanisms of **RNA splicing** are not completely understood, but it is evident that the sequences at the 5′ and 3′ ends of the exon/intron boundaries (GT and AG, respectively) are crucial for efficient RNA cleavage.

3. Translation is the process that converts the mRNA sequences derived from the coding strand of DNA into the sequences that make up amino acids. This process occurs on the ribosomes in the cytoplasm in a 5′ to 3′ direction.

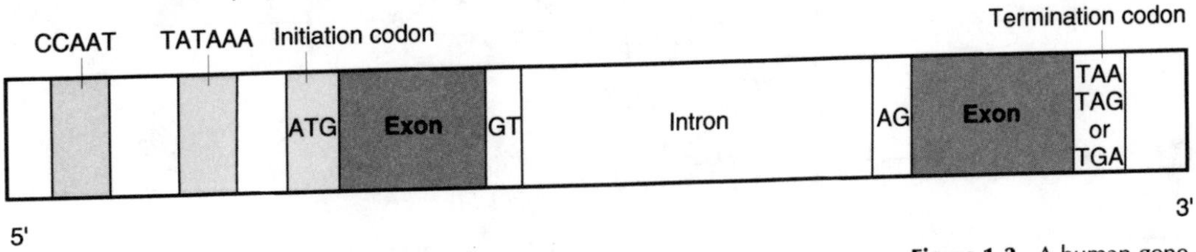

Figure 1-3. A human gene.

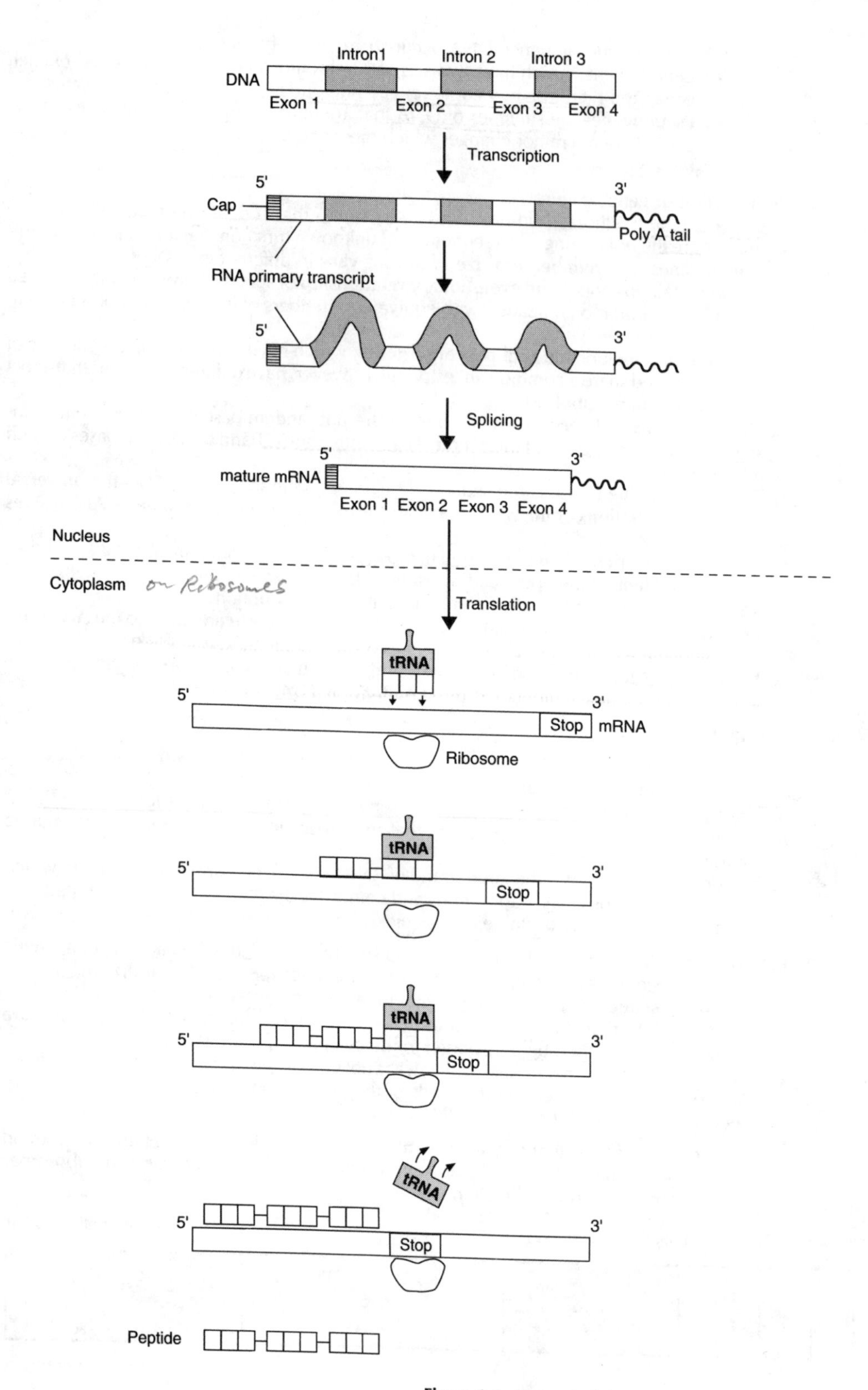

Figure 1-4. Transcription, RNA processing, and translation.

a. The mRNA moves over the surface of the ribosome, bringing successive groups of nucleotides that code for an amino acid (codons) into position.

b. **Transfer RNA (tRNA)** is an RNA molecule that brings the amino acids to the mRNA–ribosome complex. The amino acid–tRNA complexes line up next to the codons of mRNA and are joined by a peptide bond to the previously synthesized amino acid. As this happens, the tRNA is released and the mRNA moves further over the ribosome, bringing another codon into position.

c. This process is repeated in a 5′ to 3′ direction until a specific **termination codon (UAA, UAG, or UGA)** is reached. The peptide is then released, and the tRNA molecule and ribosomes are separated.

d. There are 20 naturally occurring amino acids which are the building blocks for proteins. With four different bases, and amino acids specified by groups of three bases, there are **64 possible coding triplets**.

e. Most amino acids are coded for by more than one triplet, which means **the code is degenerate**. The alternate triplets for a given amino acid often only vary by a change in the third base of the triplet (Table 1-1).

4. Levels of protein structure. Proteins are often **modified after translation** to become active. For example, insulin is produced initially as an 82–amino acid proinsulin. Subsequently, 31 amino acids are removed prior to it becoming an active hormone.

a. The **primary structure** of the protein is constituted by the number of polypeptide chains and the sequence of the amino acid residues. In some instances, proteins have nonprotein groups attached to them that are essential for functional activity. For example, the globin polypeptide is attached to heme to form hemoglobin.

b. The **secondary structure** of the protein refers to the configuration of the polypeptides due to bonding between different peptides. For example, hydrogen bonding between groups on a polypeptide chain leads to twisting of the polypeptide backbone to form an α- helix.

c. The **tertiary structure** of a protein refers to its three-dimensional form. Many proteins have helical and nonhelical regions. The three-dimensional structure represents the most favorable arrangement of the polypeptide chain for efficient and optimal activity.

Table 1-1. The Genetic Code

	Second Position				
First Position	**U**	**C**	**A**	**G**	**Third Position**
U	Phe	Ser	Tyr	Cys	U
	Phe	Ser	Tyr	Cys	C
	Leu	Ser	STOP	STOP	A
	Leu	Ser	STOP	Trp	G
C	Leu	Pro	His	Arg	U
	Leu	Pro	His	Arg	C
	Leu	Pro	Gln	Arg	A
	Leu	Pro	Gln	Arg	G
A	Ile	Thr	Asn	Ser	U
	Ile	Thr	Asn	Ser	C
	Ile	Thr	Lys	Arg	A
	Met	Thr	Lys	Arg	G
G	Val	Ala	Asp	Gly	U
	Val	Ala	Asp	Gly	C
	Val	Ala	Glu	Gly	A
	Val	Ala	Glu	Gly	G

Amino acid abbreviations: Ala = alanine; Arg = arginine; Asn = asparagine; Asp = aspartic acid; Cys = cysteine; Gln = glutamine; Glu = glutamic acid; Gly = glycine; His = histidine; Ile = isoleucine; Leu = leucine; Lys = lysine; Met = methionine; Phe = phenylalanine; Pro = proline; Ser = serine; Thr = threonine; Trp = tryptophan; Tyr = tyrosine; Val = valine.

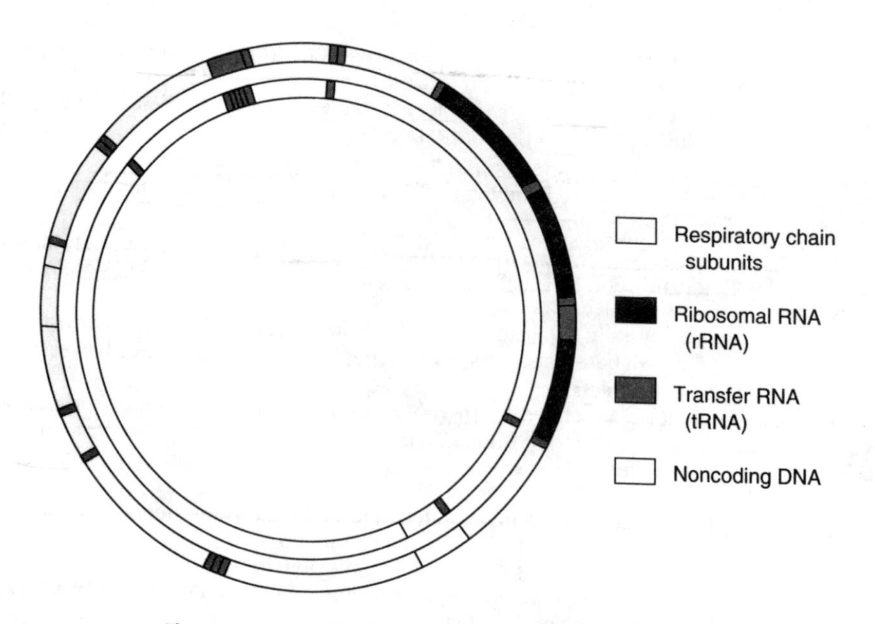

Figure 1-5. Ring structure of mitochondrial DNA (mt-DNA).

5. **Control of gene expression.** All cells have the DNA available to code for every cellular function. Yet **cells are specialized,** with specific functions due to differential expression of different genes.
   a. DNA sequences coexist in human and animal genomes in **methylated and nonmethylated forms**.
      (1) Methylation appears to have a significant repressor effect on gene expression.
      (2) Many genes that are expressed have regions of nonmethylated DNA at their 5′ end.
   b. Sequences on the mammalian X chromosome that are **expressed tend to be nonmethylated,** while sequences that are **inactive tend to be methylated**.
   c. Numerous other factors affect expression. These include:
      (1) Sequences 5′ to the gene, including TATA and CCAAT
      (2) Sequences 3′ to the gene
      (3) Occasionally, intronic sequences within the gene

6. **Mitochondrial genes.** Mitochondria are the only organelles outside the nucleus which contain their own DNA in the form of two to ten copies of double-stranded DNA, measuring about 16 kilobases (Kb) in length (Figure 1-5).
   a. **Mitochondrial DNA (mt-DNA) differs from nuclear DNA** in that it:
      (1) Is circular, rather than linear
      (2) Consists mostly of unique sequence DNA, rather than repetitive DNA
      (3) Has a slightly different genetic code
      (4) Is exclusively **transmitted** to the next generation **by mothers**
      (5) Uses the triplet **TGA** to code for **tryptophan,** rather than as a stop codon
   b. **Mitochondrial DNA codes for 13 proteins,** which are components of the mitochondrial respiratory chain and oxidative phosphorylation system, as well as 2 rRNAs and 22 tRNAs.
      (1) **Mutations in the mt-DNA coding for these proteins may result in disease.**
      (2) In such disorders, the mutation is transmitted through the ovum from an affected mother to all of her children, irrespective of sex. Examples include diseases affecting the central nervous system (CNS), such as **Leber's optic atrophy,** or diseases affecting muscle, such as **mitochondrial myopathies.**
      (3) However, only the daughters would transmit this trait in subsequent generations—an example of **nonmendelian inheritance**.

III. MITOSIS AND MEIOSIS

A. **Mitosis** is the kind of division in which a cell with 46 chromosomes produces two daughter cells, each of which also has 46 chromosomes.

1. **Somatic cells** undergo mitosis to produce genetically identical progeny.

2. **Germ cells** also undergo mitosis to increase their numbers before the final stages of development, in which meiosis occurs.

3. Mitosis is the shortest portion of the **cell cycle,** which can be arbitrarily divided into **four stages**.
   a. Just after mitosis is completed, the cell enters a phase called $\mathbf{G_1}$ during which all of the DNA is present in an unreplicated state. G_1 ends when DNA synthesis begins.
   b. DNA synthesis occurs during the **S** phase of the cell cycle. The DNA is reproduced in units called **replicons,** each of which makes a single copy of itself by **semiconservative replication** of each strand of the DNA double helix.
   c. At the completion of the S phase, each chromosome is doubled and consists of two chromatids. The stage that begins at the completion of DNA synthesis and lasts until the onset of mitosis is called $\mathbf{G_2}$.
   d. The entire **cell cycle** is thus the sequence of phases $\mathbf{G_1 \rightarrow S \rightarrow G_2 \rightarrow}$ **mitosis,** which is repeated for each replicating somatic cell.
      (1) Some terminally differentiated cells such as neurons stop dividing. They are arrested in a prolonged G_1 phase, which is sometimes called $\mathbf{G_0}$.
      (2) The part of the cell cycle between mitoses (i.e., the entire period between the onset of G_1 and the end of G_2) is called **interphase.**

4. Although **mitosis** is a continuous process, for purposes of description, it too is divided into **four stages** (Figure 1-6).
   a. At the conclusion of interphase, mitotic **prophase** begins when the chromosomes condense and become visible under the light microscope.

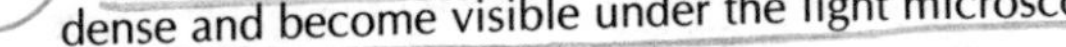

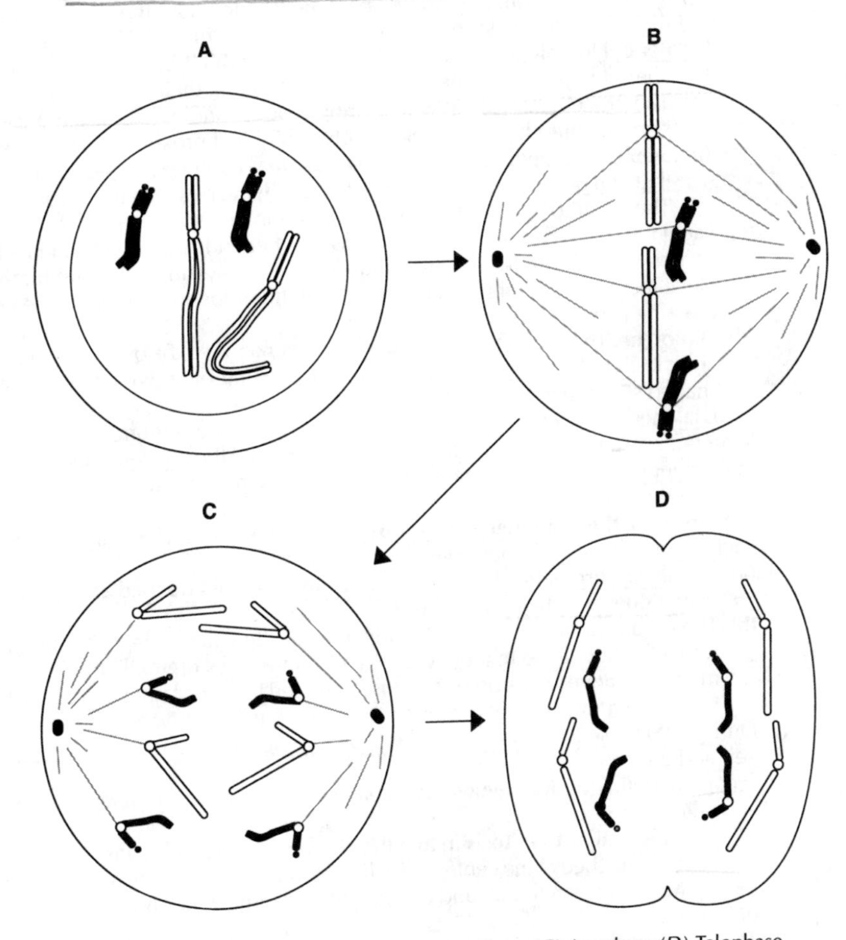

Figure 1-6. Mitosis: (*A*) Prophase (*B*) Metaphase (*C*) Anaphase (*D*) Telophase.

(1) Each chromosome consists of two parallel strands—the **sister chromatids**—which are held together by a **centromere**.

(2) The nuclear membrane disappears during prophase.

b. In **metaphase,** the chromosomes contract completely and move to the center of the cell. Spindle fibers extend from the centromere of each chromosome to two centrioles located in opposite poles of the cell.

c. In **anaphase,** the centromere of each chromosome divides and the sister chromatids separate, becoming two daughter chromosomes that begin to move to the poles of the cell.

d. In **telophase,** the daughter chromosomes reach the poles of the cell and begin to decondense. The cytoplasm divides, and nuclear membranes form again. At the completion of this process, interphase begins in the daughter cells.

B. Meiosis is the kind of cell division by which **diploid gametic precursors produce haploid gametes**. The meiotic products have 23 chromosomes rather than the 46 chromosomes that are typically present in somatic cells.

1. Meiosis is preceded by one round of DNA synthesis and consists of two special cell divisions (Figure 1-7).

a. The first meiotic division—**meiosis I**—is called **reduction division** because it reduces the chromosome number from 46 to 23.

(1) At the beginning of **meiosis I prophase (prophase I),** each chromosome has completed replication and consists of two sister chromatids attached at the centromere. For descriptive purposes, **prophase I** is divided into several sequential stages.

(a) Leptotene. The chromosomes first become visible under the light microscope as thin threads. Separate sister chromatids cannot be distinguished.

(b) Zygotene. Homologous chromosomes, one of which originates maternally and the other of which originates paternally, pair along their entire lengths.

(i) The process of pairing, which brings corresponding DNA sequences on the homologous chromosomes into close physical proximity, is called **synapsis**.

(ii) Synapsis is mediated by the **synaptonemal complex,** a specialized structure that can be seen by electron microscopy lying between the chromatin fibers. The synaptonemal complex is involved in crossing over.

(c) Pachytene. The chromosomes condense further and appear as **bivalents** because homologous chromosomes are tightly apposed in synapsis. **Recombination** takes place between individual chromatids of homologous chromosomes through **crossing over**.

(d) Diplotene. The pair of homologous chromosomes (i.e., the bivalents) begin to separate but are held together at the sites of crossing over, which appear as cross-shaped structures (**chiasmata**).

(e) Diakinesis. The chromosomes reach maximal condensation.

(2) In **metaphase I,** the nuclear membrane disappears and the bivalents line up on a plane in the center of the cell. A spindle connects the centromeres to centrioles in opposite poles of the cell.

(3) In **anaphase I, the homologous chromosomes comprising each bivalent separate** from each other and move to opposite poles of the cell.

(a) The sister chromatids of each chromosome remain attached at the centromere, which does not divide in meiosis I.

(b) The chromosomes in each homologous pair segregate independently in meiosis I; that is, there is no tendency for the chromosomes orginally obtained from the mother to move to one pole and those obtained from the father to move to the other pole. The movement of the chromosomes from each bivalent is random.

(4) During **telophase I,** the two haploid sets of chromosomes reach opposite poles of the cell and the cytoplasm divides.

b. The second meiotic division, **meiosis II,** is preceded by a brief interphase in which **DNA synthesis does not occur**.

(1) Meiosis II is similar to mitosis in that the chromosomes, each consisting of two sister chromatids attached at the centromere, line up on a central plane in the cell, a spindle forms, the centromeres split, and the daughter chromosomes move to opposite poles of the cell.

(2) The essential difference is that **in meiosis II there are only 23 chromosomes** in the original cell and in each of the daughter cells, while in mitosis there are 46.

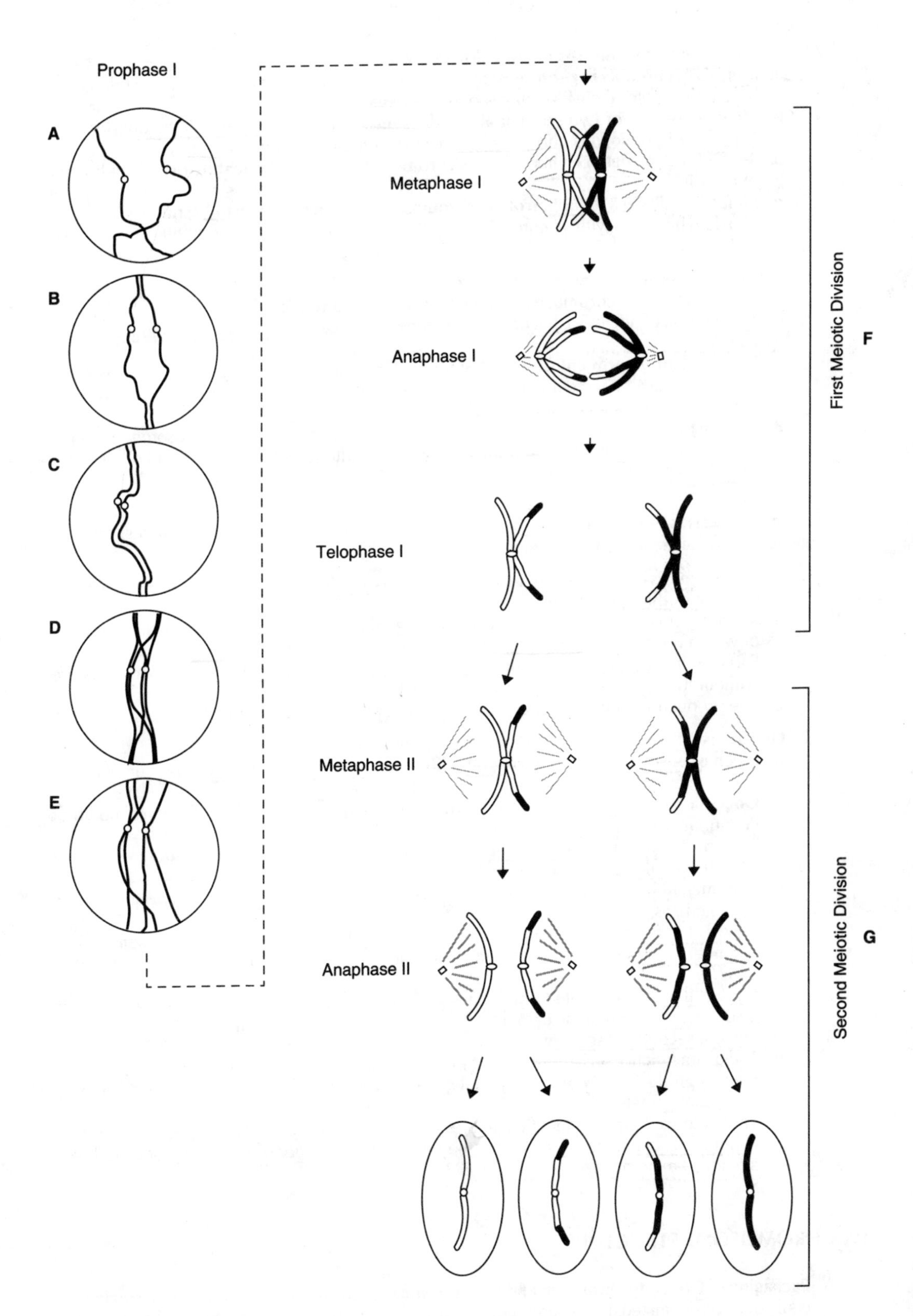

Figure 1-7. Meiosis. (*A*) Leptotene of prophase I. (*B*) Zygotene of prophase I. (*C*) Pachytene of prophase I. (*D*) Diplotene of prophase I. (*E*) Diakinesis. (*F*) Remainder of first meiotic division. (*G*) Second meiotic division.

2. **Meiosis differs from mitosis** in the following ways.
   a. **Meiosis occurs only in germ cells;** mitosis occurs in all somatic cells as well as in germ cells before they enter their final stages of development.
   b. **Meiosis consists of two sequential cell divisions;** mitosis occurs in a single division.
   c. **Pairing of homologous chromosomes occurs in meiosis** but not in mitosis.
   d. **Recombination between homologous chromosomes is a regular feature of meiosis** but not of mitosis.
   e. **Meiosis results in a reduction of the chromosome number from 46 to 23;** mitosis produces two daughter cells with 46 chromosomes, the same number that was present in the original cell.

3. The **genetic consequences of meiosis** are:
   a. **Reduction of the chromosome number from diploid to haploid**
   b. **Segregation of alleles** so that only one member of the original gene pair is included in each gamete
   c. **Independent assortment of homologous chromosomes** so that a gamete contains some chromosomes that were inherited from the mother and other chromosomes that were inherited from the father
   d. **Crossing over** between homologous chromosomes so that every chromosome in the gamete is likely to contain some genes that were inherited from the mother and other genes that were inherited from the father

C. **Gametogenesis** is the formation of the ova or sperm. In males, the process is called **spermatogenesis;** in females, it is called **oogenesis.**

1. **Spermatogenesis** begins to occur in the seminiferous tubules of the testes at the time of puberty and continues throughout life.
   a. Mitotic germ cell precursors called **spermatogonia** produce **primary spermatocytes,** each of which undergoes meiosis I to produce two **secondary spermatocytes**.
   b. Each secondary spermatocyte undergoes meiosis II to form two **spermatids,** which mature without further division to form sperm.
   c. The production of mature sperm from a spermatogonium takes about **75 days**.

2. **Oogenesis** differs in several important ways from spermatogenesis.
   a. **Much of oogenesis occurs during fetal life.** Mitotic division of germ cell precursors in females is restricted to the first few months after conception.
   b. **Oogonia** produce **primary oocytes,** which initiate meiosis I by about the third month of gestation. No primary oocytes are formed after birth.
   c. The primary oocytes become suspended in prophase I in a stage called **dictyotene** until after the female reaches sexual maturity.
   d. After menarche, individual ovarian follicles mature, releasing their primary oocytes from dictyotene as ovulation occurs.
   e. Meiosis I in females results in unequal division of the cytoplasm, with most going to the **secondary oocyte** and relatively little going to the other meiotic product, the **first polar body**.
   f. Meiosis II is not completed until after fertilization has occurred. Meiosis II also results in an unequal division of the cytoplasm, with most going to the **ovum** and relatively little going to a **second polar body**.
   g. The ovum functions as a gamete, but the polar bodies normally do not. Thus, **each female meiosis produces only one functional gamete,** whereas each male meiosis produces four functional gametes.
   h. The prolonged arrest of female gametes in prophase I, which lasts from fetal life until the ovum is ovulated, may account for the increased risk of nondisjunction associated with advanced maternal age (see Ch 2 II B 3 b).

IV. CHROMOSOME STRUCTURE

A. **Packaging of DNA into chromatin and chromosomes.** Each chromosome contains a single molecule of DNA organized into several orders of packaging to construct a metaphase chromosome. The length of a metaphase chromosome is about 0.0001 times the length of its DNA.

1. **Levels of packaging and organization of chromatin** (Figure 1-8)
   a. DNA in chromosomes is associated with DNA binding proteins; the DNA—protein complex is called **chromatin**. These proteins may mediate gene transcription or DNA replication or be the structural proteins that organize the DNA.
   b. **Histones** are the structural proteins of chromatin and are the most abundant proteins in the nucleus. Five histones—H1, H2A, H2B, H3, and H4—provide the framework for the fundamental chromatin packaging unit, the **nucleosome**.
   c. Each **nucleosome** consists of a tightly bound package of eight histones (two each of H2A, H2B, H3, and H4) with a DNA helix wound twice around the surface.
      (1) Each nucleosome core binds 146 base pairs of the DNA helix, and the cores are linked by about 60 bases into a bead-like structure.
      (2) The width of the nucleosome package is 11 nanometers (nm) compared to 2 nm for a DNA double helix.
      (3) Nucleosomes are packed tightly together with the aid of histone H1 to form a 30-nm–wide fiber, which is the basic chromatin unit seen in interphase nuclei when using electron microscopy.
   d. The 30-nm fiber is further packaged into a system of **looped domains**.
      (1) The loops, which contain from 20,000 to 100,000 base pairs of DNA, are **formed by nonhistone proteins binding specific sites along the 30-nm fiber**.

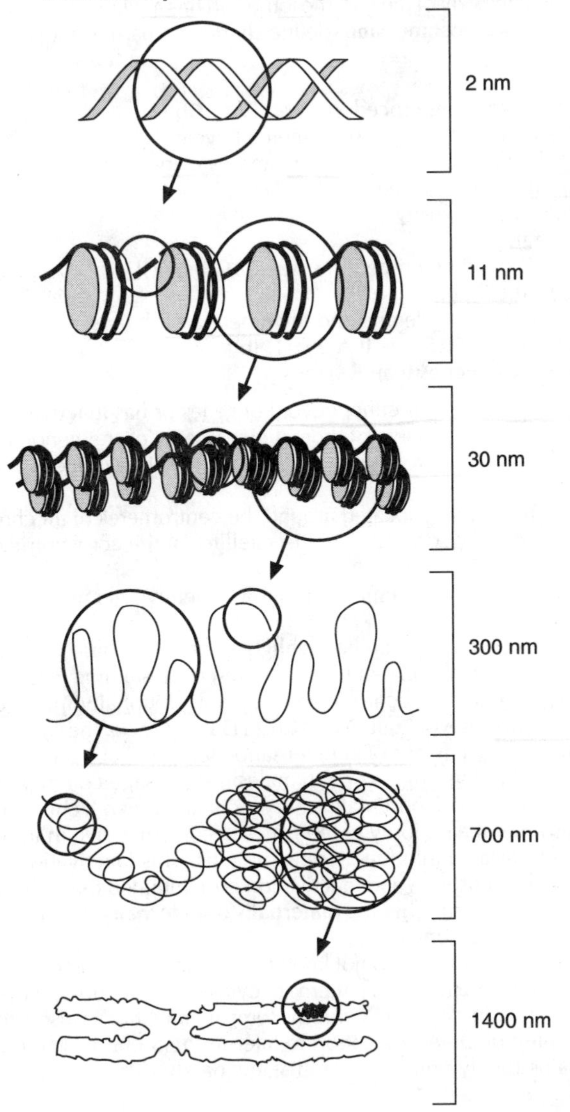

Figure 1-8. Packaging of DNA into chromatin and chromosomes.

(2) Evidence suggests that the loops house individual units of DNA transcription and replication. Therefore, they have both a structural and a functional significance.

2. **Higher order organization**

a. **Prophase and metaphase chromosomes.** Chromatin in metaphase chromosomes is highly condensed compared to interphase chromatin.

(1) The process of condensation during prophase of mitosis is thought to proceed by coiling and compaction of the looped domain structure. In the condensed state, nearby loops are held together by protein interactions that form a network or scaffold.

(2) The chromatin network is arranged in a helical fashion along the axis of the chromosome.

(3) The width of an individual chromatid of a metaphase chromosome is about 700 nm, reflecting a high degree of compression of the 30-nm fiber.

b. **Chromosome bands.** Prophase and metaphase chromosomes exhibit alternating light and dark bands under appropriate staining conditions (see IV C 1 b; Figure 1-9). These bands reflect the differential folding of clusters of looped domains and also define regions of the genome that have different properties and function.

B. **Special features of chromosomes**

1. **Euchromatin** forms the main body of the chromosome and has a relatively high density of coding regions or genes. The **chromosome bands** define alternating partitions of euchromatin with differing properties.

a. **R bands** are areas that:

(1) **Stain light** with G banding procedures [see IV C 1 b (1)]

(2) Replicate DNA early in the S phase of the cell cycle

(3) Have a relatively high content of guanine and cytosine

(4) Have the majority of **SINES** [see II A 2 c (1) (a)]

(5) Have the highest gene density

b. **G bands** are areas that:

(1) **Stain dark** with G banding procedures [see IV C 1 b (1)]

(2) Replicate late in the S phase

(3) Have a higher content of adenine and thymine than R bands

(4) Have the majority of **LINES** [see II A 2 c (1) (b)]

(5) Have relatively fewer genes than R bands

2. **Heterochromatin** is chromatin that is either devoid of genes or has inactive genes (i.e., gene transcription has been turned off). Heterochromatic segments of the genome remain more condensed in interphase than euchromatin and replicate very late in the S phase of the cell cycle.

a. **Constitutive heterochromatin** is located around the centromeres of all chromosomes, in the long arm of the Y chromosome, and in the satellites of the acrocentric chromosomes [see IV C 2 a (1) (c) (iii)].

(1) These areas contain various families of highly repetitive elements of DNA with no known function.

(2) Constitutive heterochromatin can be highlighted on metaphase chromosomes using banding procedures such as C banding or fluorescence staining [see IV C 1 b (9) (c)].

b. **Facultative heterochromatin** is euchromatin in a transcriptionally inactive state. In humans, the clearest example is lyonization (see Ch 3 I D 2 c): One of the two X chromosomes in females is inactivated as a means of compensation for the double dosage of X chromosome genes, compared to the single copy in males (i.e., **dosage compensation**).

(1) Early in the development of the female embryo, one of the two X chromosomes in each cell is altered so that the majority of the genes become inactive in transcription, and thus cannot be translated into functioning proteins. This inactivation process is permanent for that chromosome, and inactivation is maintained through subsequent cell divisions. The choice of **whether the maternally or paternally derived X chromosome becomes inactive is random.**

(2) The mechanism of inactivation is not known, but one of the outcomes of the process is modification of the methylation patterns of cytosine nucleotides in the areas around genes that regulate transcription. X chromosome inactivation is also accompanied by a change in timing of DNA replication to later in the S phase. The inactive X chromosome remains highly condensed and stains darkly during interphase, forming a **Barr body**.

3. **Centromeres** appear as a constricted or unreplicated area of the metaphase chromosome. Normally, each chromosome has one centromere. During cell division, the centromere is the last segment of the chromosome to replicate and separate.
   a. **Structure.** The centromere consists of specific DNA sequences that bind proteins, which can be identified by their antigenic properties.
   b. **Function.** Centromeric proteins participate with spindle fiber microtubules to facilitate the separation of sister chromatids or chromosomes at anaphase of mitosis or meiosis [see III A 4 c, B 1 a (3)].

4. **Telomeres** are DNA sequences found at the ends of chromosomes, which are required to maintain chromosome stability. Chromosomes without telomeres that tend to recombine with other chromatin segments are generally subject to breakage, fusion, and eventual loss.
   a. **Structure.** The terminal segments of all chromosomes have a similar sequence (TTAGGG), which is present in several thousand copies.
   b. **Function.** Telomere sequences facilitate DNA replication at the ends of chromosomes.

C. **Chromosome analysis and classification**

1. **Cytogenetic studies.** Diagnostic cytogenetic analysis can be performed with **metaphase** or **prometaphase chromosomes** obtained from rapidly dividing cells in tissue culture, or, in some cases, directly from tissues with high mitotic activity.
   a. **Preparation of cells**
      (1) **Chromosome cultures from lymphocytes.** For routine cytogenetic analysis, short-term blood cultures are established from quiescent peripheral lymphocytes.
         (a) A sample of heparinized blood is incubated for 3 days in a medium containing balanced salts, essential ingredients, and phytohemagglutinin (PHA), which stimulates cell division.
         (b) Shortly before termination of the culture, cells are arrested in metaphase with **colchicine,** which inhibits spindle formation and thereby prevents the onset of anaphase.
            (i) The medium is replaced with a hypotonic salt solution, which causes the cells to swell.
            (ii) Swollen cells facilitate chromosome spreading when the cells are fixed and spread onto microscope slides to dry.
      (2) **Cell cultures from other cell types** (e.g., fibroblasts, amniocytes, chorionic villus cells) require 1 to 2 weeks to accumulate sufficient cells for cytogenetic analysis.
         (a) The cells adhere to the culture dish surface and grow as a monolayer. When growth is sufficient, the cells are harvested in situ (i.e., on the dish surface) or put into suspension and harvested as are cultures from lymphocytes [see IV C 1 a (1) (b)].
         (b) **Fibroblast cultures** can be subcultured and maintained for weeks or months. They also can be frozen in liquid nitrogen and stored for years for future analysis.
         (c) **Direct preparation of cells** for chromosome analysis is performed routinely on bone marrow cells from leukemia patients. Chorionic villus cells also can be harvested directly. In both cases, the cells are incubated in culture medium for at least 2 hours and are harvested as a suspension. This technique takes advantage of the high mitotic rate in these tissues.
   b. **Techniques**
      (1) **G banding,** the most common procedure, uses **Giemsa** to stain chromosome bands.
         (a) Air-dried slides with well-dispersed metaphase chromosomes are briefly immersed in a solution of trypsin and are washed and stained in Giemsa.
         (b) Viewed under the light microscope, metaphases have alternating dark- and light-staining bands (Figure 1-9). The dark bands by convention are called **G bands;** the light bands are **R bands.**
         (c) In an average metaphase preparation, about 400 dark and light bands can be resolved in a haploid set of chromosomes.
      (2) **High-resolution banding** procedures capture chromosomes in **prophase** or **prometaphase,** when they are less condensed than in metaphase. This technique is helpful for showing small changes in chromosome structure (e.g., small deletions).
         (a) The procedure involves partial synchronization of mitosis in the culture. This, coupled with very brief colchicine treatment, increases the frequency of cells with very long prometaphase chromosomes.

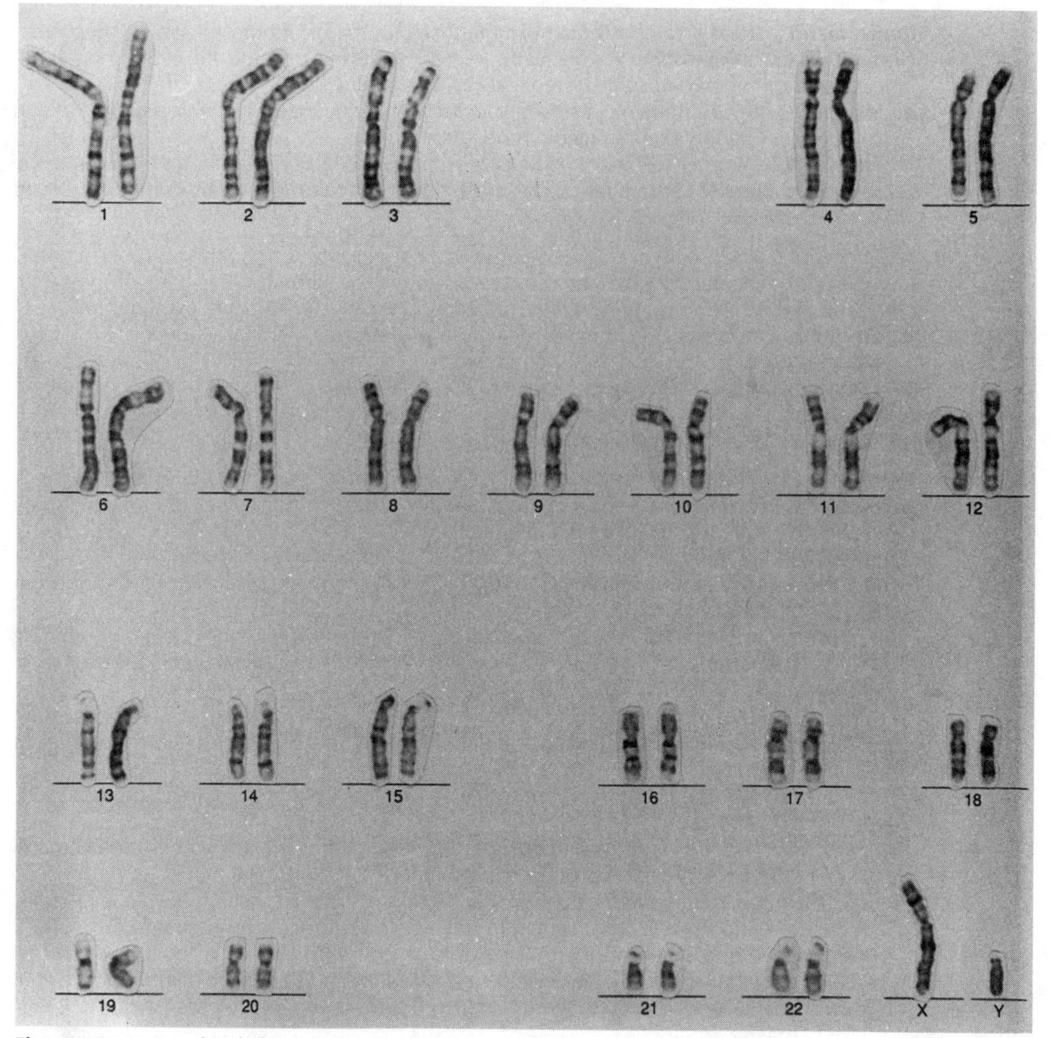

Figure 1-9. A normal male karyotype stained with the trypsin/Giemsa banding procedure. With this method, the G bands are dark-staining and alternate with light-staining R bands.

(b) **G banding** reveals that each band seen in metaphase chromosomes can be resolved into sub-bands, allowing for resolution of 800 or more bands in the best preparations. Analysis of prophase chromosomes with higher resolution than 850 to 900 bands, although possible, is not generally practical technically.

(3) **Q banding** uses **quinacrine** to stain chromosomes, which are viewed with fluorescence microscopy. Alternating bands of bright and dull fluorescence correspond to those seen by G banding for most areas of the chromosomes. Bright Q bands are equivalent to dark-staining G bands. Areas lacking this similarity are called **variable bands** [see IV C 2 a (3)].

(4) **R banding,** or **reverse banding,** uses Giemsa or fluorescent dyes to produce the reverse of the pattern seen in G banding or Q banding. R banding with Giemsa is produced by heating the chromosomes in salts prior to staining. Fluorescent dyes with a high binding affinity to GC-rich DNA (e.g., olivomycin) produce R banding.

(5) **C banding** requires heating in an alkali solution and staining with Giemsa. The treatment removes most non–C-band DNA from the chromosomes. C bands are areas of **constitutive heterochromatin** [see IV C 2 a (3)] located adjacent to the centromeres of all chromosomes and in the long arm of the Y chromosome (Yq).

(6) **Replication banding** methods are based on the fact that G bands replicate late and R bands replicate early in the S phase of the cell cycle. This technique is useful for distinguishing early-replicating from late-replicating X chromosomes. The procedure involves incorporation of the thymidine analog **bromodeoxyuridine (BrdU)** into DNA during only part of the S phase. A Giemsa staining procedure allows differentiation of BrdU-substituted DNA.

(7) **NOR staining** reveals transcriptionally active **nucleolar organizer regions,** which are sites of the ribosomal RNA (rRNA) genes on the short arms of chromosomes 13, 14, 15, 21, and 22. Chromosomes are stained with silver nitrate. Active NORs appear as black–silver deposits; other regions of the chromosomes are pale-staining.

(8) **Fragile sites** on human chromosomes can be visualized only under specific culture conditions. For example, to detect the folate-sensitive fragile site on the X chromosome associated with **X-linked mental retardation** (see Ch 2 IV A 1), cells must be grown in medium with no folic acid or be treated with methotrexate or fluorodeoxyuridine (FdUR) for several hours before harvesting the chromosomes. The lack of folic acid or the presence of methotrexate or FdUR causes a reduction in intracellular concentrations of thymidine monophosphate, which induces folate-sensitive fragile sites.

(9) **In situ hybridization to metaphase and interphase cells.** Recently developed methods for detecting specific segments of human chromosomes, such as chromosome-specific centromeric heterochromatin, individual chromosomes, or single-copy genes, are becoming available to clinical cytogenetic laboratories.

(a) These procedures involve **in situ hybridization** of DNA probes labeled with reporter molecules to metaphase chromosomes or to interphase chromatin.

(b) Common reporter molecules are biotin, digoxigenin, and dinitrophenyl, which can be detected using a variety of methods that couple the reporter molecule either to a fluorescent signal or other dye.

(c) These signals can be detected using fluorescence in situ hybridization (FISH) or conventional microscopy.

2. **Chromosome classification** is based on the International System for Human Cytogenetic Nomenclature (ISCN) from 1985.

a. **The normal human karyotype** (see Figure 1-9) consists of 23 pairs of chromosomes: **22 homologous pairs of autosomes and a pair of sex chromosomes**. The autosomes, by convention, are divided into seven groups: A (chromosomes 1 to 3), B (chromosomes 4 and 5), C (chromosomes 6 to 12), D (chromosomes 13 to 15), E (chromosomes 16 to 18), F (chromosomes 19 and 20), and G (chromosomes 21 and 22). The sex chromosomes are XX or XY. The standard ISCN idiogram of banded human chromosomes is shown in Figure 1-10.

(1) **Chromosome features**

(a) **Chromatids.** Metaphase chromosomes are divided longitudinally into two **sister chromatids** [see III A 4 a (1)].

(b) **Centromere.** The chromatids are held together at the centromere, or **primary constriction,** which delineates the chromosome into a **short arm (p)** and a **long arm (q)**.

(c) **Centromere position.** Chromosomes are divided into three groups based on centromere location.

(i) **Metacentric chromosomes,** such as chromosome 1, have a central centromere.

(ii) **Submetacentric chromosomes,** such as chromosome 6, have a centromere that is displaced from the center.

(iii) **Acrocentric chromosomes,** such as chromosome 13, have a centromere near one end.

(d) **Satellites** are small segments of chromatin distal to the secondary constriction on the short arm (p) of the acrocentric chromosomes (numbers 13, 14, 15, 21, and 22).

(2) **Bands.** Each chromosome has alternating dark and light bands. The majority of bands have constant morphology between chromosomes and between normal individuals.

(a) **Each band is numbered.** The numbering on each arm starts with 11 or 11.1 at the centromere and proceeds to the telomeres.

(b) When resolved into sub-bands with high-resolution banding, new bands are numbered as a subset. For example, band 18q22 may be divided into 18q22.1, 18q22.2, and 18q22.3 (Figure 1-11).

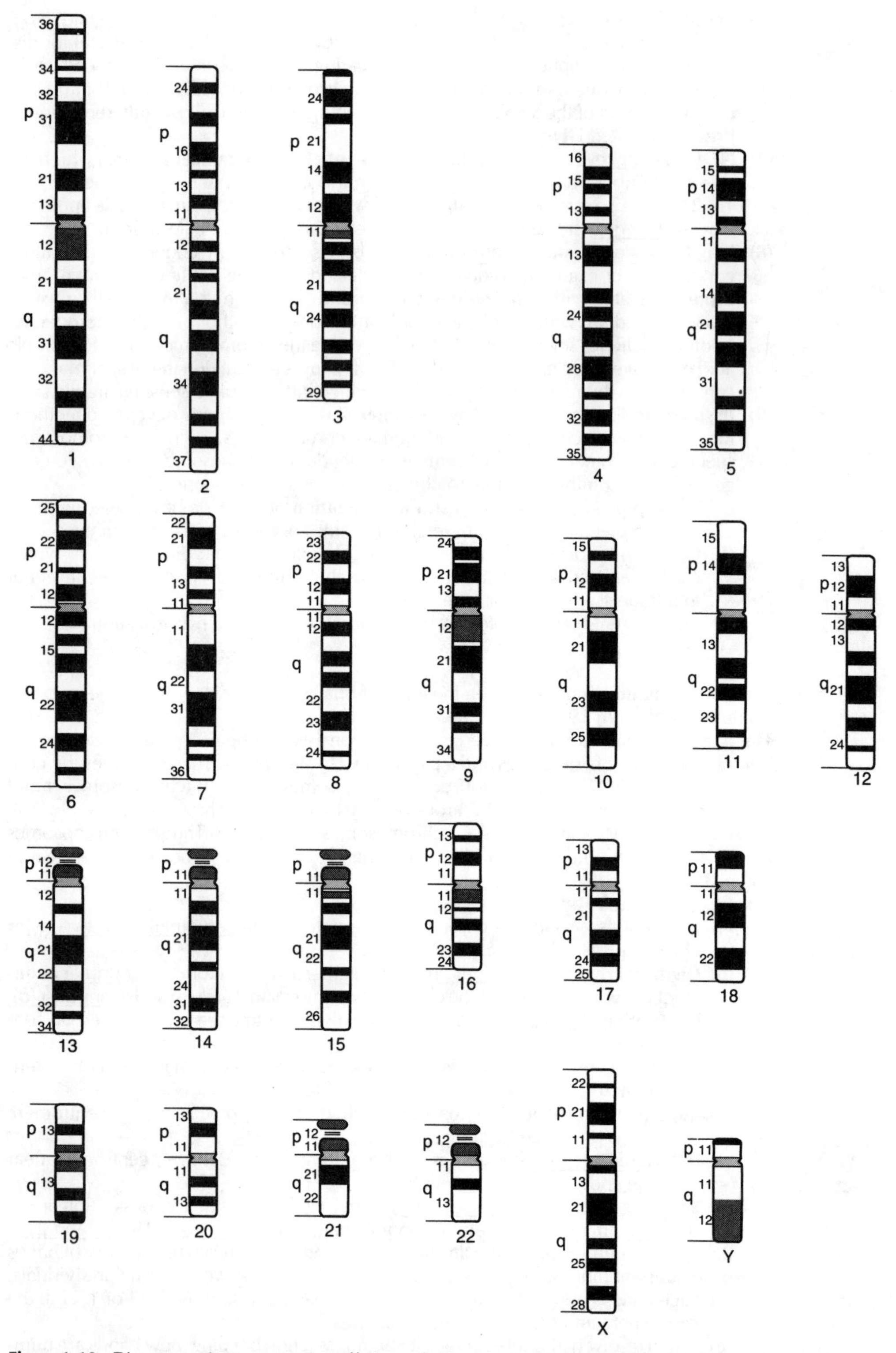

Figure 1-10. Diagrammatic representation of human chromosomes, showing specific morphologic features and band numbering system. The partially *shaded areas* denote heteromorphic, or variable, regions. Short arms of the chromosome are indicated by p; long arms of the chromosome are indicated by q. [Based on the International System for Human Cytogenetic Nomenclature (ISCN), 1985.]

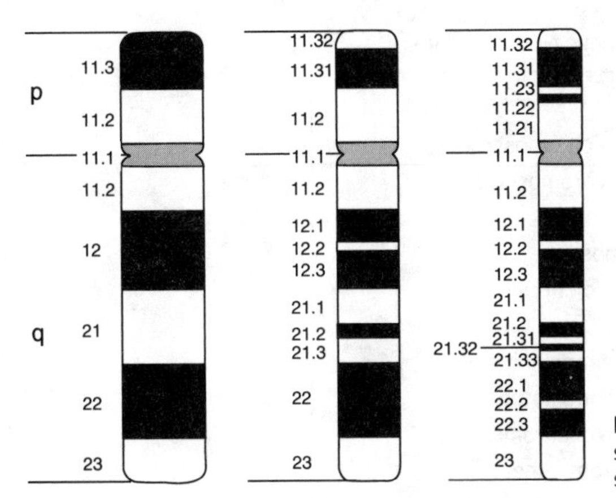

Figure 1-11. Diagrammatic representation of chromosome 18, showing standard to high resolution of chromosome bands. The level of resolution for the chromosome on the *left* is 400 bands, for the *middle* chromosome is 550 bands, and for the *right* chromosome is 850 bands.

(3) The **level of resolution** reflects the detail that can be discerned in a given preparation. In general, the higher the level of resolution (i.e., the higher the "band level"), the better is the ability to discern detail.

(4) Chromosome heteromorphic regions, or variable regions, are evident on many chromosomes (depicted as shaded areas in Figure 1-10). These are areas of **heterochromatin** (see IV B 2); loss or addition to these areas has no clinical significance.

(a) **C-band variation** is most evident in the centromeric regions of chromosomes 1, 9, and 16 and in the long arm of chromosome Y (Yq). The variation is in band size.

(b) **Q-band variation** appears on the short arms (p) and satellites of chromosomes 13, 14, 15, 21, and 22; in the centromeric region of chromosomes 3 and 4; and in Yq. The variation is in both size and intensity of fluorescence.

b. Nomenclature for normal and abnormal karyotypes. ISCN formulas can be written to describe any chromosome complement. Box 1-1 provides a list of some abbreviations used in human cytogenetic nomenclature.

(1) **Basic formula.** In describing the karyotype, the first item is the number of chromosomes; second is the sex chromosome constitution. The normal female karyotype is 46,XX; the normal male karyotype is 46,XY. Autosomes are specified only if there is an abnormality.

(2) **Formulas for karyotypes with abnormal numbers of chromosomes.** A plus or minus sign before a chromosome number indicates that the entire chromosome is extra (trisomy) or missing (monosomy). Triploidy (three copies of each chromosome) or tetraploidy (four copies of each chromosome) is evident from the chromosome number. Mosaicism (i.e., in an individual who has two or more populations of cells that differ in karyotype) is indicated by a diagonal line. Examples include the following.

(a) **45,X** indicates 45 chromosomes with one X chromosome.

(b) **49,XXXXY** indicates 49 chromosomes with four X chromosomes and one Y chromosome.

(c) **47,XY,+21** indicates 47 chromosomes, XY sex chromosomes, and an extra chromosome 21 (trisomy 21).

(d) **46,XY,+18,-21** indicates 46 chromosomes, XY sex chromosomes, trisomy 18, and a missing chromosome 21 (monosomy 21).

(e) **70,XXY,+22** indicates triploidy, XXY sex chromosomes, and an extra chromosome 22 (tetrasomy 22).

(f) **45,X/46,XX/47,XXX** indicates mosaicism for three cell lines.

(3) **Formulas for karyotypes with structurally altered chromosomes** may or may not use the band numbering system to define chromosomal rearrangements according to the points of breakage and reunion. Examples include the following.

(a) **46,X,i(Xq)** indicates a female karyotype of 46 chromosomes, one normal X chromosome, and an isochromosome (see Ch 2 III C) of Xq.

(b) **46,XX,-13,+t(13q21q)** indicates a female karyotype with 46 chromosomes and

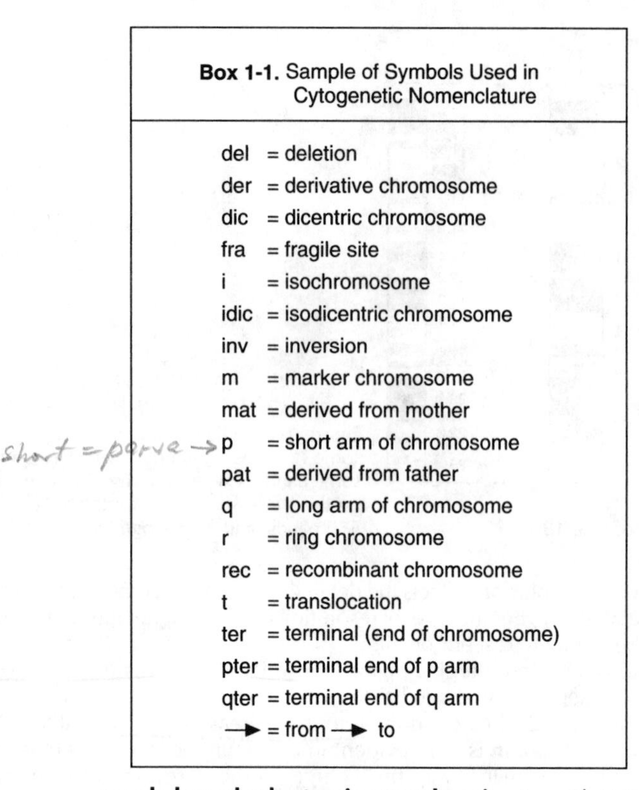

Box 1-1. Sample of Symbols Used in Cytogenetic Nomenclature

del	= deletion
der	= derivative chromosome
dic	= dicentric chromosome
fra	= fragile site
i	= isochromosome
idic	= isodicentric chromosome
inv	= inversion
m	= marker chromosome
mat	= derived from mother
p	= short arm of chromosome
pat	= derived from father
q	= long arm of chromosome
r	= ring chromosome
rec	= recombinant chromosome
t	= translocation
ter	= terminal (end of chromosome)
pter	= terminal end of p arm
qter	= terminal end of q arm
→	= from → to

an **unbalanced robertsonian translocation** (see Ch 2 III E 3 a) between chromosomes 13 and 21, with the long arm of chromosome 21 present in triplicate. The breakpoints are not indicated, but the translocation chromosome is a fusion of the long arms of chromosomes 13 and 21 near the centromeres. The short arms of both chromosomes were lost after formation of the translocation.

(c) 45,XX,t(14;21)(p11;q11) indicates a female karyotype with 45 chromosomes, a **balanced robertsonian translocation** (see Ch 2 III E 3 b) between chromosomes 14 and 21, and breakpoints at p11 on 14 and q11 on 21.

- **(i)** The term **balanced** in human cytogenetics usually refers to karyotypes with structural alterations or rearrangements with no gain or loss of essential chromosome material.
- **(ii)** Although the robertsonian class of translocations between the D and G group chromosomes are described as balanced because they present a clinically normal phenotype, chromosome material is deleted from the short arms of the chromosomes involved in the translocation.
- **(iii)** The short arms of these chromosomes contain multiple copies of rRNA genes as well as heterochromatin.
- **(iv)** Loss of heterochromatin and some copies of rRNA genes is compatible with a normal phenotype.

(d) 46,XY,r(4)(p16q34) indicates a male karyotype of 46 chromosomes with one chromosome 4 in the form of a ring. The breakpoints are at p16 and q34.

(e) 46,XY,t(2;12)(p24;q15) indicates a male karyotype of 46 chromosomes with a **balanced reciprocal translocation** (see Ch 2 III F) between chromosomes 2 and 12. The breakpoints are at band p24 on chromosome 2 and on band q15 on chromosome 12.

(f) 46,XX,-2,+der(2),t(2;12)(p24;q15)pat indicates an unbalanced female karyotype of 46 chromosomes with a derivative chromosome 2 from a balanced reciprocal translocation between chromosomes 2 and 12. There is only one normal chromosome 2, but two normal chromosomes 12. The karyotype has trisomy for segment 12q15 to 12q telomere (12q15qter) and monosomy for segment 2p24 to 2p telomere (2p24pter). The derivative chromosome is from the father.

STUDY QUESTIONS

Directions: Each of the numbered items or incomplete statements in this section is followed by answers or by completions of the statement. Select the **one** lettered answer or completion that is **best** in each case.

1. Of the following types of genetic disease, which one occurs most frequently?

(A) Autosomal dominant
(B) Autosomal recessive
(C) X-linked
(D) Chromosome abnormality
(E) Multifactorial

2. The overall incidence among livebirths of genetic disease and other congenital anomalies apparent by age 25 is approximately

(A) 0.5%
(B) 1%
(C) 8%
(D) 15%

3. The proportion of adult hospital admissions that occur for diseases that are due mostly or entirely to genetic factors is approximately

(A) 1%
(B) 10%
(C) 30%
(D) 50%
(E) 75%

4. High-resolution chromosome banding involves which one of the following? It

(A) permits identification of individual single-copy genes
(B) uses chromosomes in the mid-metaphase stage of cell division
(C) permits the resolution of 350 to 400 dark and light bands per haploid chromosome set
(D) is used to demonstrate fragile sites
(E) allows demonstration of small alterations of chromosome structure

5. Which one of the following is the correct order of meiotic events?

(A) Separation of sister chromatids, recombination, formation of the synaptonemal complex, separation of homologous chromosomes
(B) Separation of homologous chromosomes, formation of the synaptonemal complex, recombination, separation of sister chromatids
(C) Recombination, formation of the synaptonemal complex, separation of sister chromatids, separation of homologous chromosomes
(D) Formation of the synaptonemal complex, recombination, separation of sister chromatids, separation of homologous chromosomes
(E) Formation of the synaptonemal complex, recombination, separation of homologous chromosomes, separation of sister chromatids

6. Gametogenesis in males differs from that in females because of which one of the following factors?

(A) Synaptonemal complexes are only formed in females
(B) Mitotic division of germ cell precursors occurs only in males
(C) Dictyotene occurs during meiosis I in females but not in males
(D) Meiosis in males begins in the fetus, while female meiosis does not begin until puberty
(E) Oocytes do not complete mitosis until after fertilization, while spermatocytes complete mitosis before mature sperm are formed

---

1-E 4-E
2-C 5-E
3-B 6-C

Directions: The group of questions below consists of lettered choices followed by several numbered items. For each numbered item select the **one** lettered choice with which it is **most** closely associated. Each lettered choice may be used once, more than once, or not at all.

Questions 7–11

Match each of the following descriptions of cytogenetic features with the most appropriate structure.

(A) Chromatid
(B) Centromere
(C) Variable band
(D) Fragile site
(E) Satellite

7. A feature not visualized on routine chromosome preparations

8. One of the two replicated copies of the chromosome that separate during meiosis

9. The primary constriction that divides the long arm of the chromosome from the short arm

10. A structure that normally occurs only on acrocentric chromosomes

11. A structure surrounded by heterochromatin that is stained during C banding

---

7-D
8-A
9-B
10-E
11-B

ANSWERS AND EXPLANATIONS

1. The answer is E *[I B, C].*
Multifactorial disorders are the most frequently occurring type of genetic disease, with an incidence of about 46/1000 by age 25. Single gene disorders as a group occur less than 1/10 as often as multifactorial and chromosome abnormalities with an incidence less than 1/25 as great.

2. The answer is C *[I C].*
Although most individual genetic diseases are rare, as a group, genetic diseases and congenital anomalies that manifest by age 25 are quite common, affecting some 79/1000 livebirths.

3. The answer is B *[I D 2].*
Approximately 10% of adult hospital admissions are for diseases that are caused largely or entirely by genetic factors. Some 30% to 50% of admissions to pediatric hospitals are for genetic conditions or congenital anomalies.

4. The answer is E *[IV A 2 b, C 1 b (2)].*
High-resolution chromosome banding is done on cells captured in prophase or prometaphase. The chromosomes in such preparations are more extended, and regions that comprise single bands in standard G-banded preparations can be resolved into sub-bands, permitting the identification of smaller cytogenetic alterations. High-resolution banding generally permits the identification of 800 or more bands in the haploid karyotype, but this resolution is still far too poor to distinguish single-copy genes. Demonstration of fragile sites requires the application of a different special technique.

5. The answer is E *[III B 1].*
In meiosis I, DNA replication is followed by pairing of homologous chromosomes and formation of the synaptonemal complex. Recombination occurs between the homologues, which subsequently begin to separate, forming chiasmata (cross overs). In the first meiotic metaphase, one chromosome from each homologous pair goes to each daughter cell. The sister chromatids separate in the second meiotic division.

6. The answer is C *[III C 2].* prophase I
Oocytes are arrested in the dictyotene of meiosis I until just prior to ovulation. Meiosis in males begins at puberty. Formation of synaptonemal complexes and recombination occur during meiosis in both males and females. Mitotic division of germ cell precursors prior to the onset of meiosis takes place in both oogenesis and spermatogenesis.

7–11. The answers are 7-D *[IV C 1 b (8),* **8-A** *[IV C 2 a (1) (a)],* **9-B** *[IV C 2 a (1) (b)],* **10-E** *[IV C 2 a (1) (d)],* **11-B** *[IV B 2 a (2), C 2 a (3)].*
Fragile sites can be visualized only under special culture conditions such as folate deprivation. Mitotic chromosomes contain parallel sister chromatids that separate during cell division. The centromere is the primary constriction that separates the p (short) and q (long) of the chromosome. The centromere is surrounded by heterochromatin that stains specifically with the C-banding technique. Satellites occur at the tips of the short arms of the acrocentric chromosomes (numbers 13, 14, 15, 21, and 22).

(P)

2 Chromosome Anomalies

Fred J. Dill and Barbara C. McGillivray

I. GENERAL CLINICAL FEATURES

A. Chromosome anomalies (i.e., abnormalities, aberrations) may be either of two general types—numerical or structural.

1. **Numerical anomalies** result in either **polyploidy,** the addition of complete haploid sets of chromosomes, or **aneuploidy,** the addition or loss of one, or rarely, two chromosomes.

2. **Structural anomalies** are rearrangements of genetic material within or between chromosomes. These may be either **genetically balanced,** when there is no change in the amount of essential genetic material, or **unbalanced,** if there is a gain or loss of essential chromosome segments.

B. Frequency of chromosome anomalies

1. Fifteen percent of recognized pregnancies result in early spontaneous abortion, and almost fifty percent of these have chromosome anomalies. This implies that **over 5% of recognized pregnancies are chromosomally abnormal**.

2. Surveys indicate that 1/200, or **0.5%, of newborn infants have a chromosome abnormality**. Table 2-1 lists several chromosome abnormalities.
   a. Sex chromosome aneuploidy is responsible for one-third (33%) of chromosome abnormalities.
   b. Balanced autosomal structural rearrangements account for one-third (33%) of chromosome abnormalities.
   c. Autosomal aneuploidy causes one-fourth (25%) of chromosome abnormalities.
   d. Unbalanced structural anomalies cause one-twelfth (8%) of chromosome abnormalities.

3. Only the balanced structural anomalies and some sex chromosome anomalies are compatible with a normal phenotype.

C. Clinical spectrum of chromosome anomalies. A chromosome abnormality may be suspected in the following clinical situations:

1. **Couples presenting with infertility**
   a. Two to four percent of infertile couples have an autosomal rearrangement; either person of the couple may have an altered number of sex chromosomes.
   b. Chromosome abnormalities can result in nonproduction of sperm or ova or implantation failure.

2. **Miscarriage** (spontaneous abortion)
   a. **Aneuploidy** is commonly seen in miscarriages.
   b. Miscarriage may be associated with an empty sac, a growth-disorganized embryo, or an embryo with malformations (first trimester losses).
   c. The fetus may have growth retardation and malformations.

3. **Abnormal livebirths**
   a. **Common clinical features** include malformations, developmental delay, and poor physical growth.
   b. The occurrence is often secondary to aneuploidy or unbalanced structural rearrangements.

Table 2-1. Types of Chromosome Abnormalities

Chromosome Abnormality	Resultant Aberration	Presentation
Triploidy	Three copies of all chromosomes	Usually results in miscarriage
Monosomy		
Autosomal	Single copy of one autosome	Lethal in early pregnancy
X chromosome	Single sex chromosome	Usually lethal during pregnancy, may present during infancy, childhood, or adolescence
Trisomy		
Autosomal	Three copies of an autosome	Lethal during gestation for all autosomes except 13, 18, and 21, which present at birth
Sex chromosome	Extra sex chromosome	Usually presents in late childhood or adulthood
Deletion		
Autosomal	Partial monosomy of one autosome	Usually presents at birth or during early childhood
X chromosome	Partial monosomy of an X chromosome	Variable presentation from birth to adulthood
Duplication		
Autosomal	Partial trisomy	Presents from birth to early childhood

II. NUMERICAL ALTERATIONS

A. Polyploidy

1. **Two polyploid conditions occur in humans: triploidy,** with 69 chromosomes with XXX, XXY, or XYY sex chromosome complements; and **tetraploidy,** with 92 chromosomes and either XXXX or XXYY sex chromosome complements.

2. **Clinical examples**
   - a. **Triploidy represents 20% of chromosomally abnormal spontaneous abortions** and is occasionally encountered later in pregnancy.
     - (1) **First trimester.** The placenta has focal trophoblastic hyperplasia and hydatidiform changes to the chorionic villi (partial hydatidiform mole). Unlike the true molar pregnancy, a small embryo is usually present.
     - (2) **Second to third trimesters.** On ultrasound, growth retardation and progressive oligohydramnios are recognized. The fetus has a relatively large head, congenital heart lesions, and syndactyly. The placenta is small.
     - (3) **Livebirths.** Rarely, liveborn infants survive for a brief period with clinical features like those in the fetus.
   - b. **Tetraploidy represents 6% of early chromosomally abnormal abortions.**
     - (1) Most tetraploids are lost in the first trimester.
     - (2) In the rare instance of an ongoing pregnancy, the fetus has marked growth retardation, microcephaly, and multiple malformations.

3. **The pathogenesis of triploidy and tetraploidy differ.**
   - a. **Triploidy** results from a **failure of meiosis in a germ cell,** or more commonly from a **fertilization error** such as dispermy. Meiotic failure could result in fertilization of a diploid egg with a haploid sperm or fertilization of a haploid egg with a diploid sperm.
   - b. **Tetraploidy** is a consequence of a **failure of the first cleavage division,** resulting in a doubling of the chromosome number immediately after fertilization.

4. **Recurrence risks.** Neither triploidy nor tetraploidy has an increased risk of recurrence.

B. Trisomy

1. **Trisomy is the presence of three copies of a chromosome rather than the normal two copies.** Trisomies for each of the autosomes (nonsex chromosomes), except chromosome 1, have been recorded.

a. **Most trisomic embryos are lost early in pregnancy.** Trisomy is the most common finding in chromosomally abnormal embryos studied after miscarriage.

b. **Viability of embryos with specific trisomies** is dependent on both the size of the genetic imbalance and the genetic content of the specific chromosomes involved. With rare exceptions, only autosomal trisomies 13, 18, and 21 survive to term and are seen in the liveborn population.

c. **Trisomy for sex chromosomes** such as XXX or XXY have less deleterious effects on development, and most fetuses survive to the newborn period.

2. Clinical examples

a. **Trisomy 21.** Approximately 1/680 liveborn infants has Down syndrome, but the frequency is doubled at 10 to 12 weeks of pregnancy and is a common finding in spontaneous miscarriages.

(1) **Prenatal.** The **abnormalities observed on ultrasound** are nuchal thickening or cystic hygroma, proceeding to fetal hydrops when severe, duodenal stenosis or atresia ("double bubble" sign), and short femur lengths (a less reliable sign). **Low maternal serum α-fetoprotein levels** increase the possibility that a fetus is affected with Down syndrome (see Ch 10 III B 6).

(2) **Infancy.** The infant has a characteristic face with a flat nasal bridge, epicanthic folds, Brushfield spots, protruding tongue, small ears, and a flat occiput. Investigations may reveal hyperbilirubinemia, rare leukemoid reactions, or cardiac lesions (most common are atrial septal defects, ventricular septal defects, and atrioventricular canal).

(3) **Childhood and adulthood.** All affected children have mental retardation, which is usually moderate. Other features include short stature, autoimmune abnormalities, and hearing loss. Older adults are more likely to develop an Alzheimer type of presenile dementia. The life span, once the individual has survived the first year, averages 50 to 60 years.

b. **Trisomy 13.** The newborn incidence is approximately 1/5000.

(1) **Prenatal.** The most severe cases usually spontaneously abort in the first trimester. During the second trimester, abnormalities observed on ultrasound include growth retardation, congenital heart lesions, midline brain and facial lesions (e.g., holoprosencephaly), and omphalocele.

(2) **Newborn.** The newborn also has midline abnormalities (i.e., cutis aplasia of the midline scalp, midline brain malformations, central or unilateral facial clefts), omphalocele, cardiac abnormalities (which are almost universal among those affected), and polydactyly. Mental retardation in this condition is profound.

c. **Trisomy 18.** The newborn incidence is approximately 1/8000.

(1) **Prenatal.** The diagnosis is usually made in spontaneous abortions or stillbirths. Abnormalities such as severe intrauterine growth retardation, congenital heart lesions, and diaphragmatic hernia are frequently detected by ultrasound.

(2) **Infancy.** The newborn has a small facies with prominent occiput, small ears, overlapping fingers, and rocker-bottom heels. Almost all have cardiac as well as other internal malformations. Liveborns are more likely to be female and are profoundly handicapped.

d. **Klinefelter syndrome (47,XXY).** The incidence of Klinefelter syndrome is 1/1000 males.

(1) The **clinical presentations** include the:

(a) Young boy with mild delay and immaturity

(b) Older boy with small, soft testes

(c) Adult male with a eunuchoid habitus, gynecomastia, and poor musculature

(d) Normal-appearing male with infertility

(2) Affected males have increased risks for breast cancer, schizophrenia, and immaturity.

(3) Progressive hyalinization and fibrosis of the seminiferous tubules usually lead to inadequate testosterone production at puberty and in adulthood, requiring supplementation of testosterone on a long-term basis.

e. A **47,XYY** trisomy occurs with the same frequency as 47,XXY (1/1000 males).

(1) Males with 47,XYY are not dysmorphic, and they are not spontaneously aborted.

(2) The diagnosis may be made coincidentally at prenatal diagnosis or with newborn screening.

(3) Clinical features include tall stature and mild social problems, but the majority of affected males are thought to be normal with normal fertility. Although initially publicized as such, the diagnosis of 47,XYY is not associated with increased aggressive tendencies or criminal behavior involving violence.

f. 47,XXX. The incidence rate is 0.8/1000 females, and most are never diagnosed.

(1) The majority of 47,XXX females have no clinical manifestations associated with the trisomy features and have normal fertility and produce normal offspring. A small number may present with radioulnar synostosis, oligomenorrhea, and premature menopause.

(2) There is an increased risk of psychiatric problems, specifically schizophrenia, in these women.

3. Pathogenesis of trisomy and associated factors

a. Nondisjunction is a failure of segregation of chromosomes or chromatids at cell division, which can occur during meiosis or mitosis (Figure 2-1).

N·B

(1) **Nondisjunction during the first division of meiosis (meiosis I)** results from failure of homologous chromosomes to segregate (see Figure 2-1B).

(2) **Nondisjunction at the second division of meiosis (meiosis II)** results from the failure of sister chromatids to segregate (see Figure 2-1C).

(3) Both events produce **gametes that are disomic or nullisomic** for specific chromosomes, and fertilization produces aneuploid zygotes, either trisomic or monosomic (see II B 1, C 1, respectively).

(4) **Mitotic nondisjunction in somatic cells,** like a meiosis II error, is the failure of sister chromatids to segregate at anaphase. This results in a trisomic cell and a monosomic cell.

(5) Nondisjunction is a common event that **appears to occur at a higher frequency in oogenesis than in spermatogenesis**. Studies using cytogenetic and DNA polymorphisms have shown that in about 90% of the cases, the extra chromosome in Down syndrome patients comes from the mother. The majority of the extra chromosomes were derived from meiosis I errors.

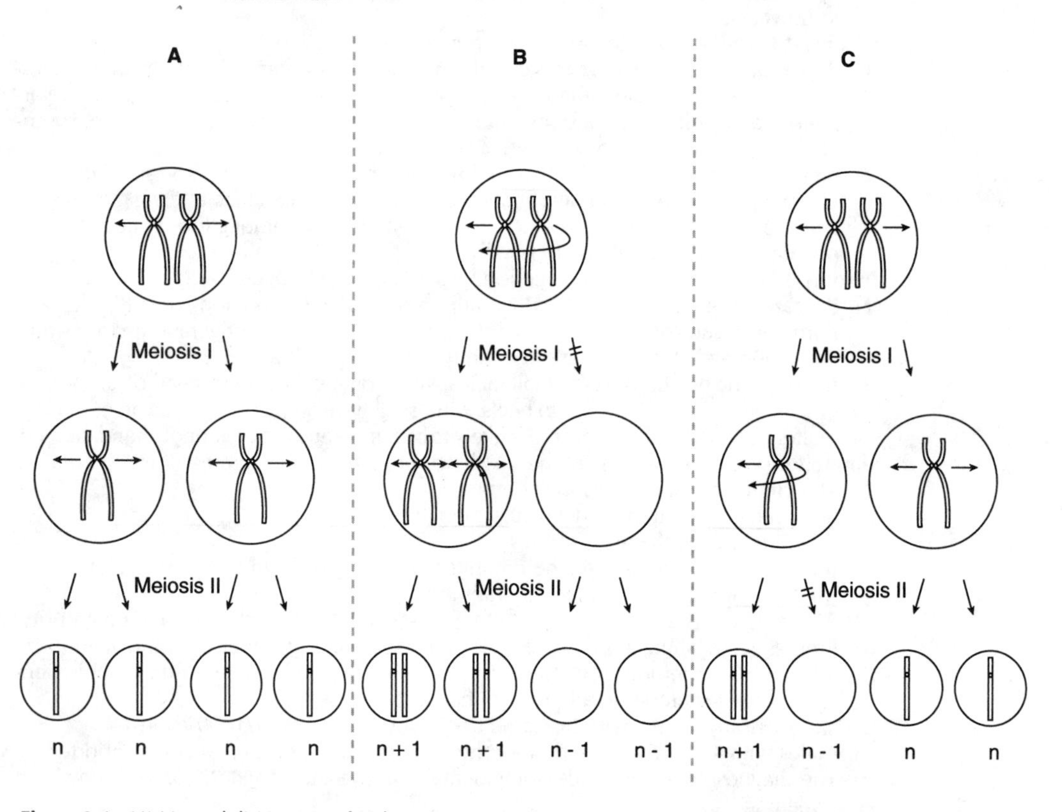

Figure 2-1. (*A*) Normal disjunction of a chromosome pair during meiosis I and II. (*B*) In meiosis I nondisjunction, it is the homologous chromosomes that fail to disjoin. (*C*) In meiosis II nondisjunction, it is the sister chromatids that fail to disjoin. Mitotic nondisjunction is equivalent to an error in meiosis II. The gametic chromosome number, which in humans is 23, is noted by *n*.

(6) **No environmental agents,** such as low-level radiation, exogenous hormones, alcohol, or other drugs, **have been shown to have a measurable influence on the rate of nondisjunction** as measured by the frequency of trisomic babies born both to women exposed to these agents and to control mothers. (See Ch 8 II D for more information regarding environmental effects in pregnancy.)

(7) There is **no established genetic etiology** for primary nondisjunction in humans.

(8) The **only clear influence** on the rate of nondisjunction **is the age of the mother**.

b. Maternal age effect (see Ch 6 II D 2). The occurrence of trisomy in livebirths and in spontaneous abortions increases with the age of the mother. Figure 2-2 indicates the rates for liveborn infants. The nature of the factors influencing chromosome segregation behavior as it relates to increasing age is not known despite considerable research effort.

c. Secondary nondisjunction occurs in existing trisomic cells during meiotic division. Fifty percent of resulting gametes are aneuploid. For example, since Down syndrome results from three copies of chromosome 21, gametes will be either disomic or monosomic for chromosome 21. Secondary nondisjunction causes individuals to have gonadal mosaicism for autosomal trisomy [see II D 2 a].

4. Recurrence risks. The occurrence rate of initial liveborn trisomy is correlated with maternal age (Table 2-2).

a. Trisomies for chromosomes 13, 18, and 21. The recurrence risk for trisomy is 1% to 2% for the mother whose first trisomic infant was born when she was under the age of 30 years. If the mother is over age 30 when the infant is born, the risk will be the same as her age-associated risk (e.g., if she was 37 at the birth of the affected infant, and is 39 during her next pregnancy, her risk will be that of a 39 year old).

b. All other trisomies (excluding those for sex chromosomes) are rare in ongoing pregnancies. The diagnosis of a trisomic spontaneous abortion does not increase the risk for a liveborn with trisomy.

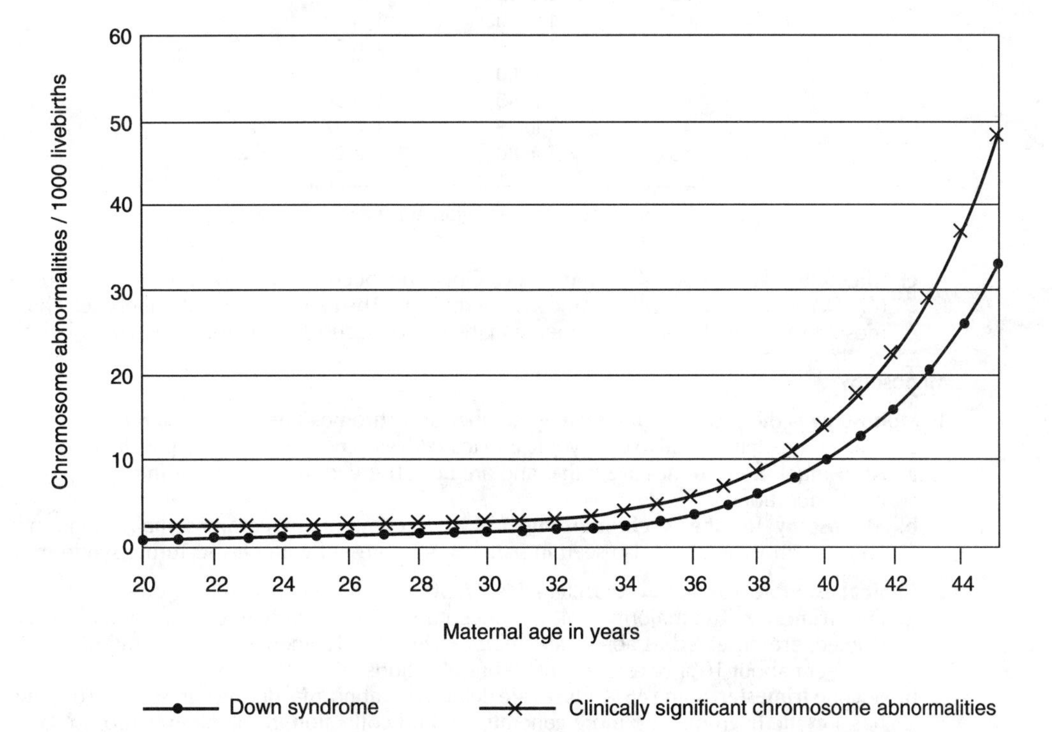

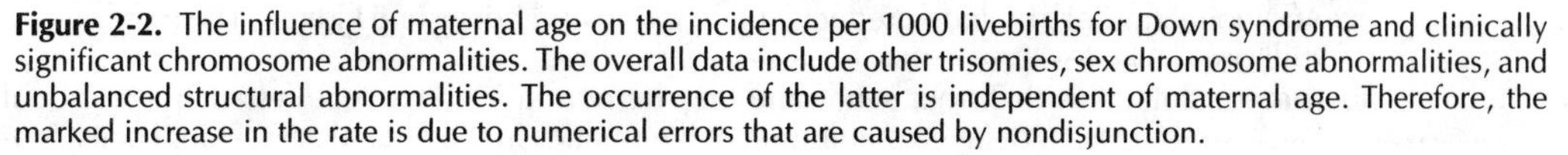

Figure 2-2. The influence of maternal age on the incidence per 1000 livebirths for Down syndrome and clinically significant chromosome abnormalities. The overall data include other trisomies, sex chromosome abnormalities, and unbalanced structural abnormalities. The occurrence of the latter is independent of maternal age. Therefore, the marked increase in the rate is due to numerical errors that are caused by nondisjunction.

Table 2-2. Chromosome Abnormalities in Newborns

Maternal Age at Birth	Risk for Newborn with: Down Syndrome	Risk for Newborn with: A Significant Chromosome Abnormality*
20	1 in 1420	1 in 500
21	1 in 1420	1 in 500
22	1 in 1330	1 in 500
23	1 in 1330	1 in 490
24	1 in 1250	1 in 480
25	1 in 1250	1 in 480
26	1 in 1210	1 in 480
27	1 in 1250	1 in 480
28	1 in 1210	1 in 460
29	1 in 1180	1 in 440
30	1 in 1140	1 in 420
31	1 in 1000	1 in 390
32	1 in 830	1 in 330
33	1 in 630	1 in 290
34	1 in 490	1 in 250
35	1 in 360	1 in 190
36	1 in 282	1 in 160
37	1 in 220	1 in 130
38	1 in 170	1 in 110
39	1 in 130	1 in 88
40	1 in 100	1 in 70
41	1 in 80	1 in 55
42	1 in 60	1 in 43
43	1 in 48	1 in 34
44	1 in 38	1 in 27
45	1 in 30	1 in 20

* Excludes unbalanced translocation and XXX.

c. **Klinefelter syndrome, XYY, and XXX.** Only the occurrence of Klinefelter syndrome (47,XXY) is associated with advanced maternal age. The recurrence rate for these sex chromosome abnormalities is not higher than the rate of occurrence in the general population.

C. Monosomy

1. **Monosomy is the presence of only one member of a chromosome pair in a karyotype** and is generally more detrimental to embryonic and fetal development than is the equivalent trisomy.
   a. **Autosomal monosomies are lethal** and are not observed in livebirths or in early spontaneous abortions.
   b. **Monosomy for the X chromosome (45,X)** is common in chromosomally abnormal aborted embryos but is also seen in patients who present with classic Turner syndrome.
2. **Clinical example.** Among liveborn females, 1/2500 will have classic 45,X Turner syndrome.
   a. **First trimester.** The majority of 45,X conceptuses are lost early in pregnancy and, when studied, are observed as abnormal embryos with developmental disorganization. These represent about 10% of early spontaneous abortions.
   b. **Second trimester.** Some 45,X fetuses are detected as abnormal during ultrasound. The fetus has a cystic hygroma or a more generalized fluid collection, resulting in hydrops fetalis. Additional findings include preductal coarctation and horseshoe kidney. The majority are stillborn.
   c. **Infancy.** Infants with Turner syndrome may be normal or have features such as residual neck webbing from delayed maturation of lymphatics in the nuchal area, shield chest, coarctation, and edema of the hands and feet.

d. Childhood. The presenting features during childhood are short stature or a cardiac murmur. Teenagers present with primary or secondary amenorrhea or lack of secondary sex characteristics because of streak ovaries. Turner syndrome females are mentally normal but may have spatial perceptual abnormalities.

e. Adulthood. Rarely, women with Turner syndrome who retain functioning ovaries are fertile and have an increased risk for offspring with chromosome anomalies.

f. Note that many females with the Turner phenotype may have karyotypes other than 45,X, such as X chromosome mosaicism, Xq isochromosome, or Xp deletion. The karyotype in spontaneous abortions and fetal hydrops, however, is likely to be 45,X.

3. Pathogenesis. Monosomy may result from nondisjunction or chromosome lag. A chromosome may lag at anaphase and be excluded from a new nucleus. In males, lag of the Y chromosome at meiosis is thought to be a common cause of X chromosome monosomy.

4. Recurrence risks for 45,X or variations do not increase with advancing maternal age, and there is no increased risk after a pregnancy with a miscarriage, stillbirth, or infant with Turner syndrome.

D. Mosaicism

1. Mosaicism is the presence of two or more cell lines with different karyotypes in a patient.

a. A normal diploid line commonly exists with an abnormal cell line. The abnormal line may have a numerical or a structural anomaly.

b. Note that a specific cell line may be represented in all tissues or may be confined to single or multiple tissues.

2. Clinical features. The range of features for any mosaic situation depends on the specific chromosome involved and the proportions of normal and abnormal cell lines.

a. Autosomal examples

(1) Mosaic trisomy 8. Children are usually well grown and are mentally retarded with mild facial dysmorphism. The fingers have flexion deformities, and there are deep creases on the palms and soles. Lymphocyte chromosomes are usually normal, while fibroblast cultures demonstrate mosaic trisomy 8. Since nonmosaic trisomy 8 has not been described in a liveborn, the condition is likely to be lethal.

(2) Mosaicism for trisomies 13, 18, and 21 occurs in some patients. About 2% of those with trisomy 21 have normal and trisomic cells. Clinical features are usually similar to those in nonmosaic cases, but milder forms do occur.

b. Sex chromosome example—mosaic Turner syndrome. The clinical presentation of mosaic Turner syndrome is variable and depends on the proportions of the normal and aneuploid cell lines.

c. Undiagnosed low-level mosaicism. A parent with low-level mosaicism or gonadal mosaicism may be clinically normal. The condition may come to attention only when two or more offspring have a similar chromosome abnormality.

d. Confined chorionic mosaicism. In approximately 2% of chorionic villus sampling preparations, mosaicism confined to the chorion is found. It is most likely to be seen in direct or short-term cultures and may be present in a normal fetus [see Ch 10 III B 3 d (4)].

3. Pathogenesis. Mosaicism results from nondisjunction, chromosome lag, or mitotic instability of a structurally altered chromosome.

a. Early development. If the nondisjunction or chromosome lag occurs very early in development, such as in the embryo cleavage stage or in the inner cell mass of a blastocyst, mosaicism may be detected in multiple tissues in the newborn or adult.

b. Late development. If these events take place later in the development of the embryo or fetus, they are confined to specific cell lines and are unlikely to be detected.

c. Abnormal cell lines that are confined to the chorion are derived from events during cleavage in cells that are destined to become placental tissue only.

d. Note that for many cases of autosomal mosaicism and for the majority of those with trisomy 21 who were tested, the zygote was trisomic. The mosaic condition resulted from chromosome lag of the extra chromosome 21 early in development, which produced the normal line.

4. Recurrence risks for offspring with mosaicism are low (equivalent to the population risk). However, parents of children with autosomal mosaicism have the same risks for future autosomal mosaic offspring as those with nonmosaic trisomic offspring.

5. **Implications for offspring of mosaic individuals.** All mosaic individuals, including those with undiagnosed low-level mosaicism, have an elevated risk of having nonmosaic chromosomally abnormal offspring because of secondary nondisjunction (see II B 3 c). The risk depends on the proportion of trisomic germ-line cells to normal germ-line cells.

III. STRUCTURAL ALTERATIONS

III. STRUCTURAL ALTERATIONS result from breakage and fusion of chromosome segments in novel ways. Many new structures, such as acentric fragments of chromosomes and dicentric chromosomes, are very unstable at cell division. These can result in major chromosome loss, leading to cell death. A variety of stable chromosome alterations occur in humans.

A. **Deletions** represent a loss of chromatin from a chromosome.

1. **Classification**
   - a. **Terminal deletions** arise from one break. The acentric fragments that are formed are lost at the next cell division.
   - b. **Interstitial deletions** arise from two breaks, fusion at the break sites, and loss of the interstitial acentric fragment.
   - c. **Ring chromosomes** arise from breaks on either side of the centromere and fusion at the breakpoints on the centric segment. Segments distal to the breaks are lost so that individuals with chromosome rings have deletions from both the long and short arms of the chromosome involved.
   - d. Deletions may also occur as a result of segregation of a familial inversion or translocation (see III D–F).
2. **Clinical examples.** Deletions of many segments of the karyotype are documented. A cytogenetically detectable deletion in euchromatin (see Ch 1 IV B 1) results in an abnormal phenotype. Three well-characterized deletion syndromes are described below. In each, the individual syndrome gestalt is secondary to the loss of a critical area of the chromosome involved.
   - a. **5p- (Cri du chat, or cat cry syndrome)**
     - (1) **Features.** This was the first autosomal deletion to be described. The infants have round facies, a cat-like cry, which disappears with time, and cardiac defects. Infants with cat cry syndrome are also mentally retarded.
     - (2) Most children have a de novo deletion of a variable amount of the short arm of chromosome 5 (5p), with the critical area being 5p15.
   - b. **4p- (Wolf-Hirschhorn syndrome)**
     - (1) **Features.** The infants have prominent foreheads and broad nasal root ("Greek warrior helmet"). The philtrum is short, and the mouth is downturned. Cardiac defects are common, and all affected infants are severely mentally retarded and exhibit growth failure.
     - (2) Ten to fifteen percent of the deletions are associated with familial translocations. The essential region involved is 4p16.
   - c. **Ring chromosome 14.** The clinical features are dependent on the breakpoints, but commonly observed breakpoints are at 14p11 and 14q32. Children may have mild dysmorphic features; seizures are frequent, and the occurrence of mental retardation is variable.
3. **Microdeletion syndromes.** These result from small deletions that require high-resolution banding for cytogenetic diagnosis.
   - a. **Prader-Willi syndrome.** The clinical features of Prader-Willi syndrome evolve after birth.
     - (1) **Infancy.** The infant has profound hypotonia and poor feeding for the first year. The differential diagnosis includes sepsis, a metabolic abnormality, or a myopathy.
     - (2) **Early childhood.** In the second to the third year, the child develops an insatiable appetite and truncal obesity; the hands and feet are small. The face has bifrontal narrowing and almond-shaped eyes; the hair color is fair. Most children have developmental delay.
     - (3) **Late childhood.** Obesity increases without therapy. Behavioral problems and rage episodes develop with age.
     - (4) **Chromosomes.** About 60% of children have deletions or rearrangements involving 15q11–13; some have deletions detectable only by molecular methods (see Ch 4 I; II).
   - b. **Miller-Dieker syndrome (agyria, lissencephaly).** Lissencephaly, or smooth brain, may be found in a variety of conditions. Miller-Dieker syndrome refers to lissencephaly and dysmorphic features (e.g., microcephaly, a high forehead, and furrowing of the central forehead), which are associated with a deletion of 17p13.3 (Figure 2-3). Death occurs early.
4. The **recurrence risk** is negligible for these deletion syndromes unless a parent has a chromosome rearrangement. Therefore, parents must be investigated before counseling the family.

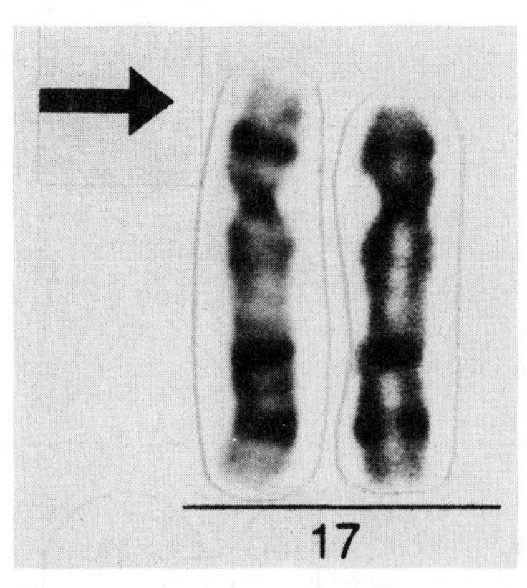

Figure 2-3. A pair of G-banded chromosomes 17 from a patient with Miller-Dieker syndrome. The normal chromosome 17 is on the *left,* and the chromosome with the p13.3 deletion is on the *right*. The breakpoint is indicated by the *arrow*.

B. Duplications represent a gain of chromosome material and result in trisomy for segments of chromosomes.

1. **Duplicated segments** may be arranged as direct tandem repeats or as inverted repeats of chromosome segments.

2. **Duplications of chromosome segments** may arise from familial rearrangements such as translocations or inversions (see III D–F).

3. The **risk of recurrence** is not high unless the duplication is part of a familial chromosome rearrangement.

C. Isochromosomes. One of the chromosome arms (p or q) is duplicated, and all material from the other arm is lost, so that the arm on one side of the centromere is a mirror image of the other.

1. **Clinical examples.** Isochromosomes are generally not seen in autosomes because they create a lethal situation. Exceptions include the following.
   a. In the situation of a long-arm (q) isochromosome of the acrocentric chromosomes [e.g., 45,XX,i(21q)], loss of short-arm (p) material does not lead to an abnormal phenotype (see III E 1 a).
   b. **Isochromosome Xq** is seen in approximately 20% of females with Turner syndrome and is the equivalent of monosomy Xp. The clinical picture is identical to classic Turner syndrome (see II C 2).

2. **Pathogenesis.** Isochromosomes may arise at mitosis or meiosis from a centromere division error that is at right angles to the normal separation. Long-arm isochromosomes have, by definition, deletions of the short arm, and short-arm isochromosomes have deletions of the long arm (Figure 2-4).

3. **Recurrence.** If a parent has an isochromosome for an autosome such as i(21q) the risk for trisomic offspring is 100%, as there can be only disomic and nullisomic gametes produced at meiosis. Fertilized nullisomic gametes will be lethal monosomies.

D. Inversions are a reversal of the order of chromatin between two breaks on the chromosome. They are encountered as new mutations or may be present in multiple generations in families.

1. **Types of inversions**
   a. **Pericentric inversions** are the **result of breaks and rearrangements on both sides of the centromere** (Figure 2-5). An inverted chromosome often has a strikingly different morphology than the parental chromosome.
   b. **Paracentric inversions** are the **result of breaks and rearrangements on the same side of the centromere**. These inverted chromosomes have similar overall morphology to the parental chromosomes, but the band order is changed.

2. **Meiotic segregation**
   a. **Pericentric inversions**
      (1) Since carriers of inversion usually have one normal and one inverted chromosome, the pairing constraints at prophase of meiosis necessitate the **formation of a loop** (Figure 2-6).

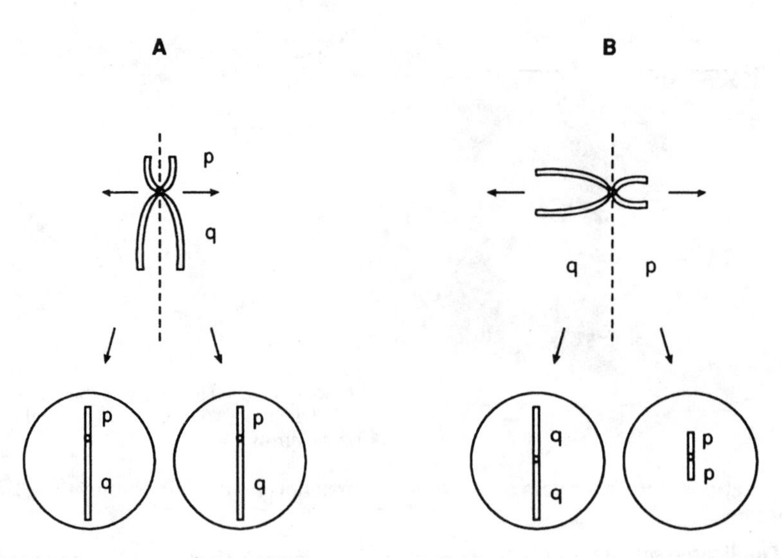

Figure 2-4. (*A*) Normal centromere division. (*B*) Centromere misdivision, creating a long-arm isochromosome on the *left* and a short-arm isochromosome on the *right*.

- **(a)** A **recombination event** within the loop structure produces chromosomes with duplications and deletions of segments outside the breakpoints that formed the inversion.
- **(b)** Chromosome strands not involved in the recombination are similar to the parental chromosome.

(2) Embryos receiving a recombined chromosome have unbalanced karyotypes. Survival to term depends on both the size of duplications and deletions and the chromosome involved.

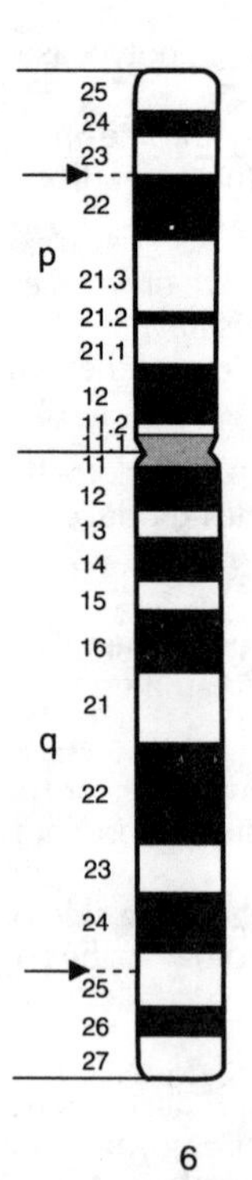

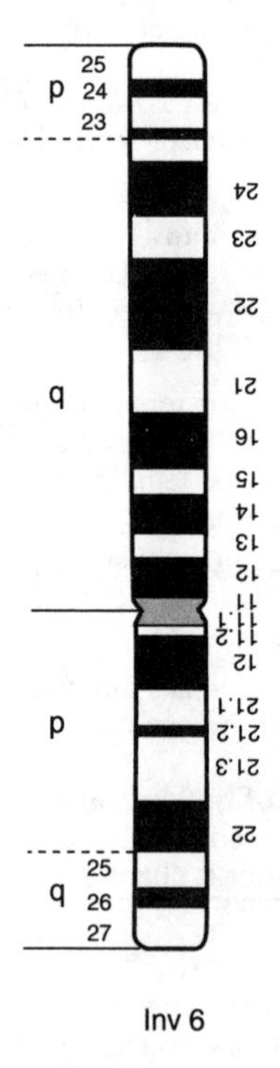

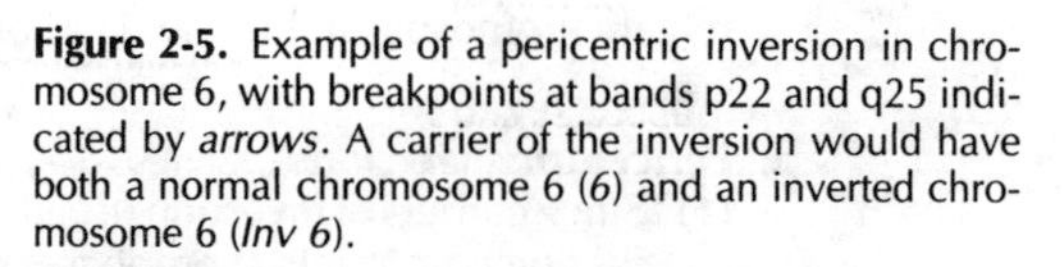

Figure 2-5. Example of a pericentric inversion in chromosome 6, with breakpoints at bands p22 and q25 indicated by *arrows*. A carrier of the inversion would have both a normal chromosome 6 (*6*) and an inverted chromosome 6 (*Inv 6*).

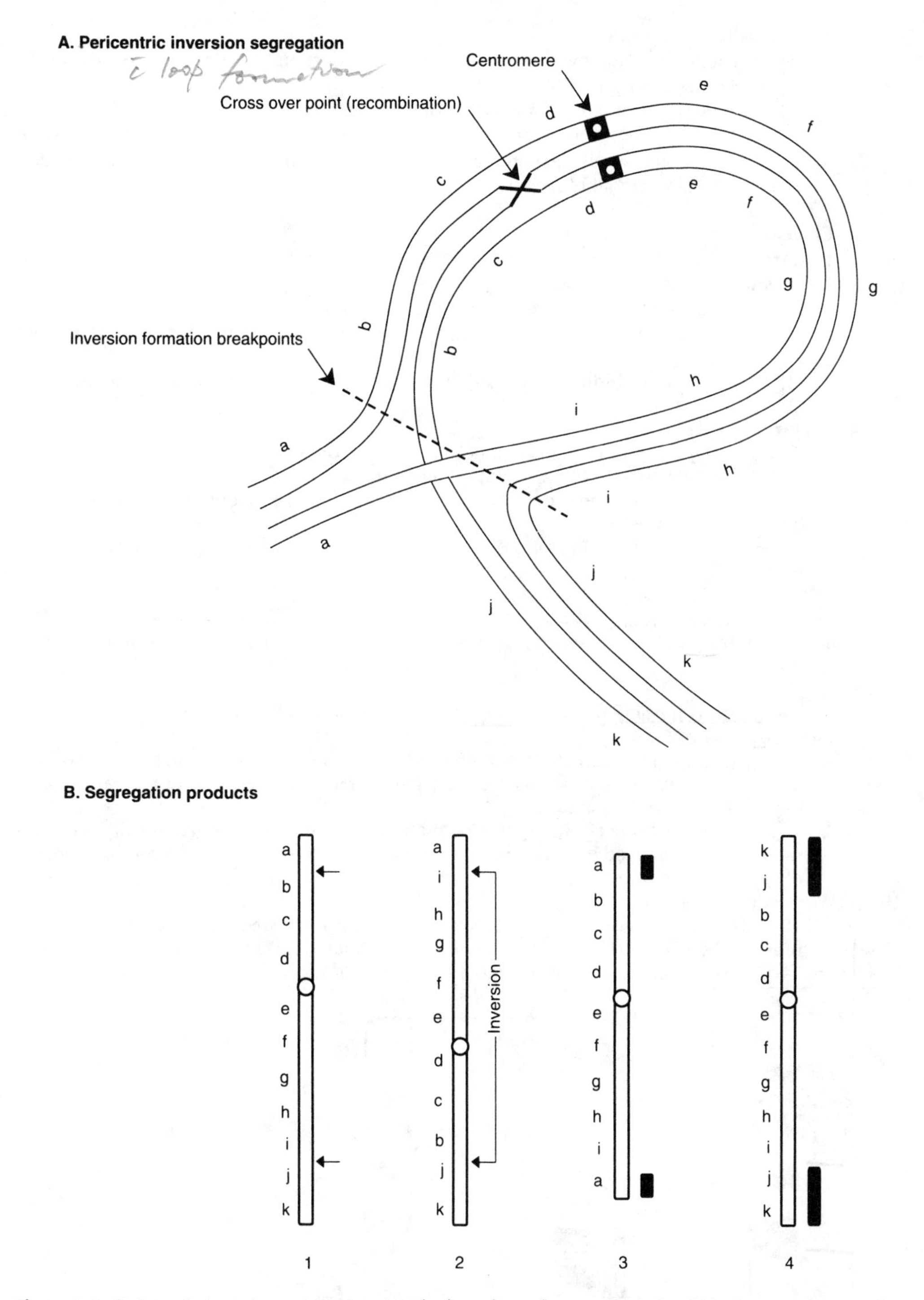

Figure 2-6. Pericentric inversion segregation. (*A*) The loop formed to accommodate homologous pairing of the inverted segment is shown here. Note that the centromere is within the loop. This example shows a single cross over event within the loop. (*B*) With a single cross over, two segregation products (*1* and *2*) will be like the parental chromosomes, normal or inverted, and two (*3* and *4*) will have duplications and deficiencies (indicated by the *black bars*). In *3*, segment *a* is duplicated and segment *jk* is absent, and, in *4*, the situation is reversed. Both products, *3* and *4*, will produce an abnormal embryo, fetus, or newborn. Note that if a cross over does not take place within the inversion loop, all products will be like the parental chromosome.

b. Paracentric inversions

(1) As with pericentric inversions, meiotic pairing constraints lead to **loop formation** between the junctions of the inverted segment.

(2) A **single recombination event** within the loop produces **unstable recombinant chromosomes** (either dicentric or acentric), which are almost never seen in the children of carriers. Usually only the parental chromosomes (normal or inverted) are transmitted to clinically normal offspring.

3. Clinical features

a. Carriers of either pericentric or paracentric inversions **are normal**.

b. Diagnosis is based on the following:

(1) Coincidental finding at prenatal diagnostic testing

(2) Spontaneous abortions, with unbalanced products having large duplications and deletions

(3) Stillbirths or livebirths, with small duplication and deletion products of the inversion

4. Recurrence risks

a. If the ascertainment is through an abnormal stillbirth or livebirth, there is an overall 5% to 10% risk for recurrence of an abnormal livebirth.

b. If the ascertainment is through recurrent miscarriage, the risk is low for an abnormal livebirth.

c. If the finding is coincidental at prenatal diagnosis, the risk of recurrence is 1% to 3%.

d. Each family needs to be considered individually.

E. Robertsonian translocations (centric fusions) result from fusion of whole arms of acrocentric chromosomes (i.e., chromosomes 13–15, 21, and 22) [Figure 2-7].

1. The **breakpoints forming these rearrangements** are at or near the centromeres of both chromosomes involved.

a. Only the long-arm fusion chromosome product is usually recovered (the short-arm fragment, or telomere, is lost), so that **heterozygous carriers for this class of rearrangements have only 45 chromosomes.**

b. These **carriers of robertsonian translocations are normal,** and the translocation is considered to be balanced because short-arm material on acrocentric chromosomes consists of genetically inert heterochromatin and ribosomal RNA (rRNA) genes, which occur in multiple copies on other acrocentric chromosomes.

c. The **robertsonian translocation 45,XX or XY,t(13q14q) is the most common translocation found in humans** and has an incidence rate of about 1/1500 individuals. Translocation 14q21q is also common and is frequently a cause of familial Down syndrome.

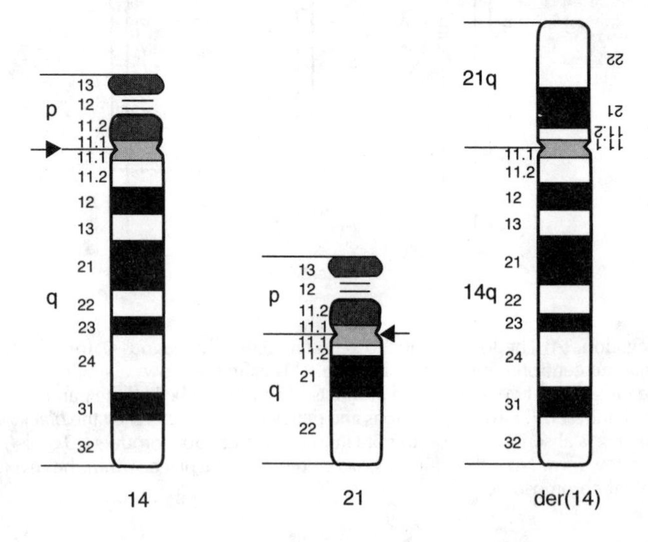

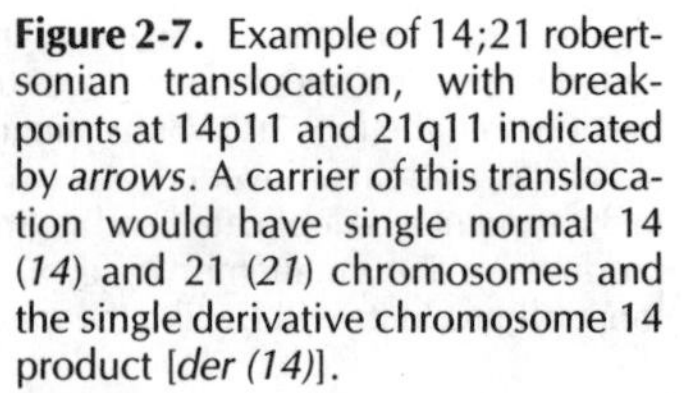

Figure 2-7. Example of 14;21 robertsonian translocation, with breakpoints at 14p11 and 21q11 indicated by *arrows*. A carrier of this translocation would have single normal 14 (*14*) and 21 (*21*) chromosomes and the single derivative chromosome 14 product [*der (14)*].

2. **Consequences of meiosis in carriers.** Homologous segments of the two chromosomes involved pair at meiosis. With robertsonian translocations, this usually involves three chromosomes since the short-arm translocation product is lost. Segregation of the chromosome will produce gametes with both genetically balanced and unbalanced products (Figure 2-8).

3. **Clinical example—t(14q21q).** In a translocation involving chromosomes 14 and 21, the carriers are clinically normal.
   a. At conception, the **unbalanced products** result in essentially trisomy 21 (likely to go to term), trisomy 14 (early miscarriage), or monosomy of either chromosome 14 or 21 (early miscarriage).
   b. Offspring with **balanced products** have completely normal chromosomes or, like the parent, carry the translocation.
   c. Therefore, the **family with this translocation** could present with multiple family members having Down syndrome or with recurrent miscarriages.
   d. **Karyotyping** of couples who have had recurrent miscarriages should be considered (see Ch 1 IV C 2 a).

4. The **recurrence risk of robertsonian translocation carriers** of all types for abnormal offspring depends on ascertainment, the specific chromosomes involved, and the sex of the carrier parent. For example, females with the translocation between chromosomes 14 and 21 [t(14;21)] have a 10% risk for offspring with Down syndrome while the male carrier has a 2% risk (Table 2-3).

F. **Reciprocal translocation** results from breakage and exchange of segments between chromosomes. The points of exchange can be at any location along the chromosomes (Figure 2-9).

1. **Consequences of meiosis in carriers** (Figure 2-10)
   a. **Homologous segments** of the two chromosomes involved in a translocation pair to form a four-part structure (Figure 2-10A).
      (1) **Segregation of the paired segments** produce gametes with several possible chromosome constitutions, including normal chromosomes, balanced translocations, and unbalanced translocations.
      (2) **Disjunction of the four-part structure** at anaphase usually distributes two chromosomes to one cell and two to the other (2:2 disjunction); or in some instances, distribution is three and one (3:1 disjunction).
   b. **Gametes with unbalanced translocation products** from 2:2 disjunction have both duplications and deficiencies for chromosome segments involved in the translocation (see Figure 2-10B). Those from 3:1 disjunction may have duplications or deficiencies of segments of both chromosomes or be nullisomic or disomic for either chromosome.

2. **Case example 2-A. Translocation of chromosomes 6 and 7** [t(6;7)]
   a. Siblings presented to a pediatric service with identical features of growth retardation, microcephaly, blepharophimosis with long lashes, and congenital heart disease. Since the parents were from a small village, consanguinity was suspected and the children were thought initially to have a recessive disorder.
   b. Studies showed both siblings had forty-six chromosomes, with one chromosome 7 replaced with a derivative 7 from a 6;7 reciprocal translocation. They both had normal number 6 chromosomes. The karyotypes were unbalanced, with duplication (partial trisomy) 6p21 → pter and deficiency (partial monosomy) for 7q34 → qter, as a result of adjacent 1 segregation of the balanced translocation in the carrier father. The father's karyotype was 46,XY,t(6;7)(p21;q34). The family history was subsequently documented to be positive for miscarriages.

Table 2-3. Risks of Carriers for Having Liveborn Offspring with Unbalanced Karyotypes

Abnormality	Risk
Robertsonian t(14;21)	10% (female) 2% (male)
Isochromosome i(21q)	100%
Reciprocal translocations	≤ 30%–40%*
Inversions	≤ 5%–10%*

*Risks for reciprocal translocations and inversions vary among individual anomalies. Those that produce a large chromosome imbalance usually have a lower risk since they are lethal to the embryo and result in miscarriage.

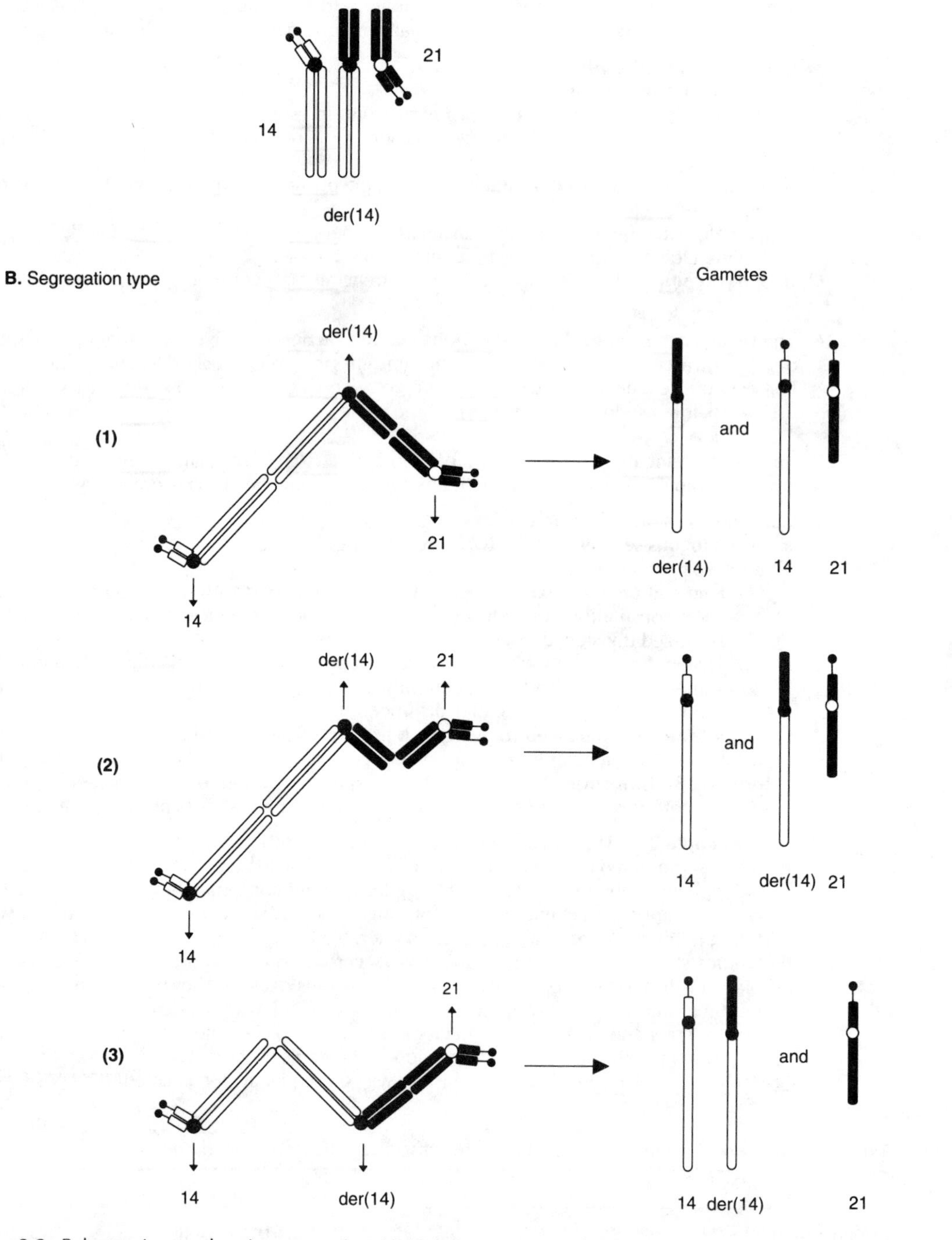

Figure 2-8. Robertsonian translocation segregation. (*A*) Meiotic pairing complex in a carrier of a t(14;21)(p11;q11) robertsonian translocation. The complex is shown as a three-part structure with homologous segments of the involved chromosomes paired. (*B*) Segregation is one of the three possibilities (*1–3*). In type *1,* the derivative [*der(14)*] chromosome segregates from chromosomes 14 and 21. Fertilization of the resulting gametes produces a balanced translocation karyotype or a normal one. Both gametes produce a normal phenotype. In segregation types *2* and *3,* all of the gametes are abnormal and will result in long-arm (q) trisomy or monosomy of either chromosome 21 or 14. Only the gametes from type *2* [*der(14) and 21*] will produce viable offspring after fertilization. These children will have Down syndrome (trisomy 21q).

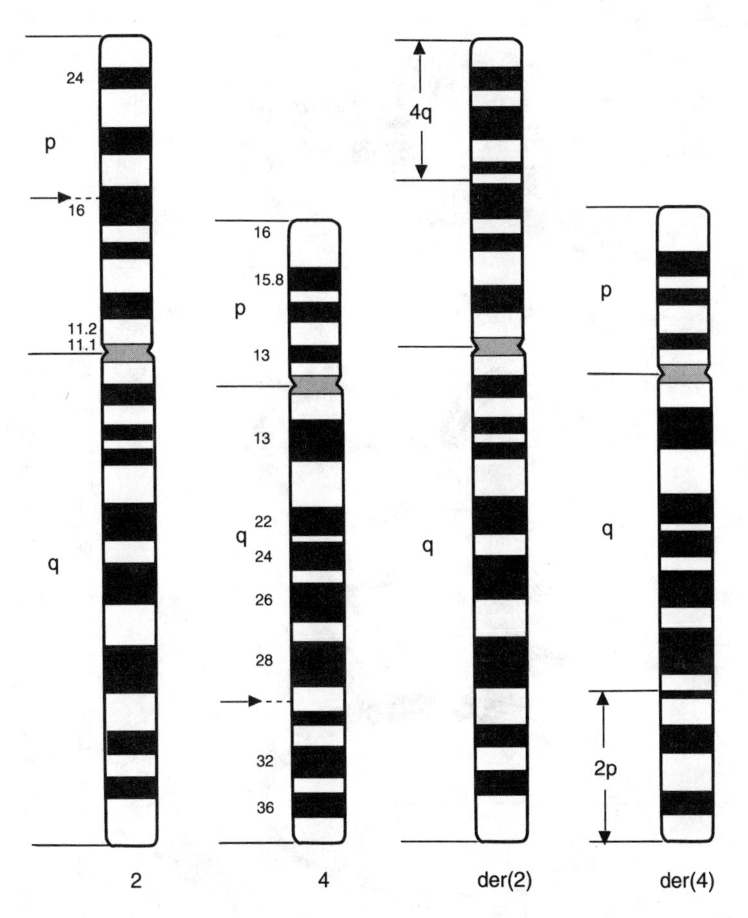

Figure 2-9. Example of a 2;4 reciprocal translocation with breakpoints at 2p16 and 4q31.1 indicated by the *arrows*. A carrier of this translocation would have single normal chromosomes 2 (*2*) and 4 (*4*) and the two derivative chromosomes [*der(2) and der(4)*].

c. In this example, one form of chromosome imbalance from adjacent I segregation is seen. The other product that would result in deficiency 6p21 → pter, duplication 7q34 → qter would likely result in early miscarriage, as the deficiency is large (more than one or two chromosome bands). Generally large deficiencies are lethal and less tolerated than duplications of the same chromosome segment.

3. The **recurrence risks of carriers** for abnormal offspring depend on previous reproductive history, ascertainment, predicted type of segregation, and the likely survival potential of zygotes with a specific chromosome imbalance. Generally, for those translocations that could result in chromosomally abnormal viable offspring, the risk for carriers is rarely greater than 20% to 30% and is likely to be less (see Table 2-3).

IV. OTHER ANOMALIES

A. Chromosome fragile sites are gaps or breaks in chromosomes that can be visualized if the cell cultures are provided with specific conditions [see Ch 1 IV C 1 b (8)]. They are genetically transmitted traits and are specific to the location on chromosomes.

1. Clinical example

a. Fragile X mental retardation

(1) Statistically, fragile X mental retardation occurs at a frequency of approximately 1/1500 males, and it occurs less frequently in females. Mental retardation is exhibited in 80% of the males and 30% of the females who carry the fragile X mutation.

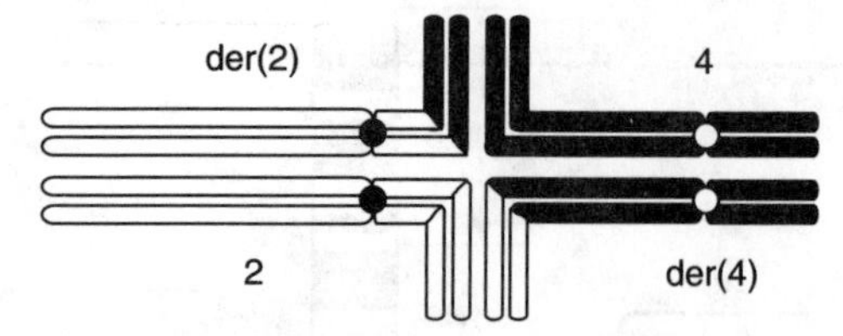

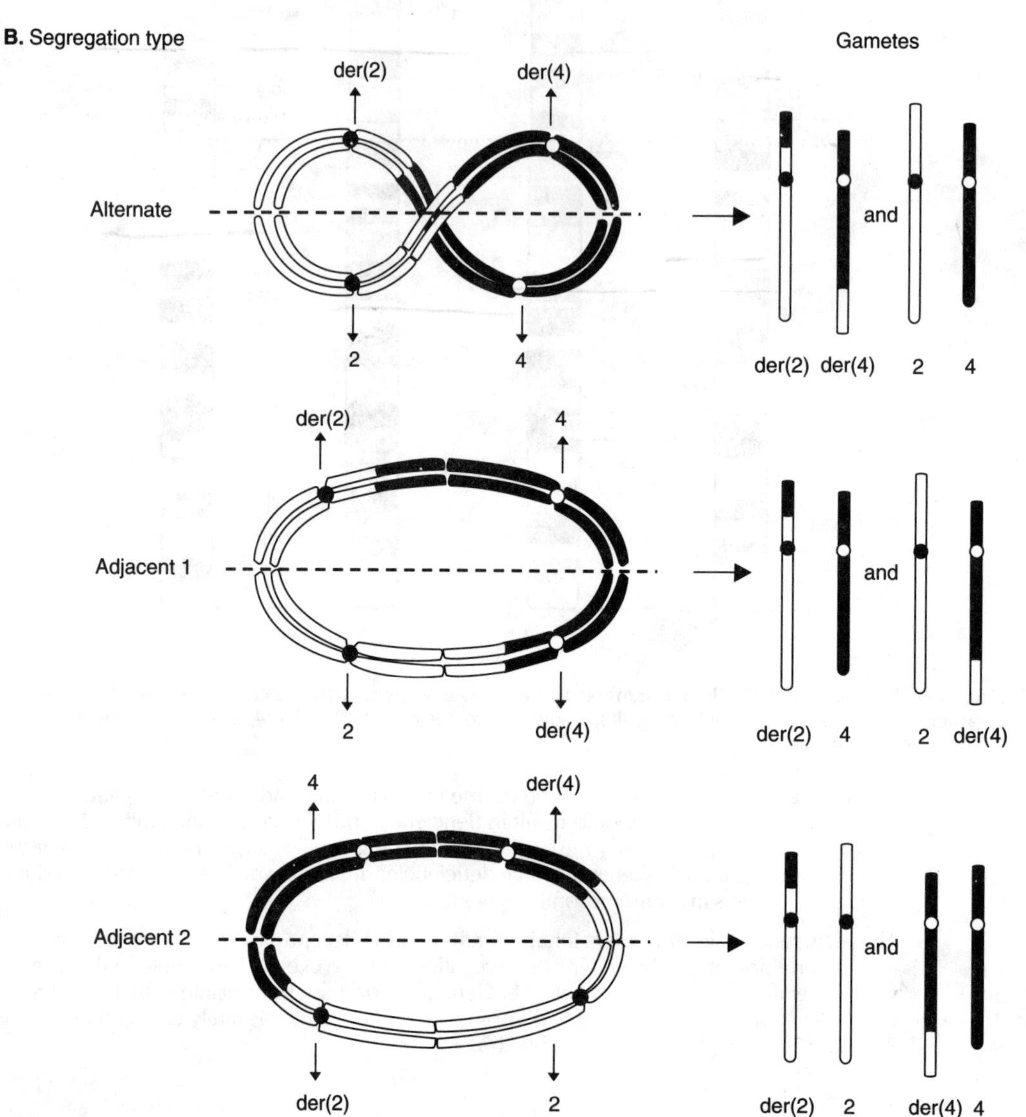

Figure 2-10. Reciprocal translocation segregation. (*A*) Meiotic pairing complex in a carrier of t(2;4)(p16;q32) reciprocal translocation. The complex shows a four-part structure with homologous segments of the involved chromosomes paired. (*B*) Segregation is one of three types (alternate or adjacent centromeres to the spindle pole at anaphase). Gametes from alternate segregation have a balanced translocation or are normal. Gametes from adjacent 1 and 2 segregation are all abnormal, with both duplications and deficiencies. For most translocations, abnormal clinical outcomes usually result from adjacent 1 segregation, as many of the unbalanced products produce viable abnormal offspring or result in recognized pregnancy loss. Most adjacent 2 segregation products have large duplications and deficiencies and would be lost prior to implantation. Note that segregation may result in movement of three chromosomes to one pole and one chromosome to the other, unlike the 2:2 segregation described in III F 1. For most translocations, this would be associated with major lethal imbalances. However, for a subset of reciprocal translocations, usually involving a small acrocentric chromosome (e.g., chromosome 21), viable 3:1 unbalanced products do occur.

(2) **Cytogenetically,** fragile X mental retardation is characterized by chromosome fragility involving Xq27.3 under cell culture conditions that lower folic acid levels or inhibit thymidine synthesis (Figure 2-11).

(3) The **clinical presentation** most often occurs in a male with speech delay and abnormal speech patterns. Microcephaly is not a feature, but the facies are narrow with large ears, prominent mandible, and close-set eyes. Macro-orchidism commonly occurs with the onset of puberty.

(4) The **gene** follows a pattern of X-linked inheritance but is responsible for **the following complications**.

(a) Females receiving the fragile X gene from a normal male (termed a transmitting male) are usually mentally normal.

(b) Females inheriting the fragile X gene from a normal mother have a 30% chance of being affected; if the gene is received from an affected mother, the risk rises to 50%.

(c) Males who receive the mutant gene from a normal mother have an 80% chance of expressing the disease phenotype and a 20% chance of being a normal transmitting male.

(d) All males receiving the mutant gene from an affected mother will express the disease phenotype.

(5) The **diagnosis** is made by analyzing the chromosome for the presence of the Xq27.3 fragile site using a folate-depleted culture medium [see Ch 1 IV C 1 b (8)].

(a) All affected males should express the fragile site in 3% to 40% of their cells.

(b) Cytogenetic diagnosis is more difficult in females. About half of affected female carriers have a positive result, but fewer normal female carriers of the mutant gene exhibit the fragile site.

(c) Molecular techniques, which search for linked DNA polymorphisms (see Ch 4 I; II), can be used in association with the chromosome analysis. Recently, a direct probe has been described and will soon be used for diagnosis.

b. Autosomal fragile sites. Although autosomal fragile sites collectively occur more frequently than the site at Xq27.3, their clinical significance is unclear. Reproduction is not thought to be affected.

B. Marker chromosomes are derived from structural rearrangements and are commonly found as extra chromosome pieces.

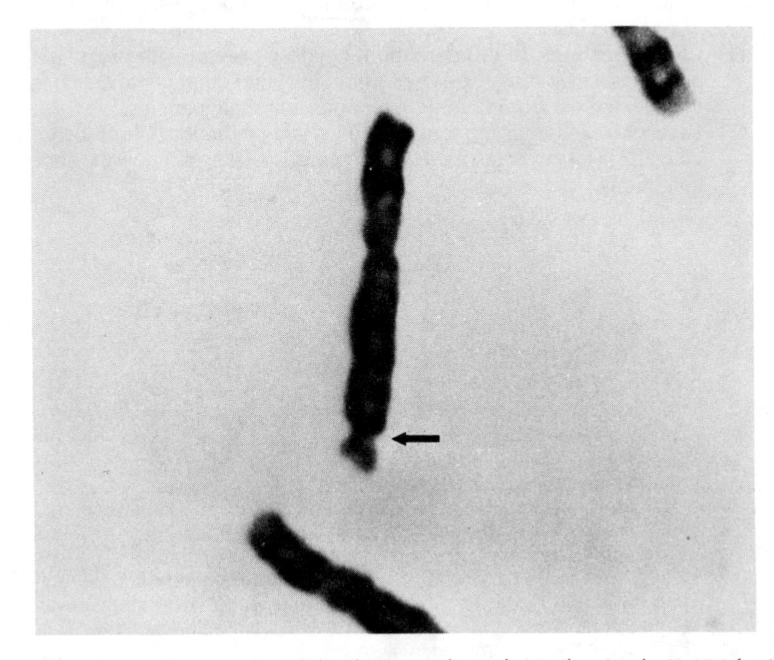

Figure 2-11. An X chromosome from a male with fragile X mental retardation showing the Xq27.3 fragile site (*arrow*).

1. **The chromosome material of origin is unknown.** Most chromosomes presenting as constitutional markers (as compared to tumor cell marker chromosomes) are commonly smaller than chromosome 22, may be in the form of a ring or a bisatellite chromosome, or may have metacentric morphology.
2. **Small markers may exhibit mitotic instability** and present in a mosaic form. Various amounts of heterochromatin or euchromatin may be present.
3. **Many markers are compatible** with normal development and normal intelligence.

C. Chromosome breakage. Chromosome breaks result in visible lesions in metaphase chromosomes and can lead to structural changes such as deletions and translocations.

1. **Induction of lesions.** Lesions resulting from chromosome breaks may be induced by a variety of sources, including faulty DNA repair or synthesis, environmental insults such as radiation, or chromosome-breaking chemicals. Cell cultures from normal individuals exhibit chromosome breakage in low frequency. Some individuals have an intrinsically high incidence.
2. **Clinical examples.** There are a number of clinical syndromes, all inherited in an autosomal recessive manner, that are characterized by hematologic abnormalities and malformations. Some of these are also associated with an increased risk for malignancy.
   a. **Fanconi anemia (Fanconi pancytopenia)**
      (1) **Clinical features**
         (a) Children may present with a variety of external and internal malformations—the most diagnostic being radial and thumb abnormalities, patchy pigmentation over the trunk, congenital heart disease, and renal malformations.
         (b) In time, children develop pancytopenia from bone marrow hypoplasia and may ultimately develop leukemia requiring a bone marrow transplant.
      (2) **Diagnosis** is confirmed with demonstration of an elevated frequency of chromosome breakage in cultured lymphocytes. The frequency of breakage is increased markedly over that in controls with the use of the alkylating agent diepoxybutane (DEB).
   b. **Bloom syndrome**
      (1) **Clinical features.** Affected children exhibit growth retardation and abnormal skin with telangiectases involving the face. All have an increased risk for developing malignancies. An increased frequency of chromosome breakage and homologous chromosome exchanges can be detected in cultured cells.
      (2) **Diagnosis** can be made by the observation of a value that is 12 times the normal value of sister chromatid exchanges in cultured lymphocytes.
   c. **Ataxia telangiectasia**
      (1) **Clinical features.** In this condition, children present with progressive ataxia and telangiectasia involving the ears, conjunctiva, and other facial areas. Survival is poor as the affected die from chronic infections and malignancies.
      (2) **Diagnosis.** Cell lines are sensitive to ionizing radiation. Chromosomes from children with ataxia telangiectasia exhibit increased breakage frequency under normal culture conditions.

STUDY QUESTIONS

Directions: Each of the numbered items or incomplete statements in this section is followed by answers or by completions of the statement. Select the **one** lettered answer or completion that is **best** in each case.

1. A family seeks their physician's advice. The woman is pregnant, and she is concerned because her brother has Down syndrome. The physician ascertains that the affected brother lives in a local group home. Of the following methods of investigating and counseling, which one would be most appropriate?

(A) Review the family history
(B) Confirm the diagnosis in the affected brother
(C) Karyotype the affected brother
(D) Karyotype the woman coming for counseling

2. A newborn female is noted to have dorsal edema of the hands and feet as well as a cardiac murmur. The clinical suspicion is Turner syndrome, and a karyotype is quickly arranged. The results show that the infant is mosaic, and 50% of the cells are 45,X and 50% of cells are 46,XY. The physician should advise the parents to take which one of the following actions with their child?

(A) The child has a Y chromosome and should be raised as a boy, which requires surgery and male hormones to help the process
(B) The child has a variant of Turner syndrome and will be short, is likely to have a congenital heart lesion, needs cardiac assessment, and should be raised as a female
(C) The child should be raised as a female but will require a gonadectomy because of the risk of malignancy

3. Of the following clinical situations, which one warrants a karyotype?

(A) A woman with one spontaneous abortion
(B) The parents of a child with trisomy 21
(C) The sister of a boy with Prader-Willi syndrome demonstrated to have the common deletion
(D) A couple with a stillbirth and three spontaneous abortions
(E) The maternal uncle of an infant with a maternally derived t(21;21)

Directions: The group of items in this section consists of lettered options followed by a set of numbered items. For each item, select the **one** lettered option that is most closely associated with it. Each lettered option may be selected once, more than once, or not at all.

Questions 4–9

For each chromosomally abnormal pregnancy, assess the recurrence risk in future pregnancies.

(A) Increased risk
(B) Decreased risk
(C) Unchanged risk

4. A spontaneous abortion with triploidy
5. A stillbirth with triploidy
6. A spontaneous abortion with 45,X
7. A spontaneous abortion with trisomy 18
8. A liveborn with trisomy 21 (mother age 23)
9. A liveborn with trisomy 21 (mother age 40)

1-A 4-C 7-C
2-C 5-C 8-A
3-D 6-C 9-C

ANSWERS AND EXPLANATIONS

1. The answer is A *[II B 2 a].*
If there is a history of recurrent miscarriages, difficulties conceiving, or other family members with Down syndrome, the physician should consider the possibility of a translocation accounting for the Down syndrome individual as well as the additional history. If the mother was over 35 at the birth of the brother affected with Down syndrome, then trisomy would be the suspected cause. Since the affected brother is available, he could either be examined or the diagnosis could be documented. However, the question of a familial translocation would not be answered without karyotyping.

The physician could arrange for blood to be sent to the local cytogenetics facility. By karyotyping the affected person first, the physician is able to offer information to the whole family. If the diagnosis is confirmed as a trisomy, the physician can reassure the family that the risk of Down syndrome is that of the population risk. If a translocation is discovered, the physician will need to karyotype the parents of the affected individual.

2. The answer is C *[II C 2].*
Although this infant does have a variant of Turner syndrome, it would be insufficient to offer the usual counseling. The parents need to be aware that a cell line with a Y chromosome is present, as this will put the gonads at increased risk for gonadoblastoma development.

The child has presented as a female and, even though there is a Y chromosome line, it is not appropriate to rear her as a male. Although surgical conversion is possible, it would be extremely difficult to surgically create a normal appearing male, and sexual function would be lacking. If an infant presents as a male (i.e., traits such as unilateral testis or hypospadias are present), consideration may be given to raising him as such. Males with mixed gonadal dysgenesis may do well, but most often have short stature, and the gonads must be carefully watched.

3. The answer is D *[III A 3 a, E, F; Tables 2-2 and 2-3].*
The couple who experienced a stillbirth and three miscarriages has a history very suggestive of translocation. In couples having three or more miscarriages, 1% to 3% have such a rearrangement.

A woman who had one miscarriage is not at an increased risk statistically to have a translocation.

The parents of a child with trisomy 21 are not necessarily candidates for karyotyping because the risk of recurrence depends on the maternal age, and the chromosomes of the parents would be expected to be normal.

If the boy with Prader-Willi syndrome has the common interstitial deletion, it is not secondary to a familial rearrangement. Therefore, a karyotype of other family members is not necessary.

A mother with a translocation resulting in t(21;21) is most likely to have chromosomally normal parents and siblings. If a parent of her own had the rearrangement, she would have had Down syndrome herself. Her brother also would be unlikely to carry the translocation.

4–9. The answers are: 4-C *[II A 2 a; Table 2-1],* **5-C** *[II A 2 a],* **6-C** *[II C 2; Table 2-1],* **7-C** *[II B 2 c],* **8-A** *[II B 2 a; Table 2-2],* **9-C** *[II B 2 a; Table 2-2].*
Triploidy is a common cause of early spontaneous abortions, is a sporadic event, and is not associated with an increased risk in future pregnancies.

A spontaneous abortion with 45,X is also a common finding in early miscarriages. Turner syndrome is not known to be associated with increased maternal age, nor is there an increased risk in subsequent pregnancies.

Data suggest that trisomies ending in early miscarriages do not increase the risk for a subsequent liveborn with a trisomy.

Data confirm that the recurrence risk of trisomy 21 to a mother at or below 30 years of age is increased to 1% to 2%. The same data show no increased risk (beyond that already existing with age) for an older mother.

3
Single Gene Alterations

Barbara McGillivray and Michael R. Hayden

I. PATTERNS OF INHERITANCE

A. Introduction. Single gene (mendelian) inheritance describes those conditions where single mutant genes have a large effect on human health. In 1866, Mendel termed "those characteristics that are transmitted entire[ly], or almost unchanged by hybridization," as dominant; "those [that] become latent in the process," were termed recessive. Genes are recognized by the physical characteristics or traits determined by them. Pedigree analysis is essential to determine the mode of inheritance of many traits.

1. **Definitions.** The question for most single gene inheritance is whether the specific phenotype is seen in the heterozygote or only in the homozygote.
   - a. **Gene.** This term describes the hereditary factor that interacts with the environment to determine a trait.
   - b. **Alleles** are alternative forms of both normal and abnormal genes.
   - c. A **locus** is the physical location of a gene on a chromosome. Since human chromosomes are paired, individuals have two alleles at each locus.
   - d. **Genotype** describes the genetic constitution of an individual, which is the specific allelic makeup of an individual.
   - e. **Phenotype** describes the end result of both the genetic and environmental factors giving the clinical picture or the observed expression of the gene.
   - f. **Homozygous** describes the condition of having identical alleles at one locus, which can be either normal or abnormal.
   - g. **Heterozygous** describes the condition of having two different alleles at one locus and usually refers to having one normal and one abnormal or mutant allele. Heterozygous individuals are also referred to as **carriers**.
   - h. A **dominant condition** is seen in both the heterozygote and the homozygote. This implies that a single copy of the allele is enough for the condition to be expressed.
   - i. A **recessive condition** is seen only in the homozygote, which means that the allele must be present in both chromosomes. Dominant and recessive refer to the clinical conditions, not to the genes themselves.
   - j. **Autosomal** refers to the autosomes, which are the nonsex chromosomes. An autosomal condition results from alleles on the autosomes.
   - k. **X- or Y-linked** refers to genes having loci on either the X or Y chromosome. With X-linked alleles, both recessive and dominant inheritance may be seen. The term **sex-linked** is also used to represent X-linked inheritance.
   - l. A **compound heterozygote** is an individual with two different mutant alleles at one locus. β-Thalassemia may be caused by inheritance of independent mutations of the normal β-globin gene from each parent. Many autosomal recessive conditions are seen in a compound heterozygote, not in a true homozygote.
   - m. A **double heterozygote** is an individual with two mutant alleles that are each at a different locus. The offspring of two parents, each with autosomal recessive deafness at different loci, will have normally hearing children who are double heterozygotes.
2. **Pedigree symbols and construction.** When the family history is taken, the pedigree is a shorthand or graphic representation of the details. Individuals are characterized according to their sex, their generation, and their biological relationship to each other. In Chapter 10 I B, the steps involved in constructing pedigrees are explained. Figure 10-1 illustrates the most commonly used symbols, with an example pedigree.

B. Autosomal dominant inheritance. Over half of the presently described traits are inherited in a dominant fashion: about one-third as recessive and one-tenth as X-linked. Dominant implies that the disease allele need be present only in single form (as in the heterozygote) to result in the phenotype.

1. The **criteria** of autosomal dominant inheritance include the following.
   - **a.** The transmission of the trait continues from generation to generation without skipping. In a typical dominant pedigree (Figure 3-1), there can be many affected members in each generation.
   - **b.** Except for a new mutation, every affected child has an affected parent.
   - **c.** In the mating of an affected heterozygote to a normal homozygote, the segregation frequency is one-half, which means with each pregnancy the risk is 50% that the child will inherit the mutant allele from the affected parent.
   - **d.** The two sexes are affected in equal numbers. Since the loci are on the autosomes (i.e., nonsex chromosomes), a sex differential is not seen.

2. **Features.** A number of complications are important to consider in autosomal dominant inheritance to provide the most accurate counseling. Each of the following may alter the idealized dominant pedigree.
   - **a. New mutations.** An isolated case of a dominant disorder may be the result of a new mutation, but other causes, such as genetic heterogeneity and paternity, should also be considered.
     - **(1)** A new mutation is seen more often with diseases that are so severe that people who are affected by them are less likely to reproduce than normal. The majority of cases of achondroplasia are the result of new mutations.
     - **(2)** Increased paternal age may be associated with an increase in new mutations at some loci.
     - **(3)** The rate of new mutations is available for some dominant conditions and is useful for counseling.
   - **b. Reduced penetrance. Penetrance** is often expressed as the frequency of the number of individuals who have a gene and show the trait. Since penetrance is an all-or-nothing phenomenon, if the frequency of expression of a genotype is less than 100%, **reduced penetrance** exists. **Nonpenetrance** refers to the situation where the mutant allele is inherited but not expressed.
     - **(1)** In a particular disorder, penetrance would be calculated by determining the total number of individuals expressing the disease divided by the total number of individuals inheriting the gene.

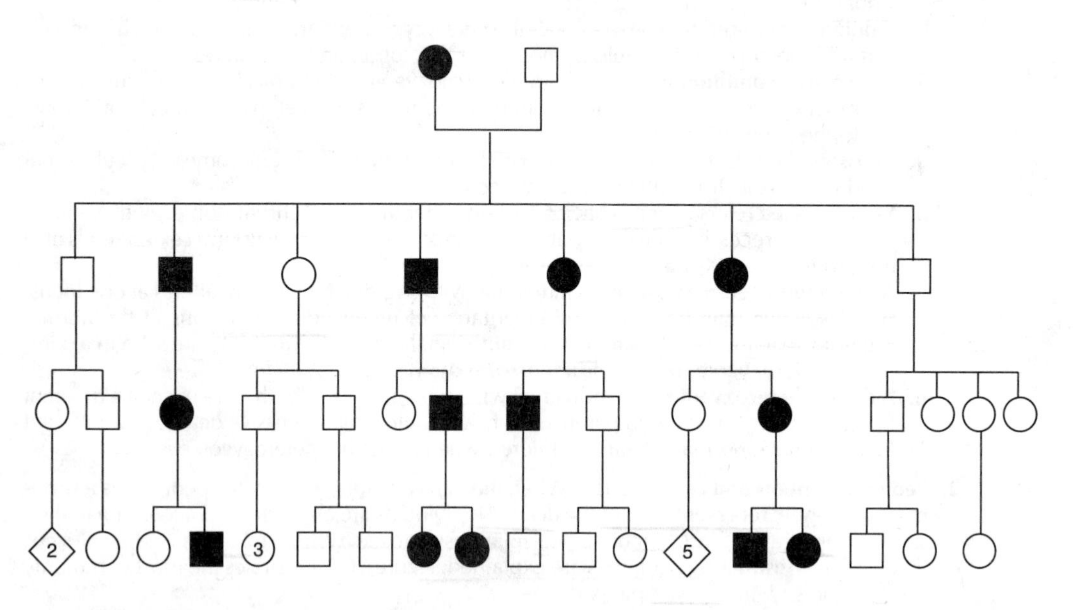

Figure 3-1. Pedigree typical of dominant inheritance.

(2) Penetrance is a clinical term and may be affected by factors including age (e.g., Huntington's disease) or better detection methods (e.g., magnetic resonance imaging for signs of tuberous sclerosis).
(3) In disorders such as polydactyly with a known reduced penetrance, an apparent singleton individual may have an affected parent.

c. Variable expressivity. Expressivity refers to how or to what degree a particular allele is expressed in an individual.
(1) Variable expression is **common in dominant disorders,** which means that all family members should be examined carefully, with the features of the condition kept in mind.
(2) Affected individuals who are capable of reproducing are generally less severely affected. In their potential children, more severe expression may be seen (i.e., they may express the full spectrum of the syndrome), and such severe expression may be unexpected by the parents.

d. Observer fallibility. Before counseling a family, the geneticist should consider how the diagnosis was made and whether the diagnosing individual has sufficient expertise to make a reliable genetic diagnosis. The danger lies in basing the counseling on another physician's or the family's diagnosis. As much information as possible should be obtained to corroborate a diagnosis.

e. Variation in age of onset. Adult polycystic kidney disease (PKD) is inherited as a dominant trait, but the cysts appear with time. Before the advent of molecular testing (see Ch 5) , the diagnosis could not be reliably ruled out in a young person. Affected individuals may die before any signs of the condition are seen.

f. Genetic heterogeneity. Different mutations, at the same locus or at different loci, result in a similar clinical picture. Heterogeneity causes difficulties in counseling two unrelated, but similarly affected, individuals (e.g., parents with sensorineural deafness). In addition, the conditions may be inherited in different manners (e.g., retinitis pigmentosa may be inherited as either an autosomal dominant or recessive trait).

g. Phenocopy is an environmentally caused phenotype producing a clinical picture similar to that produced by a specific genotype. For example, an infant may present in a similar clinical fashion with warfarin embryopathy or Conradi syndrome (a single gene condition).

h. Variation in severity dependent on sex. The severity of a dominant condition may depend on the sex of the affected parent. In Huntington's disease, those with early severe disease are more likely to have an affected father; in myotonic dystrophy, those with early onset are usually born to affected mothers.

3. Clinical examples

a. Huntington's disease is an autosomal dominant adult-onset disease involving selective cell death of caudate nuclei. The clinical presentation includes a choreic movement disorder, mood disturbances, and progressive loss of mental activity. The course from onset to death is usually 15 years and begins in the fourth decade. (There is less frequent juvenile onset when the affected parent is the father.) The gene location is the short arm of chromosome 4 (4p), but the gene has not been identified.

b. Marfan syndrome is a disorder of connective tissue with variable expressivity involving the skeletal system (e.g., abnormal proportions, scoliosis, arachnodactyly), cardiovascular system (e.g., mitral and atrial valve prolapse, aortic dilatation leading to aortic dissection), and ocular system (e.g., lens dislocation). The diagnosis is made on clinical grounds, with involvement in at least two systems. The abnormality may involve fibrillin. A tentative gene location is the long arm of chromosome 15 (15q).

C. Autosomal recessive inheritance. The phenotype is usually observed only in the homozygote, and the typical pedigree demonstrates affected male and female siblings with normal parents and offspring. Recessive inheritance is suspected when parents are consanguineous and is considered proven when enzyme levels are low or absent in affected individuals and are at half-normal values in both parents.

1. Criteria

a. If the trait is rare, parents and relatives, other than siblings, are usually normal.
b. If the recessive genes are alleles, all children of two affected parents are affected.
c. In the mating of two normal heterozygotes, the segregation frequency with each pregnancy is 25% homozygous normal, 50% heterozygous, and 25% homozygous affected.
d. Both sexes are affected in equal numbers.
e. If the trait is rare, the possibility of parental consanguinity is increased.

2. **Features.** A number of features are important to keep in mind before counseling families regarding autosomal recessive inheritance.
   a. **Heterozygote frequency.** Affected individuals are almost always the offspring of normal heterozygotes (not homozygotes). The relationship of different phenotypes of one locus is described by the Hardy-Weinberg law (see Ch 6 I B 4). The homozygote frequency (q^2) is usually known, and the heterozygote frequency can be calculated from 2pq. The trait is maintained in the much more common heterozygote. Frequencies range from 1 in 10 individuals for hemochromatosis to greater than 1 in 200 people for rare disorders. Calculating the carrier frequency is important for designing screening programs.
   b. **Carrier state.** Carriers are assumed to be unaffected, but some may show half-normal enzyme levels (e.g., the level of hexosaminidase A in Tay-Sachs disease heterozygotes) or minimal clinical features. The majority of carriers are clinically normal, however.
   c. **Consanguinity.** A description of alkaptonuria notes that 60% of affected children had parents who were cousins. Consanguineous individuals have a proportion of their genes in common by descent. First cousins have one-eighth of their genes in common, while siblings have one-half of their genes in common. Consanguinity may come up in counseling in several situations:
      (1) Couples concerned that they are related by descent
      (2) Religious isolates, such as the Amish in North America
      (3) Geographic isolates, such as the Lac St. Jean area in Canada or areas of the Appalachians
   d. **Autosomal recessive phenotypes in the local population.** It is important to know an area's population when counseling, as there may be particular racial groups in which a disease allele is carried more frequently because of the founder effect (see II C 3) or heterozygote advantage. Knowledge of the local population allows appropriate screening and counseling. Examples of this include:
      (1) Sickle cell anemia in blacks
      (2) Tay-Sachs disease in Ashkenazi Jews
      (3) Cystic fibrosis in northern European Caucasians
   e. **Heterogeneity.** Clinically similar individuals may have genetic causes for their condition but have mutant alleles at different loci. Albinism, sensorineural deafness, and congenital hypothyroidism may all be recessively inherited but have different loci involved in different families. If two affected parents with deafness have different loci involved, their children will have normal hearing but will be **double heterozygotes**.

3. **Clinical examples**
   a. **Galactosemia**
      (1) **Clinical aspects.** Affected infants present with cataracts or hepatomegaly and may die of sepsis. Abnormal or absent galactose-1-phosphate uridyl transferase results in an inability to use lactose. Female survivors have ovarian failure, perhaps due to intrauterine damage.
      (2) **Treatment** involves dietary restriction of lactose-containing products. Untreated survivors suffer mental deficiency.
      (3) **Genetic background.** The condition is heterogeneous, with a number of allelic variants described, such as the Duarte allele.
   b. **Homocystinuria**
      (1) **Clinical aspects.** Individuals with homocystinuria may be mistakenly diagnosed as having Marfan syndrome since they have an abnormal body habitus (long limbs) and may have dislocated lenses. Two-thirds of people with homocystinuria are mentally handicapped or may have psychiatric problems. Deficient cystathionine β-synthetase results in increased levels of homocystine and methionine, the latter being thought to cause the problems in the condition.
      (2) **Treatment** of homocystinuria involves a low methionine diet. Sudden death is common in untreated patients because of thromboembolic phenomena. Individuals with metabolic defects other than cystathionine β-synthase deficiency, which are less common, are less likely to respond to dietary manipulation.
      (3) **Genetic background.** Homocystinuria is an autosomal recessive amino acid disease that results from an inability to convert homocysteine to cystathionine.
   c. **Cystic fibrosis**
      (1) **Clinical aspects.** About 1/5000 infants in northern European populations have cystic fibrosis and present with chronic respiratory infections, malabsorption, or failure to

thrive. Secretions of the gut and lung are thick, and sweat is altered, with high sodium and chloride levels. The chronic pulmonary disease begins before 1 year of age and is the cause of 90% of deaths.

(2) **Treatment.** Presently, the best treatment includes regular chest physiotherapy to loosen secretions, antibiotics, and replacement of digestive enzymes.

(3) **Genetic background.** The gene has been localized to the short arm of chromosome 7 (7p), and multiple alleles are described. This can potentially allow screening of at-risk populations as well as prenatal diagnosis.

D. **X-linked inheritance** can be either recessive or dominant, although with a recessive X-linked condition, the phenotype is usually only observed in the male. X-linked inheritance is suspected when several male relatives in the female line of a family are affected. Since males have only one X chromosome, they are **hemizygous,** not heterozygous, for X-linked genes.

1. **Criteria**
   a. If the trait is rare, parents and relatives (except maternal uncles and other male relatives in the female line) are usually normal.
   b. Hemizygous affected males do not transmit the trait to children of either sex, but all daughters are obligate heterozygotes (i.e., they do not show a mutation clinically, but have been determined to be carrying the mutant allele).
   c. Female heterozygous carriers are clinically normal but will transmit the trait to sons 50% of the time. Daughters are normal heterozygotes 50% of the time and normal homozygotes 50% of the time.
   d. Most often, affected daughters are produced only by matings of heterozygous females with affected males.
   e. Except for those with new mutations, every affected male is born of a heterozygous female.

2. **Features.** Some specific features of X-linked inheritance that cause problems are important to keep in mind.
   a. **The sporadic case.** If a singleton male is affected, the question is whether the condition is the result of a new mutation or if the mother is heterozygous.
      (1) **Heterozygous females** may be detected by subtle clinical features, intermediate enzyme levels, or molecular methods. In approximately two-thirds of cases, the mother is a heterozygote and in one-third of cases, the son has a new mutation.
      (2) **Bayesian methods** (see Ch 10 II C) utilizing other information about the family can further qualify the risk of X-linked inheritance.
   b. **Heterogeneity.** Diseases with similar clinical features may be inherited through differing mechanisms.
      (1) **Albinism** is usually inherited as an autosomal recessive condition, with involvement of eyes, hair, and skin.
      (2) An X-linked form, **ocular albinism,** involves only the eyes, but the condition may be overlooked in the blond child.
      (3) A good clinical examination and family history are essential.
   c. **Affected females.** Dosage compensation (Lyon hypothesis; see Ch 1 IV B 2 b) occurs as a consequence of **random X chromosome inactivation** at the late blastocyst stage in females. The net effect is **hemizygosity** of the X chromosome. Females who are heterozygous for X-linked conditions have clones of cells in which **one allele or the other is active**. X-linked conditions affecting females can be explained by the following mechanisms.
      (1) **Random inactivation.** By chance, the normal X chromosome is inactivated more often than the abnormal, resulting in a situation similar to that in a hemizygous affected male, with the majority of cells expressing the mutant X chromosome.
      (2) **Heterozygous mother and an affected father.** The risk for affected female offspring (i.e, homozygous for the mutant allele) would be 25% with each pregnancy.
      (3) **45,X Turner syndrome.** A female exhibits hemizygosity for X-linked genes, as does a male.
      (4) **46,XY females with sex reversal syndromes** (e.g., androgen insensitivity) can also inherit an X-linked disorder.
      (5) An **affected male** can produce an affected female offspring if there is a new mutation in the X chromosome from the normal mother.
      (6) A **heterozygous female** can produce an affected female offspring if there is a new mutation in the X chromosome from the normal father.

(7) **Nonrandom X inactivation.** In Turner syndrome that is caused by an isochromosome and a mutant allele on the normal X chromosome; the isochromosome is nonrandomly inactivated (see Ch 2 II C 2; III C 1 b). Nonrandom X inactivation may also be seen in X;autosome translocations, where the X chromosome involved in the translocation is not inactivated. If the X chromosome has a mutant allele or the translocation itself produces a mutation, the female will be affected.

d. X-linked lethal conditions. If the condition is lethal in males, any affected males would die before or at birth. Heterozygous females would be normal, but half of their male offspring would not survive. The end result would be a sex ratio for the condition of two females to every male.

3. Clinical examples

a. Duchenne muscular dystrophy (DMD)

(1) **Clinical aspects.** The incidence of DMD is about 1/3600 males, with onset in the first year of life. Characteristics include progressive weakness and muscle wasting, hypertrophy of calf muscles, and finally death, with cardiac or respiratory failure in the late teens to twenties.

(2) **Genetic background.** Gene localization pinpointing Xp21 was accomplished by studying affected females with X;autosome translocations and affected males with X chromosome deletions. The normal gene product was identified as the muscle protein, dystrophin, which perhaps acts as a fiber protector. About 60% of affected males have detectable deletions, allowing direct molecular techniques [see Ch 5 II A 2, 3 a (3)] for analysis, and the remainder of families require restriction fragment length polymorphisms (RFLP) analysis [see Ch 4 I B 1 a (1)].

b. Fragile X syndrome

(1) **Clinical aspects.** The **most common cause of inherited mental retardation** is fragile X syndrome, which occurs in 1/1500 males. Mental retardation is more common in males, suggesting the existence of X-linked causes. Affected males have large ears, a prominent chin, megalotestes, and mild-to-moderate mental retardation. As children, they usually present with speech delay or autistic features. Unlike most X-linked syndromes, females are affected in one-third of cases (which are usually milder).

(2) **Genetic background.** The diagnosis is made with specialized chromosomal culturing methods, which deplete folate and lead to expression of a fragile site at Xq27.3 in a proportion of the cells. Twenty percent of males inheriting the gene are both cytogenetically and clinically normal (transmitting males), and their daughters are clinically normal. The fragile X mutation produces instability of a repetitive DNA sequence that can increase in length during mitosis or meiosis.

E. Mitochondrial inheritance. Human cells have hundreds of mitochondria dispersed throughout the cytoplasm, each containing a number of circular DNA molecules. Mutations involving the mitochondrial DNA are now known to account for a small number of genetic conditions.

1. Criteria

a. Each mitochondrion contains a number of copies of the circular genome. Mitochondrial enzymes are usually encoded by nuclear genes, but some of the cytochrome oxidase enzymes are derived solely from mitochondria.

b. Almost all mitochondrial DNA is maternally inherited. Pedigrees with mitochondrial inheritance should show that all children of an affected mother are affected and all children of an affected father are normal (Figure 3-2).

c. Specific tissues have different proportions of mitochondria. Rich areas include striated and cardiac muscle, the kidney, and the central nervous system (CNS). Therefore, mitochondrial diseases are commonly myopathies, neurological syndromes, and cardiomyopathies.

2. Clinical examples

a. Kearns-Sayre syndrome is a sporadic condition with onset before age 20. Characteristics include progressive external ophthalmoplegia, retinal pigment abnormalities, heart block, and cerebellar ataxia. Studied patients exhibit deletions of muscle-derived mitochondrial DNA.

b. Leber's optic neuropathy. Optic atrophy, occasional movement disorder, and electrocardiographic (EKG) abnormalities characterize this disorder, which is transmitted exclusively by women to children of both sexes. The disorder is thought to be secondary to a mutation of one of the genes coding for subunits of complex I in the electron transport chain.

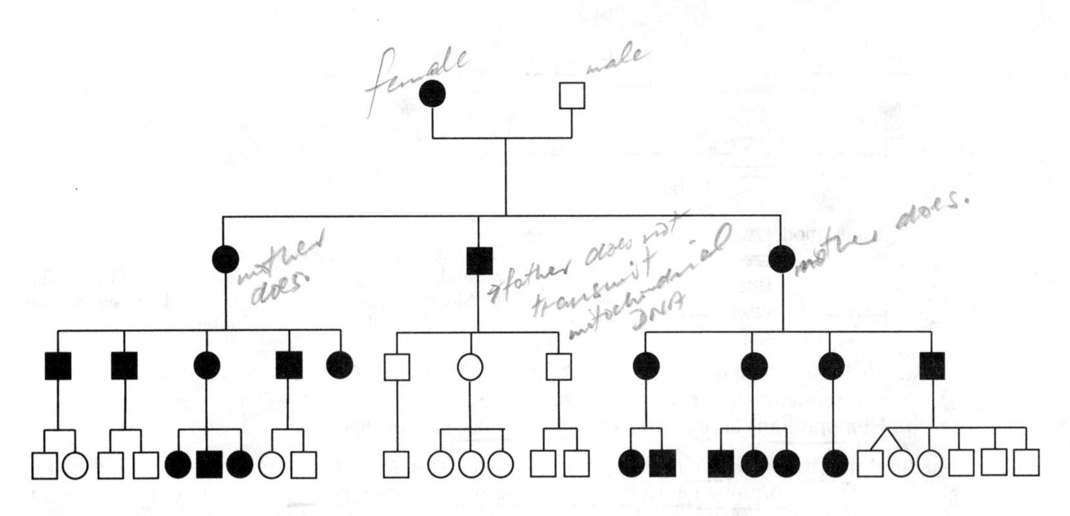

Figure 3-2. Pedigree showing typical maternal transmission of mitochondrial DNA.

II. TYPES AND MECHANISMS OF MUTATION. The development of DNA technology has allowed the delineation of mutations at the level of DNA sequence. In the human genome, there are many different types and mechanisms of mutations, including major gene rearrangements, such as deletions, insertions, duplications, and point mutations. These mutations affect transcription and translation, with altered expression or function of the protein product.

A. Major gene rearrangements

1. Deletions are easily detected by recombinant DNA technology (see Ch 5 II A) by the absence or altered size of a DNA fragment. Gene deletions are uncommon causes of mutation in the human genome except for a few diseases.

a. α-Thalassemia is caused by deletion of the α-globin gene cluster, which is four highly related sequences in close proximity to each other.

(1) Because of the sequence similarity of these genes, mispairing of chromosomes may occur during meiosis.

(2) After mispairing and crossing over, one product contains a deficiency of genetic material (Figure 3-3).

(3) Unequal crossing over in the region of the α- globin gene cluster has resulted in chromosomes that lack functional α-globin genes. These chromosomes are commonly present in Southeast Asian populations.

(4) Fetuses that lack all α-globin genes usually die of generalized edema either during the pregnancy or shortly after birth.

b. Growth hormone deficiency is another disorder which commonly results from deletions.

(1) Like the α-globin gene cluster, the growth hormone gene cluster also consists of similar DNA sequences in close proximity to each other.

(2) Mispairing and unequal crossing over in meiosis again causes deletions of the growth hormone gene, with the resultant effect of dwarfism in homozygotes.

c. Familial hypercholesterolemia (FH), which results from defects in the low-density lipoprotein (LDL) receptor gene (see Ch 5 II A 3 c), can also be caused by deletions due to unequal crossing over involving the highly repetitive *alu* sequences. Deletions occur in this situation after mispairing of homologous *alu* sequences with deletions of part of the gene for this protein.

2. Duplications of DNA sequences are common in evolution and may be caused by mispairing between homologous DNA sequences in close proximity, with duplication of genetic material that is contained within the gene (see Figure 3-3). The mechanism of duplication is similar to that seen in α-thalassemia. Duplications may alter the reading frame. FH and DMD are examples of genetic disorders caused by duplications.

3. Insertions are rare causes of mutation in the human genome. However, **transposition of DNA** is not an uncommon occurrence in the human genome although it usually does not involve the coding sequence. When transposition does occasionally disrupt a gene, defective gene expression may result.

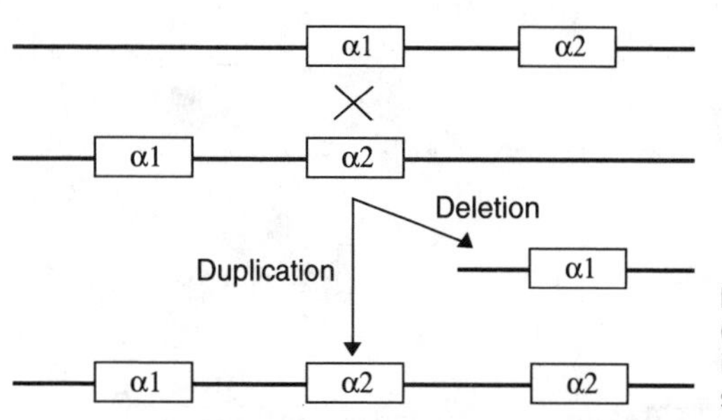

Figure 3-3. The alpha (α)-globin gene comprises 2 pairs (α1 and α2) of highly related sequences in close proximity. Mispairing during meiosis can result in one chromosome with a deletion of an α-globin gene.

a. The long interspersed repetitive sequence (LINE) [see Ch 1 II A 2 c (1) (b)] is an example of a sequence that can insert within genes.

b. **Hemophilia** is an example of a disease that has resulted from insertion of LINE.

4. **Point mutations.** Single nucleotide substitutions similar to those seen in bacteria and viruses are the most common cause of mutagenesis in the human genome. If codons are affected by a single nucleotide substitution, these point mutations are called **missense mutations**. Disorders which have clearly been shown to be due to different point mutations include β-thalassemia, cystic fibrosis, phenylketonuria (PKU), and Tay-Sachs disease.

   a. **β-Thalassemia** was the first disease to be clearly defined as a point mutation, which, in this case, affects transcription, translation, RNA cleavage, or RNA splicing and translation.

      (1) **Clinical aspects.** Children born with homozygous β-thalassemia are usually not anemic at birth due to the high levels of fetal hemoglobin. However, the disease usually manifests at 1 to 2 years of age with anemia, pallor, and failure to thrive. Hepatosplenomegaly results from extramedullary hematopoiesis. In addition, bone marrow proliferation results in frontal bossing. Sexual and physical development is retarded.

      (2) **Treatment.** Blood transfusion is the only feasible method of treatment.

   b. **Cystic fibrosis.** See Chapter 5 II B 3 for a full explanation of the clinical and genetic aspects of cystic fibrosis.

   c. **PKU**

      (1) **Genetic aspects.** PKU is inherited as an autosomal recessive trait caused by mutations in the enzyme phenylalanine hydroxylase, which converts phenylalanine to tyrosine.

         (a) In patients with PKU, phenylalanine cannot be degraded and thus accumulates, which causes damage to the CNS and mental retardation.

         (b) Mutations other than point mutations [e.g., translation mutations (see II A 6)] that cause PKU have been defined.

      (2) **Treatment.** In many instances, neonatal screening for PKU is performed so that appropriate treatment can begin early to prevent mental retardation. CNS damage and mental retardation can be avoided, to a large extent, by dietary changes to prevent phenylalanine accumulation.

      (3) **Clinical aspects.** See Chapter 7 I B 1 b for discussion of metabolic and clinical aspects of PKU.

   d. **Missense mutations** can result in the following:

      (1) **An altered codon for a particular amino acid residue** [e.g., a single nucleotide change at the sixth position of the β-globin chain of hemoglobin causes a glutamic acid to valine substitution, which results in sickle cell anemia; see Ch 5 II B 2 b (2)]

      (2) **Altered function of a protein** by either of the following mechanisms:

         (a) Directly involving crucial residues such as a catalytic site

         (b) Altering the three-dimensional structure of the protein such that its function is compromised

5. **Transcription mutations** may occur 5′ to the start codon in the DNA sequence that is critical for regulating transcription. For example, different mutations have been found in the TATA box, a region located about 30 nucleotides upstream from the initiation codon. In addition, there are residues at the 3′ end of the β-globin gene, which are also important in regulating transcription. Mutations in these distal promoter elements might cause decreased transcription with decreased production of the β-globin protein.

6. **Translation mutations.** Mutations affecting translation have frequently been seen in β-thalassemia and have also been described in other diseases such as lipoprotein lipase deficiency and PKU.

a. **Nonsense mutations** affect translation via nucleotide substitution, which produces a premature stop codon (see Ch 1 II A 3 g). These substitutions result in truncated proteins, which are frequently unstable and degraded.

b. **Frameshift mutations** are either deletions or insertions of one or a few nucleotides in the coding region of the gene. These mutations change the triplet code for all the codons that follow and thus completely alter the sequence of amino acids. In some cases, frameshift mutations lead to premature termination of the protein by creating a stop codon.

B. RNA mutations

1. **RNA cleavage** and stability mutations can result in inappropriate cleavage of the growing RNA, which causes cleavage to occur at sequences downstream. The result is RNA that is abnormally large, appears to be unstable, and is rapidly degraded.

2. **RNA splicing mutations** are critical to normal gene expression because introns must be precisely removed to produce appropriate messenger RNA (mRNA) for translation into protein. Mutations may affect splicing either by altering normal splice sites at intron and exon junctions or by creating new splice sites within introns or exons.
   - a. Mutations creating new sites are termed **cryptic splice sites**. In normal situations, RNA would not be spliced at one of these sites.
   - b. The 5′ exon/intron junction is called the **donor splice site,** and the 3′ intron/exon junction is the **acceptor splice site**.
   - c. The **5′ end** of the intron always begins with nucleotides **GT,** and the **3′ end** of the intron always has the nucleotides **AG**.
   - d. **Mutations that alter either the AG or GT splice junctions** lead to complete absence of normal splicing, which results in an unstable mRNA that is rapidly degraded.
     - (1) Persons with these mutations have **no functional protein product**.
     - (2) Many mutations that cause **β-thalassemia** affect RNA splicing by either of the following mechanisms.
       - (a) Mutations can **affect the normal donor or acceptor splice site** and prevent normal splicing of RNA, resulting in a molecule that is unstable and rapidly degraded.
       - (b) Mutations can **affect nucleotides** that are important for creating new splice sites, which results in an abnormal, unstable RNA product that is rapidly degraded.
   - e. **The end result of a mutation** is alteration of the function of a gene.
     - (1) Mutations in the coding region result in a protein with altered function, while mutations in regulatory sequences of a gene may result in either increased or decreased amounts of the protein.
     - (2) The net effect of these mutations is usually loss of normal function or, less frequently, gain of a new biological function with resultant disease.

C. Ethnic distribution of specific mutations

1. **Frequency in specific populations.** Several genetic diseases are at high frequency in certain populations. Examples include the following.
   - a. **Sickle cell anemia** is carried by 1/10 blacks.
   - b. **Tay-Sachs disease** is carried by 1/25 Ashkenazi Jews.
   - c. **FH** affects 1/100 French Canadians.

2. **Selective advantage** may explain the high frequency of a particular genetic trait within certain populations.
   - a. For example, the geographic distribution of sickle cell anemia closely parallels that of malaria, and it has now been shown that the **sickle-globin gene protects an individual against malaria**.
   - b. Therefore, persons living in Africa who have the sickle cell trait have a selective advantage, which results in a decreased frequency of malaria within that population.

3. **Founder effects** are another cause of a high frequency of specific mutations in certain populations.
   - a. The **Afrikaner population** of South Africa is descended from a few Dutch ancestors (founders) who arrived in South Africa in the middle of the seventeenth century.
   - b. One of these early settlers carried the gene mutation for **porphyria variegata**. As a result, this genetic disease is of the highest frequency in the world in this population.

c. Porphyria variegata is inherited as an autosomal dominant trait and predisposes individuals to photosensitivity and episodes of severe abdominal pain and shock following exposure to certain drugs, including anesthetics.

4. **Identical mutations occurring in more than one population** can be attributed to at least two possibilities.
   a. Mutations may have occurred prior to racial divergence and are observed in different racial groups.
   b. Mutations may have occurred independently in different population groups. Certain sequences, such as the CG dinucleotides, are particularly prone to single nucleotide substitutions. CG substitutions may produce both nonsense and missense mutations and represent mutation **hotspots** in humans.
5. **Chromosomal haplotypes.** One way to determine whether similar mutations have arisen independently or share a common origin is to determine the **pattern of RFLPs on the chromosome with the particular mutation**. This pattern has been termed a chromosomal haplotype.
   a. **Mutations that have arisen independently** often occur on different background chromosomal haplotypes.
   b. **Mutations that have a common origin** usually have the same background chromosomal haplotype.
   c. For example, **similar mutations causing β-thalassemia are present in both blacks and Chinese**. However, these same mutations occur on different β-globin gene haplotypes, which suggests independent origins of these alleles in these two population groups.

STUDY QUESTIONS

Directions: Each of the numbered items or incomplete statements in this section is followed by answers or by completions of the statement. Select the **one** lettered answer or completion that is **best** in each case.

1. X-linked ichthyosis is a skin disorder often caused by deletion of the gene for steroid sulfatase (STS), which is located between repeat sequences on the short arm of the X chromosome (Xp). The most likely reason for this deletion is the presence of which one of the following mutations?

(A) A missense mutation
(B) A nonsense mutation
(C) Recombination between repeat sequences
(D) A mutation affecting RNA splicing
(E) A high frequency of CG dinucleotides in the STS gene

2. X-linked recessive inheritance is associated with all of the following statements EXCEPT

(A) affected males are related through carrier females
(B) male-to-male transmission
(C) offspring of affected males are usually all clinically normal
(D) females are rarely affected
(E) offspring of heterozygous females are at risk to be either affected males or carrier females

3. When counseling a couple who are cousins about the risks of consanguinity, all of the following details are important EXCEPT

(A) the degree of relationship between the couple
(B) whether the couple is related through the maternal or paternal grandparents
(C) a family history of cystic fibrosis in the man's brother
(D) a family history of spina bifida in several of the cousins
(E) the ethnic origin of the couple

4. For the pedigree shown below, the most likely modes of inheritance include all of the following EXCEPT

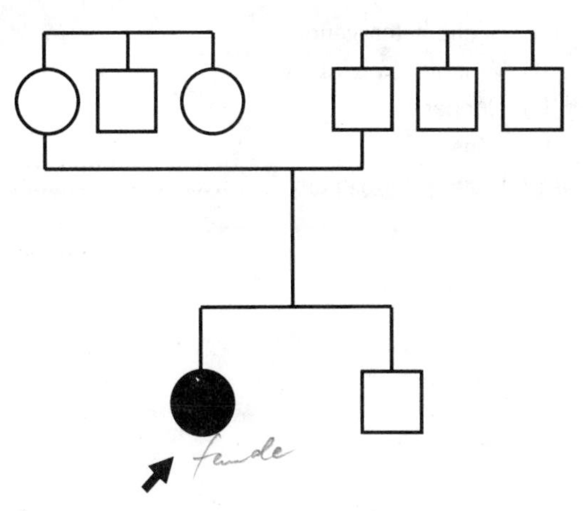

(A) autosomal recessive
(B) chromosomal
(C) autosomal dominant
(D) X-linked recessive

5. A child is diagnosed to have a type of dwarfism, and the parents are reassured that since they are normal the recurrence risk for another pregnancy is not increased. Everyone is surprised when a second child is born with features similar to the first child. All of the following explanations are valid in this instance EXCEPT

(A) the affected children have an autosomal recessive condition
(B) the affected children have a dominant condition, but there is nonpaternity
(C) the first child has a new dominant mutation
(D) the condition is autosomal dominant but has extreme variability of expression
(E) both children are male, and the condition is X-linked with the mother being a heterozygote

---

1-C
2-B
3-B
4-D
5-C

Directions: The group of items in this section consists of lettered options followed by a set of numbered items. For each item, select the **one** lettered option that is most closely associated with it. Each lettered option may be selected once, more than once, or not at all.

Questions 6–10

Match each condition described below with the term of inheritance that best defines it.

(A) Allelic heterogeneity
(B) Variable expressivity
(C) Nonpenetrance
(D) Consanguinity
(E) Locus heterogeneity

6. A man shows no detectable signs of Marfan syndrome (an autosomal dominant disorder of connective tissue), although his father and two daughters are affected

7. Retinitis pigmentosa (a type of retinal degeneration) occurs in an autosomal dominant, an autosomal recessive, and an X-linked form

8. Tay-Sachs disease is seen in high frequency in the Chicoutimi area of Quebec, where most individuals are descended from a few early French settlers

9. Duchenne muscular dystrophy (DMD) is the consequence of either a deletion involving the dystrophin gene or a point mutation of the same gene

10. In a family with myotonic dystrophy, the father has frontal balding, severe weakness, and cardiac arrhythmia; his sister has early-onset cataracts; and his child has electromyographic abnormalities

---

6-C
7-E
8-D
9-A
10-B

ANSWERS AND EXPLANATIONS

1. The answer is C *[II A 1].*
The gene for steroid sulfatase (STS) is located between repeat sequences, and mutations often arise as a result of mispairing of homologous repeat sequences with deletions of part of this protein. The presence of these repeat sequences on either sides of the gene would not, in any way, favor a missense or nonsense mutation or affect RNA splicing. There is no particular high frequency of CG dinucleotides in the STS gene.

2. The answer is B *[I D 1].*
An affected father can transmit the mutant X chromosome only to females, since the males will receive the Y chromosome as the sex chromosome. Therefore, there can be no male-to-male transmission. Males may be affected, or hemizygous, as the result of either a new mutation, or they may have inherited the mutant allele from the clinically normal carrier mother. Females are rarely affected, and when they are, it is usually the result of other mechanisms.

3. The answer is B *[I C 2 c, 3 c].*
How the couple is related (e.g., whether their fathers were brothers or other relationships of the parents) will usually make no difference in the counseling process. The closer a couple is related, the higher the risk for both autosomal recessive and multifactorially inherited disorders. If the couple are third or fourth cousins, the risk would not be increased above that for the general population; the risks would be higher for first cousins or closer relationships. Therefore, the family history of both cystic fibrosis (recessive) and spina bifida (multifactorial) would be of concern for calculating risk. The relationship of the affected people in each situation would have to be ascertained. The ethnic origins of the couple may prompt screening for recessive disorders that are more prevalent in a particular group (e.g., Tay-Sachs disease if the couple is Jewish).

4. The answer is D *[I B 1, I C 1, I D 1].*
X-linked recessive inheritance is unlikely with an affected female. The pedigree shows a single affected female, which could be the result of autosomal recessive inheritance where both carriers are normal heterozygotes. Autosomal dominant inheritance is possible if the child has a new mutation. A chromosome abnormality, either familial or de novo, could be manifested in the single affected child.

5. The answer is C *[I B 2 a].*
If the first child is affected by a new dominant mutation (advanced paternal age may be a clue), the second child would be expected to be normal. There are many types of dwarfism, and each of the inheritance patterns may be seen. Making the precise diagnosis may require careful examination of the child, radiological studies, and literature review. If the condition is autosomal recessive, then it would be expected that the parents are clinically normal and that there would be a 25% recurrence risk. If the condition is autosomal dominant, several mechanisms could explain the two affected siblings. Variable expression may mean that an affected parent is only mildly short or may only have unusual proportions. Gonadal mosaicism may be another mechanism for recurrence of dominant conditions and has been described for type II osteogenesis imperfecta.

6–10. The answers are: 6-C *[I B 2 b],* **7-E** *[I B 2 f, C 2 e, D 2 b],* **8-D** *[I C 2 c],* **9-A** *[I B 2 f, C 2 e, D 2 b],* **10-B** *[I B 2 c].*
A disorder is said to be nonpenetrant when there is no clinical evidence of a mutant allele in an individual known to have inherited the gene. When the gene is expressed, the form of expression may be highly variable, with some family members being severely affected and others having few signs of the disorder. This is common with autosomal disorders and is called variable expression. Different mutations of either the same gene (i.e., at the same locus) or different genes may give a similar clinical picture. Consanguineous individuals have a proportion of their genes in common by inheritance from a common ancestor. In some villages originated by a few settlers, disease alleles may be in higher frequencies, and a particular recessive disorder may be more common in that community.

4
Markers and Mapping

J. M. Friedman

I. POLYMORPHISMS AND GENETIC MARKERS

A. Definitions

1. **Genetic polymorphism** is the occurrence in a population of two or more alleles at a locus in frequencies greater than can be maintained by mutation. **Polymorphisms** are genetic differences which provide variation within a species.
   a. In practice, it is difficult to know what frequency of an allele can be maintained by mutation so an operational definition of polymorphism is often used: **A polymorphism is said to exist if the most common allele at a locus has a frequency of less than 99%** (see Ch 6 I B 6 b for an explanation of gene frequency).
   b. Polymorphisms are common, particularly in **noncoding regions of DNA**.

2. A **genetic marker** is a simply inherited genetic trait with readily recognizable alleles. Genetic markers are useful in family and population studies. The marker may or may not be part of an expressed gene.

B. Polymorphisms.
All polymorphisms ultimately reflect alterations of DNA sequence, but the presence of such alterations may be demonstrated in various ways—via DNA technology; altered proteins, enzymes, or antigens; or abnormal physical features.

1. **Detecting polymorphisms.** Polymorphisms may be defined by their method of detection.
   a. **DNA polymorphisms** are identified by direct detection of altered DNA sequence.
      (1) **RFLPs (restriction fragment length polymorphisms)** are inherited variations in DNA sequence that result in the gain or loss of a site recognized by a restriction endonuclease (e.g., *EcoRI* or *TaqI*) or in alteration of the number of nucleotides between such sites (Figure 4-1).
         (a) RFLPs are transmitted as simple codominant traits.
         (b) RFLPs are usually detected by Southern blotting (see Ch 5 I A).
         (c) RFLPs are widely used in **gene linkage and mapping studies** (see II and III, respectively).
      (2) **VNTRs (variable number of tandem repeats)** are a special kind of RFLP that involve variations in the number of short, repeated DNA segments between restriction sites.
         (a) VNTRs tend to be **extremely polymorphic**. So many different alleles exist at some VNTR loci that it is almost impossible for two unrelated people to exhibit the same genotype.
         (b) VNTRs are especially **valuable in forensic medicine**. **DNA fingerprinting,** which usually involves testing for VNTRs, has been used to establish paternity, determine zygosity, and identify a specific individual on the basis of a small sample of blood, semen, hair, or other tissue.
      (3) **PCR (polymerase chain reaction)** can be used to demonstrate polymorphic alterations of DNA sequence. See Chapter 5 I B for a detailed description of PCR.
      (4) **Direct DNA sequencing** could also be used, theoretically, to demonstrate DNA polymorphisms. However, this approach is slow and not generally practical considering the technology that is currently available.
   b. **Many polymorphisms** are detected by identifying **altered gene products**.
      (1) **Inherited enzyme variants** have altered protein structure.

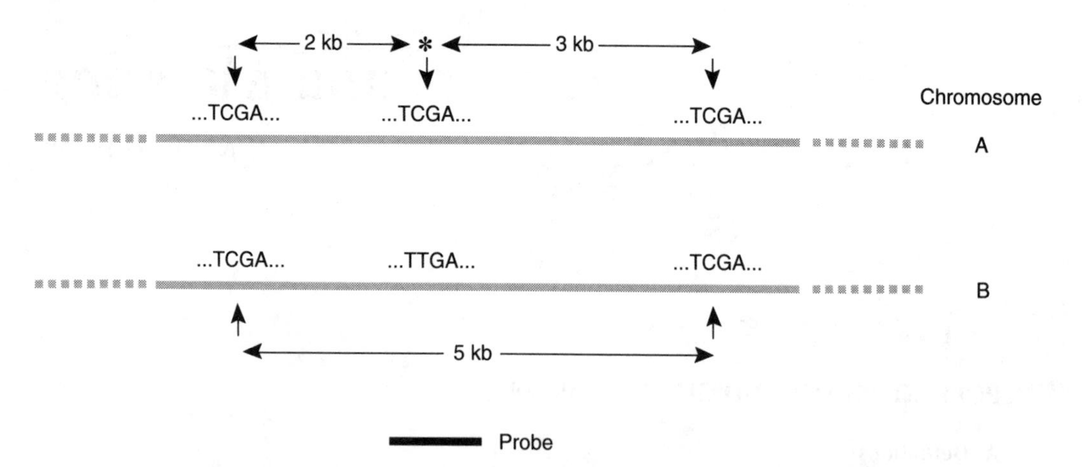

Figure 4-1. In this figure, action of the restriction enzyme *TaqI* is denoted by the asterisk. *TaqI* recognizes TCGA as a cutting site, so treatment of chromosome A produces two fragments—one measuring 2 kilobases (kb) and one measuring 3 kb. Treatment of chromosome B produces a single 5-kb fragment since *TaqI* does not recognize the sequence TTGA as a restriction site. The probe enables the corresponding piece of the chromosome to be recognized. (Reprinted with permission from Gelehrter TD, Collins FS: *Principles of Medical Genetics.* Baltimore, Williams & Wilkins, 1990, p 78.)

- **(a)** This may be demonstrated by alteration of enzyme activity, electrophoretic mobility, thermostability, or other physical properties.
- **(b)** Examples include the electrophoretic and activity variants of glucose-6-phosphate dehydrogenase (G6PD), which are common in people of African or Southern European origin.

(2) Antigenic variants may also reflect alterations of protein structure.

- **(a)** Antigenic variants are usually demonstrated by serologic tests such as red cell agglutination or antibody-mediated cytotoxicity.
- **(b)** The ABO and Rh blood groups and the human leukocyte antigen (HLA) system (see Ch 9 II D 2, 3, E 1) are important examples of polymorphic antigenic variants.

c. Some polymorphisms are observed as **alterations of physical features**. An example is the dominantly inherited postaxial polydactyly that is common in black Americans.

d. Chromosome heteromorphisms are heritable differences in chromosomal appearance. These polymorphisms may require special cytogenetic techniques for their demonstration (see Ch 1 IV C 1).

(1) The most common kinds of chromosome heteromorphisms include:

- **(a)** Variations in the size of the Y chromosome long arm
- **(b)** Variations in the size of the centromeric heterochromatin
- **(c)** Variations in satellite size and structure
- **(d)** The occurrence of fragile sites

(2) Chromosome heteromorphisms are useful in determining the origin of **chromosomal rearrangements**.

2. Clinical importance of polymorphisms. Most polymorphisms produce no clinical phenotype. Regardless of whether or not a clinical effect is produced, polymorphisms are useful as **genetic markers**.

a. Some disease genes occur with polymorphic frequencies. This is usually true when the carrier is heterozygous for relatively common autosomal recessive diseases.

(1) Examples include the heterozygous states of sickle cell disease among people of African origin, thalassemia among people of Mediterranean or Southeast Asian origin, and cystic fibrosis among people of Northern European origin.

(2) In some instances, mild disease manifestations occur in the heterozygous carriers, while severe disease is seen among homozygotes for the same gene. This is the case with familial hypercholesterolemia and β-thalassemia.

b. Genetic polymorphisms may produce disease only if exposure to certain chemicals occurs. Examples of such **pharmacogenetic predispositions** include G6PD deficiency and malignant hyperthermia (see Ch 7 II B 1).

c. Polymorphisms that determine antigenic differences are important in **blood transfusion** and **tissue typing** (see Ch 9 II D, E).

d. As mentioned, genetic polymorphisms are used in **forensic medicine.**

e. Polymorphic genetic markers can be used to determine the likelihood of associated genes that predispose to disease within families or populations.

II. LINKAGE is the occurrence of two or more genetic loci in such close physical proximity on a chromosome that they are more likely to assort (or segregate) together than independently during meiosis (Figure 4-2). Linkage occurs because crossing over does not usually take place between loci that are close to each other.

A. Concepts of genetic linkage

1. Linkage refers to loci, not to alleles.

a. For example, consider a marker locus *M* that has two allelic forms, M^a and M^b. If locus *M* is linked to locus *D* for a dominantly inherited disease, some families may carry the mutant *D* allele on the same chromosome as M^a, and others may carry the mutant *D* allele on the same chromosome as M^b. In either case, the allele at the marker locus and the *D* allele at the disease locus will be transmitted together at meiosis.

b. The process of determining whether an allele at one locus is carried on the same chromosome as a given allele at another locus is called **determining the phase of the linkage**.

(1) Thus, if a person is known to be heterozygous M^a/M^b at a marker locus and heterozygous *D/d* at a disease locus, two possibilities exist: The chromosome that carries the mutant allele at the disease locus might carry either M^a or M^b at the marker locus.

(2) Family studies are necessary to determine the phase of the linkage (i.e., to establish which of these two possibilities in fact exists).

2. The specific alleles present at linked loci in a given individual or family are irrelevant as long as they are informative for testing.

B. Measurement of genetic linkage can only take place in family studies. Meiosis, which is required to demonstrate linkage, only occurs during gametogenesis. Therefore, only family members can be studied to determine if linkage of genes has occurred.

1. The closeness of a genetic linkage is expressed in **centiMorgans** (abbreviated **cM,** not cm) or **percent recombination**.

a. Loci that are separated by crossing over in 1% of gametes are 1 cM apart.

b. Loci that are so close that they are almost never separated by crossing over are linked at a genetic distance of 0 cM.

c. Unlinked loci are separated by a genetic distance of 50 cM because a given allele at one locus has a 50% chance of being transmitted with either allele at an unlinked locus.

2. The usual statistical method of measuring linkage in family data is calculation of the **lod score**.

a. Lod is an acronym for "logarithm of the odds." The lod score is a logarithm of the **likelihood ratio**.

b. A lod score of **+3 or greater** at a recombination distance of less than 50 cM between two loci is considered to be strong evidence of linkage (1000:1 odds for linkage).

c. A lod score of **−2 or less** is taken as strong evidence that there is no linkage (100:1 odds against linkage).

C. Two related concepts distinguished from linkage

1. Linkage disequilibrium is the tendency for **certain alleles** at two linked loci to occur together more often than expected by chance.

a. For example, if marker locus *M* with alleles M^a and M^b is linked to disease locus *D* at a genetic distance of 5 cM and if it is found that the mutant allele at *D* occurs on the same chromosome as M^b more often than expected within a certain population, linkage disequilibrium is said to exist.

b. Linkage disequilibrium is **measured in a population, not in a family,** and usually varies in different populations.

2. Synteny is the occurrence of two loci together on the same chromosome. Syntenic loci may not be linked if they are far enough apart so that crossing over usually occurs between them.

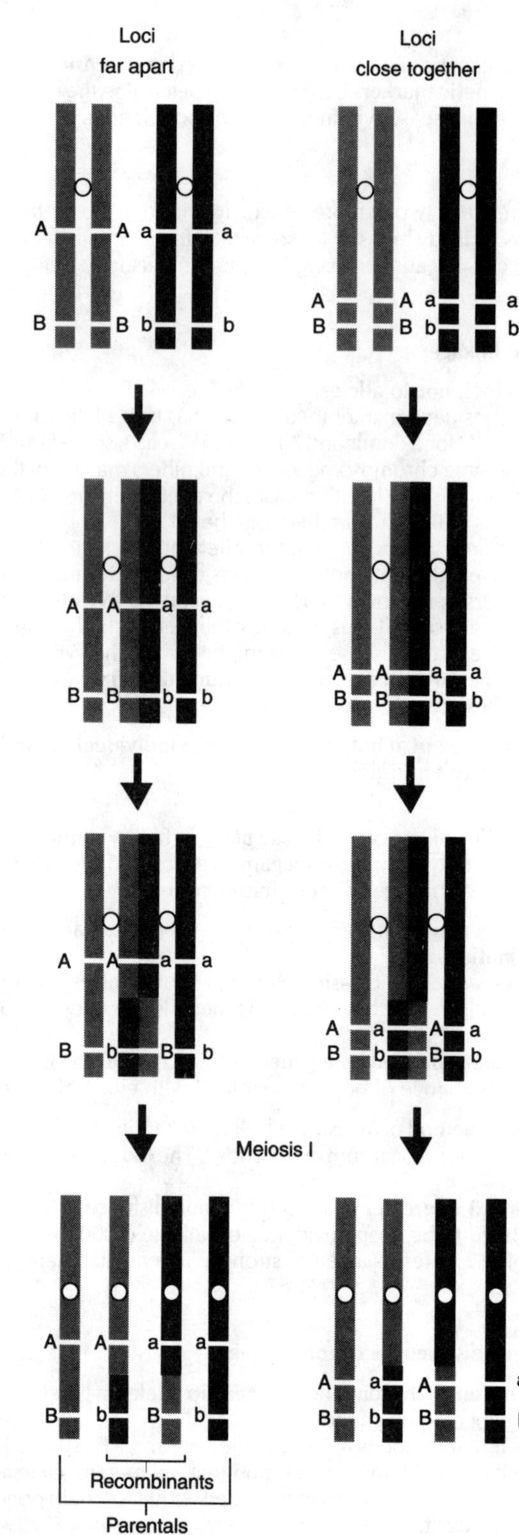

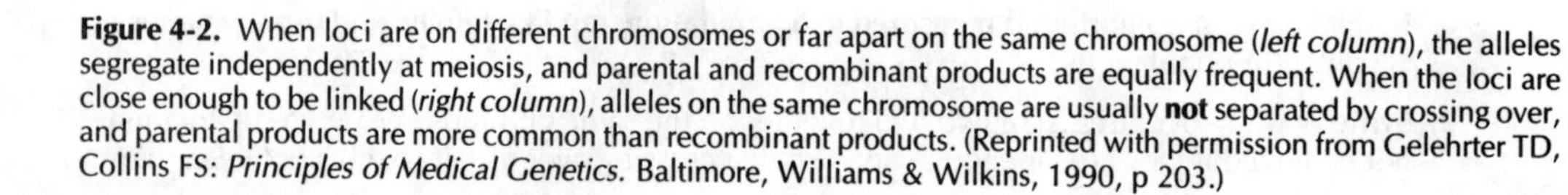
Figure 4-2. When loci are on different chromosomes or far apart on the same chromosome (*left column*), the alleles segregate independently at meiosis, and parental and recombinant products are equally frequent. When the loci are close enough to be linked (*right column*), alleles on the same chromosome are usually **not** separated by crossing over, and parental products are more common than recombinant products. (Reprinted with permission from Gelehrter TD, Collins FS: *Principles of Medical Genetics.* Baltimore, Williams & Wilkins, 1990, p 203.)

D. Clinical application of linkage

1. Linkage is clinically useful because it may permit:
   a. **More precise determination of the genotype** in an individual at an unidentified disease gene locus on the basis of readily identified linked markers (Figure 4-3).
   b. **The determination of the pattern of inheritance** or specific form of a disease that exhibits genetic heterogeneity
   c. **Gene mapping** by determining the recombination distance between two genes on a chromosome

2. **Circumstances in which linkage has been used clinically** (see Ch 5 III B 2 for further discussion and specific examples) include:
   a. **Prenatal diagnosis** of genetic diseases such as phenylketonuria (PKU)
   b. **Carrier detection** in autosomal or X-linked recessive diseases such as Duchenne muscular dystrophy
   c. **Presymptomatic diagnosis** of late-onset autosomal dominant diseases such as Huntington's disease
   d. **Elucidation of genetic factors in genetically heterogeneous conditions or multifactorial diseases** such as insulin-dependent diabetes mellitus.

3. The clinical use of linkage **always requires obtaining information on other family members** besides the one whose genotype is being determined.

4. The use of a linked marker to provide information about a disease gene **always involves some uncertainty** because recombination may occur between the loci for the marker and the disease.

III. GENE MAPPING

GENE MAPPING is the assignment of genes to specific chromosomal locations. About 2000 genetic loci have been mapped to specific human chromosomes. More than 20 additional loci are mapped each month. Despite this rapid progress, **more than 95% of the human genome remains unmapped**.

A. **Several methods of gene mapping** are available. The results of one method often complement those of another.

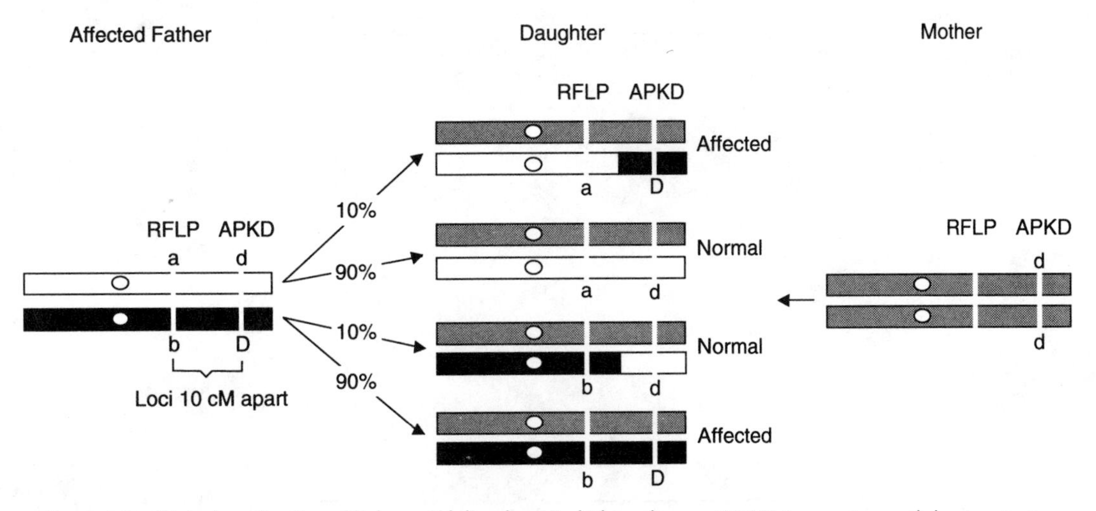

Figure 4-3. Clinical application of linkage. Adult polycystic kidney disease (APKD) is an autosomal dominant condition. The daughter has a 50% chance of inheriting the mutant *D* allele at the APKD locus from her affected father who is heterozygous *D/d*. If the father is also known to be heterozygous *a/b* at an RFLP located 10 cM from the APKD locus and if the phase of the linkage is such that the *b* allele of the RFLP locus and the mutant *D* allele at the APKD locus reside on the same chromosome, then there is a 90% chance that the daughter will inherit the mutant *D* allele at the APKD locus if she inherits her father's *b* RFLP allele. This 90% chance is calculated by subtracting the chance of recombination (10 cM = 10%) from 100%. On the other hand, there is a 90% chance that the daughter will inherit the normal *d* allele at the APKD locus if she inherits her father's *a* RFLP allele. A child who receives the father's *a* RFLP allele can only inherit the mutant *D* allele at the APKD locus if crossing over occurred; however, the probability in this case that crossing over occurred is only 10%.

1. **Family studies** are used to demonstrate **linkage** between loci. If linkage between the loci can be shown to exist, both must reside on the same chromosome. The frequency of recombination between linked markers is a measure of genetic distance and can be used to place them in linear order.
2. **Somatic cell genetic methods** are used to **demonstrate synteny** or show that an unmapped locus resides on a given chromosome. In general, somatic cell genetic methods involve the demonstration of loss of a given gene from cells that lack a specific chromosome segment and retention of the gene in cells that retain that chromosome segment.
3. **Cytogenetic techniques** such as **in situ hybridization** [see Ch 1 IV C 1 b (9)] can be used to visualize the presence of a gene directly on a specific chromosome.
4. **Gene dosage studies** are an indirect means of identifying the location of genes.
   a. These studies are based on the concept that the amount of DNA or protein product specific for a given gene is directly proportional to the number of copies of that gene present in an individual or cell.
   b. Thus, if a cell with trisomy for chromosome 21 exhibits one and one-half times the amount of a protein as a normal cell and three times as much of the protein as a cell with monosomy 21, the implication is that the gene for that protein is on chromosome 21.

B. Importance of gene mapping

1. **The gene map is the anatomy of the human genome.** Just as an understanding of anatomy is necessary to understand the function of the human body, an understanding of the gene map is a necessary prerequisite to understanding the function of the human genome.
2. **Analysis of the heterogeneity and segregation of human genetic diseases** is substantially enhanced by knowledge of the gene map.
3. Improved knowledge of genomic organization is necessary to **develop optimal strategies for gene therapy.**
4. Gene mapping **provides information regarding linkage** that is clinically useful (see II D 2).

STUDY QUESTIONS

Directions: Each of the numbered items or incomplete statements in this section is followed by answers or by completions of the statement. Select the **one** lettered answer or completion that is **best** in each case.

1. Which one of the following statements regarding genetic polymorphisms is correct?

(A) A polymorphism that occurs in noncoding regions of DNA is a genetic marker
(B) Polymorphisms resulting from alteration of the protein structure of a restriction endonuclease are restriction fragment length polymorphisms (RFLPs)
(C) Chromosome heteromorphisms are usually detected by DNA fingerprinting
(D) Genetic polymorphisms are variations in DNA sequence that commonly occur

2. Both members of a couple are carriers of β-thalassemia, an autosomal recessive disease caused by a defect in the β-globin gene. The couple has only one child who is neither affected nor a carrier of β-thalassemia. DNA analysis using a β-globin probe gives the following phenotypes for a very closely linked polymorphic restriction site [*1* and *2* are alternate alleles for this restriction fragment length polymorphism (RFLP)]:

Mother: *1,2*
Father: *1,2*
Child: *1,1*

All of the following statements regarding this family are true EXCEPT

(A) the mother's mutant β-globin allele is on the same chromosome as her *2* allele at the marker locus
(B) the father's mutant β-globin allele is on the same chromosome as his *2* allele at the marker locus
(C) testing of this RFLP by amniocentesis will permit identification of a fetus with homozygous β-thalassemia with high likelihood
(D) the mother's abnormal β-thalassemia gene must be the result of a new mutation if her mother is *1,2* at the marker locus and is neither affected with nor a carrier for β-thalassemia

3. Which one of the following statements regarding gene mapping is true?

(A) Most genes in the human genome have now been mapped
(B) Family studies are necessary for gene mapping
(C) Gene mapping is usually performed by direct DNA sequencing
(D) Gene mapping can be used to show genetic heterogeneity within a disease

4. Which one of the following statements provides the best reason that genetic polymorphisms are clinically important?

(A) They can be used as genetic markers in linkage studies
(B) They often cause chromosomal rearrangements
(C) They are usually in linkage disequilibrium with disease genes
(D) They are valuable vectors in gene therapy
(E) They cause a variable number of tandem repeats (VNTRs) that are important in tissue transplantation

---

1-D 3-D
2-D 4-A

Directions: The group of items in this section consists of lettered options followed by a set of numbered items. For each item, select the **one** lettered option that is most closely associated with it. Each lettered option may be selected once, more than once, or not at all.

Questions 5–10

For each of the descriptions listed below, select the most appropriate term.

(A) Gene mapping
(B) Linkage
(C) Linkage disequilibrium
(D) Genetic polymorphism
(E) Synteny

5. Usually identified by a lod score of +3 or greater at a recombination distance of less than 50%

6. Restriction fragment length polymorphisms (RFLPs) are a commonly used example

7. Can be performed by somatic cell genetic and cytogenetic methods

8. The occurrence of two or more alleles at a locus in frequencies greater than can be maintained by mutation

9. The tendency in a population for specific alleles at two loci to occur together more often than is expected by chance

10. The occurrence of two loci on the same chromosome regardless of how far apart they may be

---

5-B 8-D
6-D 9-C
7-A 10-E

ANSWERS AND EXPLANATIONS

1. The answer is D *[I A 1, 2, B 1 a (1)].*
Genetic polymorphisms are commonly occurring variations in DNA sequence. Genetic markers are readily identifiable, simply inherited polymorphisms found throughout the genome. Variations of DNA sequence that result in the gain or loss of a restriction endonuclease site or in alteration of the number of nucleotides between restriction sites are called restriction fragment length polymorphisms (RFLPs). Chromosome heteromorphisms are recognized under the microscope in cytogenetic studies.

2. The answer is D *[II A 1, D].*
The marker restriction fragment length polymorphism (RFLP) is closely linked to the β-globin locus, but one must determine the phase of the linkage in both parents to use the information provided in prenatal diagnosis. The child is homozygous normal at the β-globin locus, and she is homozygous for allele *1* at the marker locus. This means that in both of her parents the *1* allele at the marker locus is likely to be on the same chromosome as the normal allele at the β-globin locus and the *2* allele is likely to be on the chromosome with the mutant β-globin allele.

The information is sufficient for prenatal diagnosis: A fetus found to be *2,2* at the marker locus is likely to have homozygous β-thalassemia; a fetus found to be *1,2* at the marker locus is likely to be a heterozygous β-thalassemia carrier; and a fetus found to be *1,1* at the marker locus is likely to be homozygous normal at the β-globin locus.

The mother's father could have transmitted the chromosome with the marker *2* RFLP and the mutant β-globin allele if the mother's mother was a *1,2* heterozygote for the marker RFLP and homozygous normal at the β-globin locus. Linkage does not imply anything about which allele at one locus is associated with a particular allele at a linked locus. In some individuals, the mutant β-globin allele may be carried on the same chromosome as the marker *1* RFLP or the normal β-globin allele on the same chromosome as the marker *2* RFLP. Only after the phase of linkage is determined in an individual family can linkage be used to predict the allele present at one locus on the basis of an allele present at a linked locus.

3. The answer is D *[III A, B].*
If a genetic disease is shown to map to different loci in different families, genetic heterogeneity is established. Only a small proportion of genes in the human genome have currently been mapped. Family studies are an important method of gene mapping, but a variety of other techniques are also used. DNA sequencing currently has very limited utility in gene mapping because the distances between most loci are too great to define efficiently by sequencing.

4. The answer is A *[I B; II].*
Polymorphic genetic markers are widely used in linkage studies. Chromosome heteromorphisms can be used to identify the origin of chromosomal rearrangements but are generally not involved in their causation. An association of specific alleles at linked loci (i.e., linkage disequilibrium) occasionally occurs with disease genes but is of clinical importance in only a few cases. Polymorphisms are variants, not vectors, and their use in gene therapy is only to mark a gene that is being manipulated. Variable number of tandem repeats (VNTRs) are a highly polymorphic kind of restriction fragment length polymorphism (RFLP); they are not involved in the antigenic variation that is critical to tissue transplantation.

5–10. The answers are: 5-B *[II B 2],* **6-D** *[I B 1 a (1)],* **7-A** *[III A],* **8-D** *[I A 1],* **9-C** *[II],* **10-E** *[II C 2].*
Lod scores are the usual statistical method of measuring linkage. Loci that are separated by recombination less than 50% of the time are said to be linked. This recombination fraction corresponds to a genetic distance of 50 cM.

Common examples of genetic polymorphisms include restriction fragment length polymorphisms (RFLPs), variable number of tandem repeats (VNTRs), chromosome heteromorphisms, inherited enzyme variants, and antigenic variants of proteins.

Several methods are available for the assignment of genetic loci to specific chromosomes (i.e., for gene mapping). Family studies to demonstrate linkage are widely used, but somatic cell genetic methods are often easier to perform when there is no knowledge of the location of the gene. Cytogenetic methods of gene mapping are particularly useful for regional localization within a chromosome.

Genetic polymorphism is defined as the occurrence of two or more alleles at a locus in a frequency greater than can be maintained by mutation alone. In practice, polymorphism is often said to exist when the most common allele at a locus accounts for less than 99% of all alleles. Many genes exhibit polymorphism of their coding regions, but polymorphic DNA variation in noncoding regions is even more common.

Linkage describes the close physical proximity of two or more loci on a chromosome, but the specific alleles present at each locus are irrelevant to the identification of linkage. If a certain allele at one locus tends to be found more often than expected by chance with a certain allele at another locus linked to the first, linkage disequilibrium is said to be present. Linkage disequilibrium is specific for a given population, and the same alleles may or may not be similarly associated in a different population.

Synteny occurs when two or more loci are on the same chromosome. Syntenic loci may or may not be linked, but linked loci are always syntenic. Syntenic loci may be far enough apart on a chromosome for crossing over to occur between them. By definition, linked loci are so close that recombination usually does not occur between them.

5
DNA Diagnosis

Michael R. Hayden

I. TECHNIQUES FOR DNA ANALYSIS. In the past, the diagnosis of genetic disorders was based on recognition of a secondary product or cellular effect of a mutant gene. However, recent progress in technology and an enhanced understanding of the structure and function of human genes have provided the opportunity to investigate and diagnose genetic disease at the level of the abnormal gene itself. Several methods of recombinant DNA technology that have important application to clinical genetics are discussed in this section.

A. Southern blot

1. **Method.** This technique combines gel electrophoresis with the use of specific probes (Figure 5-1).
   - **a.** Usually DNA is obtained from white blood cells. Approximately 5–10 μg of DNA is treated with a restriction enzyme that makes sequence-specific cuts. The resulting DNA fragments are separated by gel electrophoresis.
   - **b.** The DNA is denatured to separate the strands and is transferred to a membrane, usually by placing the membrane between the nitrocellulose gel and a stack of paper towels. This allows the DNA to move out of the gel and onto the membrane.
   - **c.** A labeled DNA or RNA probe is added, which recognizes and hybridizes with the complementary DNA sequence, thus allowing the specific fragment of interest to be detected.
2. **Applications.** The Southern blot is a very common approach to DNA diagnosis; it takes advantage of the fact that sequences within a DNA probe recognize similar sequences in digested human DNA (the principle of molecular hybridization). Most genetic diseases for which DNA diagnosis is possible can be detected using Southern blot analysis. Examples of the use of this technology for specific diseases are given later in this chapter (see II A 2).
3. **Limitations.** Significant amounts of DNA are needed for the Southern blot. The technique is also labor intensive and may take 7–14 days to achieve a result. Exposure to radioactive material used in some probes is also considered a limitation.

B. Polymerase chain reaction (PCR)

1. **Method.** PCR is a powerful technique for selective and rapid **amplification** of target DNA or RNA sequences.
   - **a.** The technique is based on the enzymatic amplification of a DNA fragment that is flanked by two stretches of nucleotide primers, which hybridize to opposite strands of the sequence being investigated (Figure 5-2).
   - **b.** The technique works by using heat to separate the two strands of DNA, which allows primers to bind to their complementary sequence. The segment between the primers is amplified in the presence of *Taq* polymerase and deoxynucleotides.
   - **c.** Further increases in temperature stops the reaction, which begins again when the temperature is lowered.
   - **d.** The amount of amplified DNA increases rapidly each time the cycle occurs.
2. **Applications.** PCR is becoming extremely important and is revolutionizing approaches to both genetic analysis and detection of disease pathogens.
3. **Advantages and limitations**
   - **a. Advantages.** Only a very small amount of DNA (1 μg or less, or only a single cell) is needed for PCR. Results are achieved often in less than 1 day.

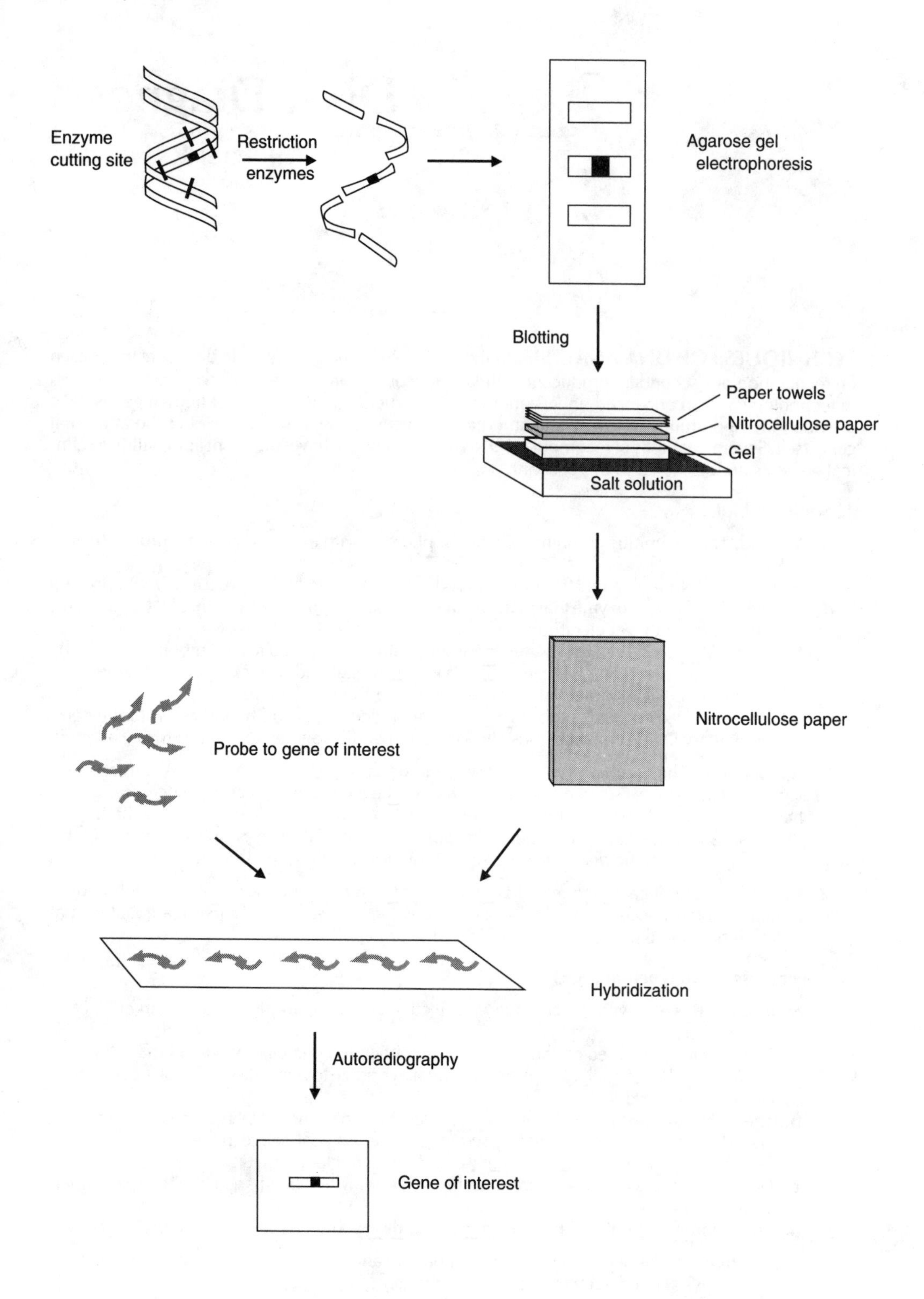

Figure 5-1. Southern blot analysis.

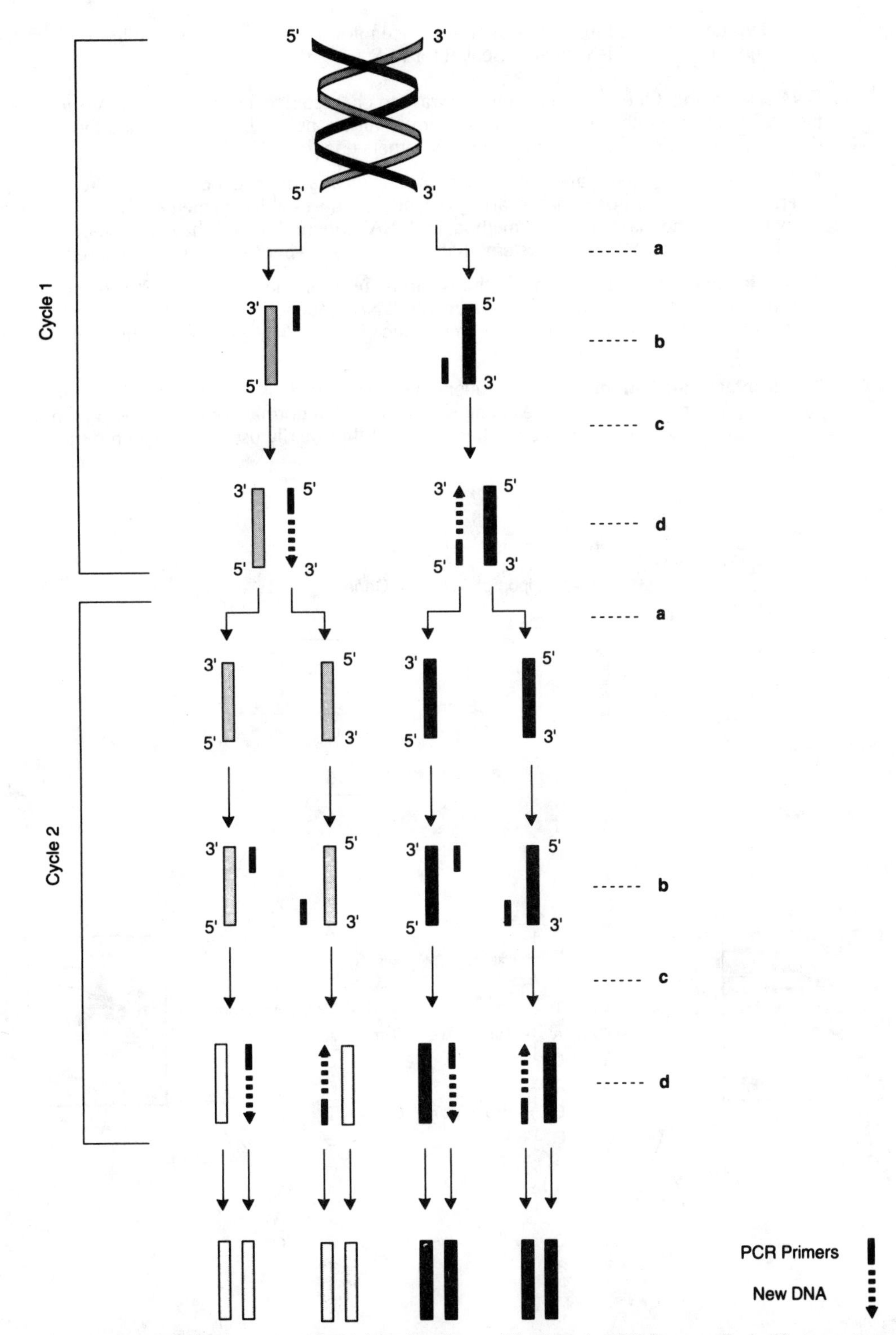

Figure 5-2. Polymerase chain reaction (PCR) is based on the amplification of a DNA fragment flanked by two primers that are complementary to opposite strands of the sequence being investigated. In each cycle, (*a*) heat denaturation separates the strands of the target DNA. (*b*) Primers are added in excess and hybridized to complementary fragments. (*c*) Deoxynucleotides and *Taq* polymerase are added while the temperature increases. (*d*) The primer is extended in the 3′ direction as new DNA is extended in the 5′ direction.

b. Limitations. One limitation is that sequence information must be available to synthesize the oligonucleotide primers essential for PCR.

C. DNA sequencing. One of the most precise ways to characterize a segment of DNA is to obtain the precise DNA sequence. In this way, abnormal variations can be identified and then used to detect such mutations in family members and others (Figure 5-3).

1. **Methods.** Two methods are used for determining DNA sequence, both of which rely on separating fragments whose lengths vary by one nucleotide. Established methods (e.g., the Sanger method and the Maxam-Gilbert method) for DNA sequence analysis have been modified and incorporated into automated systems that allow sequence information to be gathered rapidly.

2. **Applications.** DNA sequencing methods can be used on the products of PCR reactions, resulting in rapid sequencing of these fragments. DNA sequencing is not used routinely for DNA diagnosis but is likely to be of major importance in the future when the technology has improved and the cost has decreased.

3. **Advantages and limitations.** DNA sequencing allows direct assessment as to whether the gene of the patient has any sequence changes compared to a normal control. At the present time, the technology is still limited and, therefore, is still primarily used for research purposes.

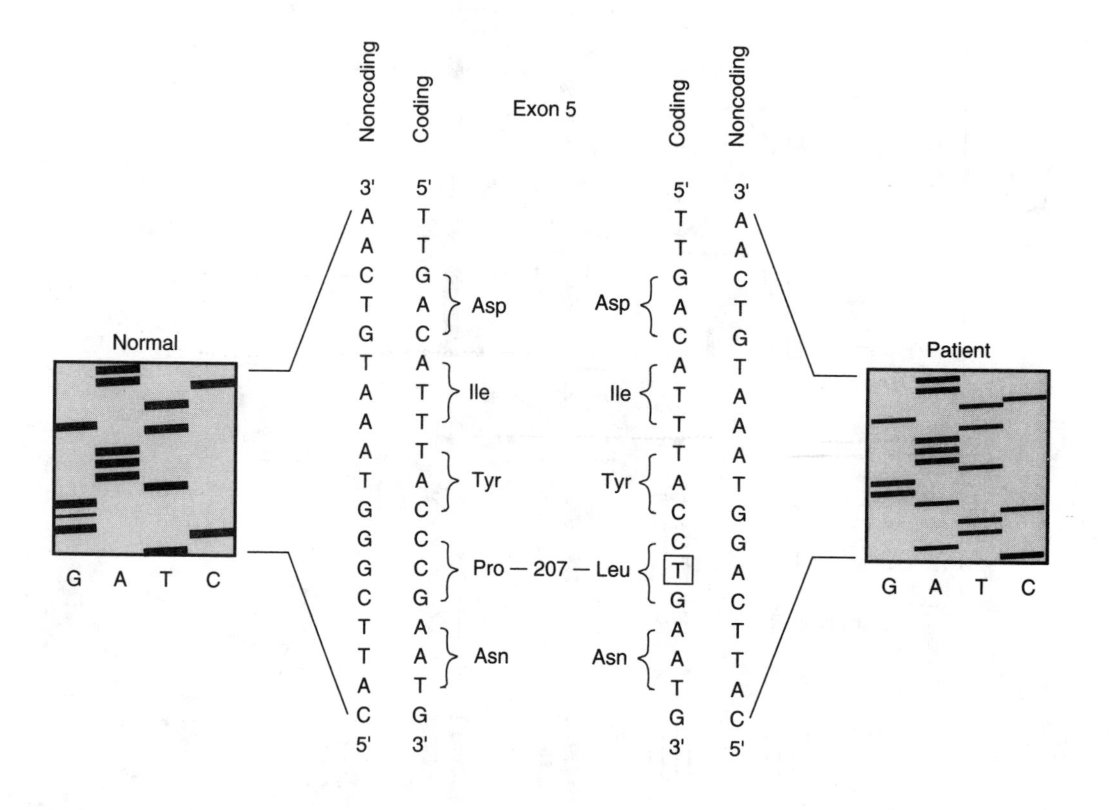

Figure 5-3. DNA sequencing allows precise identification of the abnormal variation in an abnormal patient's DNA. Note the difference in the boxed nucleotides between the abnormal patient and the normal control. (Reprinted with permission from *The New England Journal of Medicine* 324:1761–1766, 1991.)

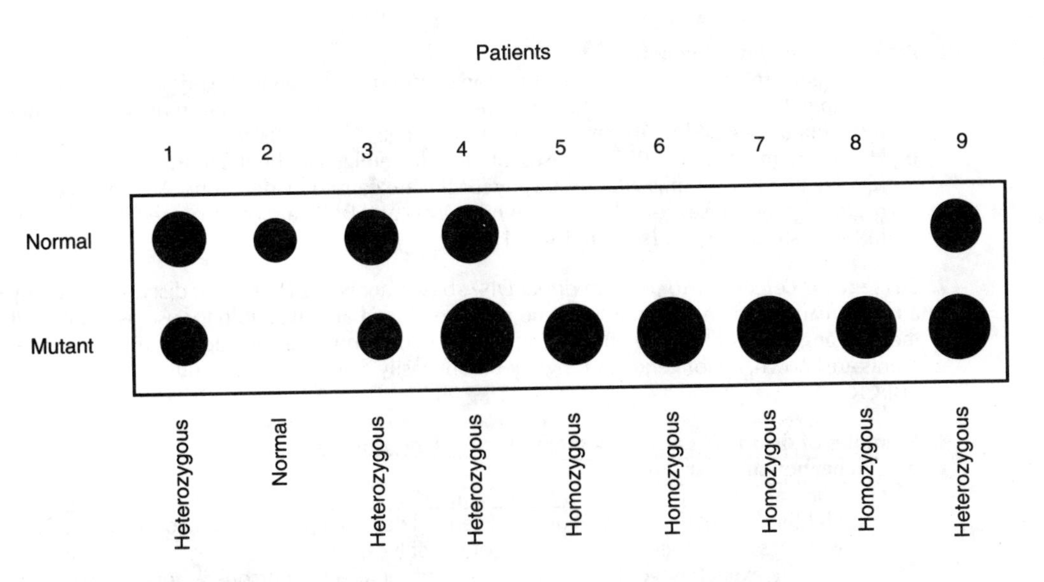

Figure 5-4. Allele-specific oligonucleotide (ASO) hybridization allows detection of both the normal gene and the abnormal gene that cause a particular disease. In this instance, the patient has a recessive disorder and the persons who are homozygous and heterozygous for the mutation can be detected. Results show that only *patient 2* is normal. *Patients 1, 3, 4,* and *9* are heterozygous carriers. *Patients 5–8* are homozygous for the mutant gene and, therefore, are affected.

D. Allele-specific oligonucleotide (ASO) probe analysis

1. **Method.** This technique uses two synthetic oligonucleotides: one specific for a normal gene and the other specific for a known genetic mutation. The ASOs for the normal and mutant gene are used as hybridization probes to determine whether a patient has two copies of a normal gene or a mutation responsible for a particular disease (Figure 5-4).
2. **Applications.** Oligonucleotide analysis is most helpful for diagnosing genetic diseases that typically arise from a single predominant mutation (e.g., sickle cell disease, α_1-antitrypsin deficiency).
3. **Advantages and limitations**
   a. This technique can be used on DNA that has been selectively amplified by PCR and allows rapid and accurate DNA-based diagnosis.
   b. ASO probe analysis is possible only with precise knowledge of the structure of the normal gene and of the molecular basis of the mutation. Also, the method is limited in that, for many diseases, there is a separate and specific mutation in each family.

E. Electrophoresis of single-stranded or denatured DNA

1. **Method.** Under appropriate conditions this technique allows the identification of differences between DNA sequences from a patient with a particular disease and from a normal control. These differences are detected by shifts in the bands that are detected.
2. **Application.** This approach is most useful for scanning DNA fragments of genes that have been amplified by PCR for variations.
3. **Advantages and limitations**
   a. This technique allows rapid screening of a particular gene for possible mutations.
   b. This technique detects normal DNA variation or polymorphisms as well as DNA sequence changes that cause disease. To distinguish between these possibilities, DNA sequencing of the altered fragment must be undertaken to identify whether the change detected is a mutation causing a specific genetic disease.

II. DETECTION OF KNOWN MUTATIONS: DIRECT DNA DIAGNOSIS.

Mutations that are directly detectable include **major gene rearrangements** (e.g., deletions, insertions, duplications) and single nucleotide substitutions, or **point mutations**.

A. Major gene rearrangements (see also Ch 3 II A)

1. **Prevalence in clinical genetics**
   a. Major gene rearrangements are uncommon causes of mutation in the human genome. For example, less than 10% of mutations causing familial hypercholesterolemia (FH) and hemophilia are caused by major gene rearrangements.
   b. However, more than 50% of cases of α-thalassemia, Duchenne muscular dystrophy (DMD), and Becker muscular dystrophy (BMD) are caused by deletions. A deletion of the α-globin gene causes α-thalassemia, while DMD and BMD are often caused by deletions in the dystrophin gene (see Ch 3 II A 1).

2. **Strategy for detection.** To perform direct DNA-based analysis of a genetic disease caused by a major rearrangement, the normal gene must be cloned and available for use as a probe. If the rearrangements are large enough, and if the gene sequence surrounding the rearrangements are known, major gene rearrangements are easily detected using Southern blot analysis or PCR.

3. **Examples of disorders caused by major gene rearrangements**
   a. **Duchenne muscular dystrophy (DMD)**
      (1) **Clinical features.** DMD is an X-linked inherited disorder, which presents during early childhood with delay in walking and general development. DMD leads to generalized weakness with pseudohypertrophy of the calves.
         (a) **Life expectancy.** The mean age of death of affected persons is approximately 17 years, and death is primarily due to cardiorespiratory problems.
         (b) **Incidence.** DMD occurs in about 1 of every 4000 male births.
      (2) **Genetic background**
         (a) **DMD** has been shown to be due to multiple different defects in the dystrophin gene, which maps to the short arm of the X chromosome.
         (b) A **frameshift mutation** (see Ch 3 II A 6 b), which interrupts the coding of the protein, is more likely to result in DMD than BMD.
      (3) **Diagnosis.** DMD can be diagnosed using DNA technology either by detecting the high frequency of deletions in the disorder or by using polymorphic DNA markers within and around the gene.
      (4) **Case 5-A. Direct DNA diagnosis of DMD.** A 6-year-old boy was recently diagnosed with DMD. The boy's maternal aunt is planning a pregnancy and wishes to know her risk of having a child with DMD (Figure 5-5). There are no other family members with DMD. In this instance, **DNA analysis** revealed that this boy had a large deletion in his gene for DMD. His mother had a similar deletion and was a carrier for DMD. DNA testing of the boy's aunt revealed that she did not have the deletion in the DMD gene. Therefore, her offspring would not be at increased risk of having DMD.
   b. **Becker muscular dystrophy (BMD)**
      (1) **Clinical features** are similar to DMD but have a much milder presentation.
      (2) **Genetic background**
         (a) BMD has been shown to be **allelic to DMD** with defects also in the dystrophin gene, which result in the milder presentation of symptoms.
         (b) BMD is most often due to a mutation that does not alter the rest of the coding sequence of the dystrophin gene.

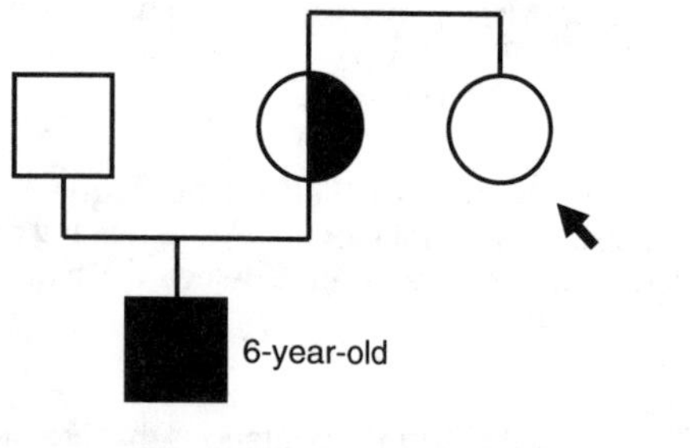

Figure 5-5. Pedigree of family in case 5-A. The proband is the person undergoing predictive testing and is indicated by an arrow. The proband's aunt does not carry the deletion for Duchenne muscular dystrophy; therefore, her offspring are not at risk of having this disease.

(3) **Diagnosis.** BMD can also be diagnosed either by detecting the high frequency of deletions in the disorder or by using polymorphic DNA markers within and around the gene.

c. Familial hypercholesterolemia (FH)

(1) **Clinical features.** In people with FH, low-density lipoprotein (LDL) cholesterol is not appropriately removed from the circulation due to mutations in the LDL receptor. LDL levels are thus caused to increase, which results in blood vessel wall damage and premature atherosclerosis.

(a) **Incidence.** In the general population, FH occurs in approximately 1 in 500 people except in certain parts of the world such as Quebec, Canada, and South Africa, where FH occurs in approximately 1 in 100 people.

(b) **Life expectancy.** Screening for FH is important, as FH does result in a high frequency of premature heart attacks and early death.

(c) **Treatment** is directed to lowering cholesterol levels, which can reduce the risk for later atherosclerotic events.

(2) **Genetic background.** FH most often results from defects in the LDL receptor gene. The condition is inherited as an autosomal dominant trait.

(a) In most separate families, FH is caused by a very common **predominant mutation**.

(b) Major rearrangements do not account for a very large number of mutations underlying FH (see Ch 3 II A 1 c) except in the **French Canadian** population where a **major deletion** accounts for approximately 65% of FH patients.

(c) In the French Canadian and South African Afrikaner populations, a **single founder** was known to have introduced this defective LDL receptor gene into the population's gene pool centuries ago. Subsequent expansion of the population has caused the frequency of this gene to increase.

(3) **Case 5-B. Direct DNA diagnosis of FH.** A father (I-1) died of a heart attack at age 42. His oldest son (II-1), a 35-year-old, nonsmoking man of French Canadian descent with FH has a history of myocardial infarction at age 33. The younger son (II-2) also has moderate hypercholesterolemia and wishes to know whether this is environmentally modulated by diet or is due to FH. The latter diagnosis would carry a serious prognosis. Physical examination was not helpful. Knowing that approximately 65% of French Canadian patients with FH have a major deletion, a simple way to assess the cause for the younger son's hypercholesterolemia would be to look for a deletion. This information would be helpful particularly in view of the fact that the older son (II-1) had previously been shown to have this rearrangement. **DNA analysis** revealed that the younger son (II-2) has a 10 kilobase (10 Kb) deletion in the LDL receptor gene causing FH (Figure 5-6). Because of the younger son's family history and the current diagnosis, more vigorous cholesterol-lowering therapy would be indicated.

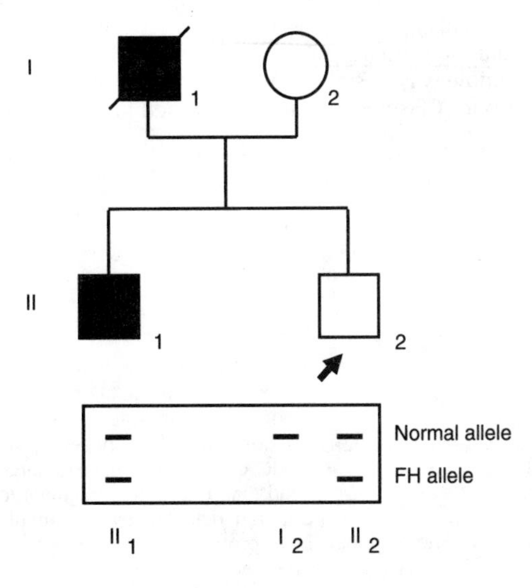

Figure 5-6. Pedigree of family in case 5-B. The proband is the younger son who requested predictive testing (*arrow*). The proband has inherited a pattern of fragments in the low-density lipoprotein receptor gene that is similar to his affected brother's pattern. The proband is also affected with familial hypercholesterolemia.

B. Point mutations

1. **Prevalence in clinical genetics.** Point mutations represent very common types of mutations underlying genetic disease (see Ch 3 II A 4).

2. **Point mutations** are also called **missense mutations** (see Ch 3 II A 4).
   a. **Strategy for detection.** Direct DNA diagnosis requires that the normal gene be cloned and available for use as a probe. Southern blot analysis or PCR amplification and digestion with the appropriate enzyme may reveal an altered pattern of restriction fragments between normal and affected persons.
   b. **Examples of disorders caused by point mutations that frequently alter restriction sites**
      (1) **Tay-Sachs disease**
         (a) **Clinical features.** Tay-Sachs disease is characterized by excessive accumulation of gangliosides in the neurons of patients due to defective lysosomal degradation of these compounds.
            (i) **Incidence.** Infantile Tay-Sachs disease occurs with highest frequency in Ashkenazi Jewish populations. The carrier rate for Tay-Sachs disease in this population is approximately 1 in 35.
            (ii) **Life expectancy.** Children with infantile Tay-Sachs disease often present at about 5 months of age with motor weakness; progressive motor and mental deterioration result in death between the second and fourth years of life.
         (b) **Genetic background.** Detailed molecular analysis has revealed that many different mutations cause Tay-Sachs disease. Molecular heterogeneity occurs even in the Ashkenazi Jewish population, but there is a predominant mutation in this population that alters a restriction site.
      (2) **Sickle cell anemia.** Similar point mutations that alter restriction sites can aid in the diagnosis of sickle cell anemia.
         (a) In this common hereditary disease in black Americans, the mutation in the β-globin gene changes the amino acid glutamine to valine by changing the codon GAG to GTG. This also abolishes a particular enzyme cutting site.
         (b) Persons with sickle cell anemia have an abnormally large fragment detected after hybridization with β-globin to DNA digested with the appropriate enzyme (Figure 5-7) due to absence of the restriction site caused by the mutation.

3. **Point mutations may not directly alter restriction sites.** An example of a disease with a predominant mutation in this category is **cystic fibrosis (CF).**
   a. **Clinical features.** Intestinal obstruction and infection are the most life-threatening manifestations of CF.
      (1) **Incidence.** CF affects Caucasian populations with an incidence of approximately 1 in every 2500 births. Approximately 1 in 20 Caucasians are carriers of CF.
      (2) **Life expectancy.** Average life expectancy for a person with CF is 20 years.
   b. **Genetic background.** CF is a common autosomal recessive disease. The gene for CF has now been cloned and over 80 different mutations have been defined.
      (1) Approximately 70% of Caucasian patients with CF have been shown to have a 3-base pair (3 bp) deletion that does not alter a restriction site.
      (2) ASO probes provide a rapid and simple way to screen for common CF mutations in the general population (see III B 2 b for CF screening among affected families).

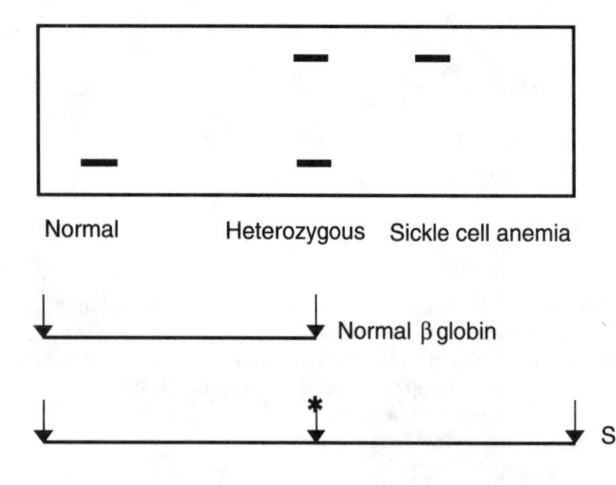

Figure 5-7. The mutation causing sickle cell anemia abolishes a restriction site. Thus, patients with sickle cell anemia have a larger band than the normal fragment for β-globin detected by a particular DNA probe.

III. DETECTION OF UNKNOWN MUTATIONS: INDIRECT DNA DIAGNOSIS. Most of the approximately 4000 inherited diseases are caused by unknown mutations. In some instances, these disorders can be diagnosed indirectly, using markers for the mutant gene. These "tags" are variations in DNA that do not code for a protein but have been demonstrated to be close to or linked to the gene of interest.

A. Limitations of indirect DNA diagnosis

1. **Numerous family members are needed** to determine the form of the marker that segregates with the mutant gene in that family.

2. **The distance between the marker and the mutant gene** limits the accuracy of indirect DNA diagnosis.
   a. For each distance of 1 million bp between the marker and gene, there is a 1% chance of recombination during each meiosis, which can render the results incorrect.
   b. For example, as shown in Figure 5-8, if it has been shown that in a particular family the autosomal dominant disease is segregating with marker A, any offspring who inherits marker A from an affected person whose DNA has markers AB is highly likely to have also inherited the mutant gene. The accuracy of the assessment largely depends on the distance between the marker and the mutant gene.

3. **Genetic heterogeneity.** When using an indirect approach to DNA diagnosis, it is important to know the likelihood of genetic heterogeneity (i.e., whether mutations underlying a particular disease may occur at different loci in different families). **Polycystic kidney disease (PKD)** is an example of a disease that demonstrates genetic heterogeneity.
   a. The gene underlying PKD has not been identified. Most affected people have a mutation on chromosome 16, although recent findings indicate that some people with PKD have the causative mutation on another chromosome.
   b. If a DNA marker from chromosome 16 is used in all instances to determine the risk of someone having the mutant gene, the results of DNA testing will likely be incorrect in families who have the causative gene on another chromosome.
   c. The discovery of genetic heterogeneity has made DNA diagnosis of PKD difficult. In each case, the likelihood that the mutant gene is on chromosome 16 must be determined before DNA analysis can be used to assess risk for the disease.

B. Special applications for indirect DNA diagnosis

1. **Diseases involving a known gene** [see II A 3 a (1); case 5-A]
   a. When the causative gene is identified but the specific mutation in a particular family is unknown, variations in noncoding regions within the gene that alter a restriction site can help mark the abnormal gene and trace its inheritance.In this instance, blood from family members is needed.
   b. The accuracy rate will be very high, because the distance between the harmless DNA variation that alters a restriction site and the DNA variation that causes the mutation is likely to be small.

2. **Use of linkage disequilibrium.** In some cases, a mutation that causes a disease occurs on a background of DNA variation outside the affected gene that might be used to indicate the presence of the abnormal gene. This background DNA variation is much less common in the general population than in people affected with a particular disease. The nonrandom association between a particular mutation and a pattern of DNA variation is termed linkage disequilibrium

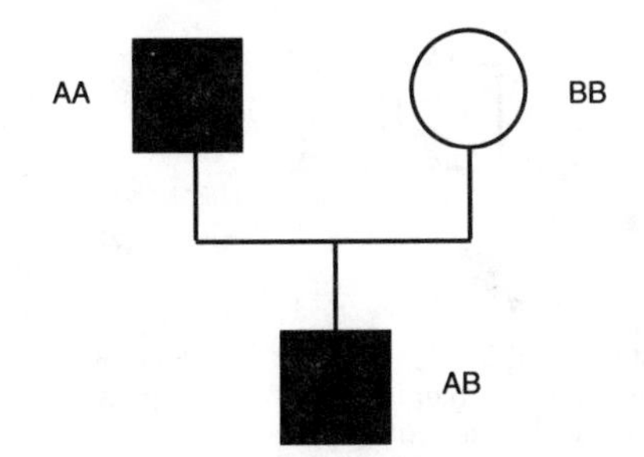

Figure 5-8. In this family, the son has inherited marker A from his father and has also inherited the gene for an autosomal dominant disease. The DNA marker is linked to the gene causing this disease, which means that the A marker co-segregates with the mutant allele.

(see Ch 4 II C 1). The demonstration of linkage disequilibrium can be important for providing information to persons at risk for a genetic illness.

a. A good example is **sickle cell anemia,** which is caused by a mutation that is associated with a particular genetic variation close to the β-globin gene. In many instances, the presence of the mutation can be inferred by the presence of a particular restriction fragment.

b. The recent finding of a strong association between the mutation underlying **CF** and a particular harmless variation in neighboring DNA in affected people has allowed linkage disequilibrium to be used in genetic counseling for CF, which is an autosomal recessive condition. The strong association may reduce the need for DNA samples from affected people in some families. Clinical implications of this finding are demonstrated by the following example.

(1) A couple had a child with CF who died. The mother is pregnant again and wishes to know whether or not the fetus carries the CF mutation, which, in this example, is assumed to be strongly associated with a B variation in nearby noncoding DNA. No blood is available from the affected child.

(2) In this instance, if the fetus has inherited two B variations in DNA near the CF gene from each parent, the risk of that child developing the disease is very high. Since CF is an autosomal recessive disease, two abnormal genes cause the disease state.

C. Case examples of indirect DNA diagnosis

1. Case 5-C. Indirect DNA diagnosis of Huntington's disease (HD). The proband is a 30-year-old man (III-1) who requests predictive testing for HD, an autosomal dominant genetic disorder. His father (II-1) died of the disease, which was confirmed at autopsy, and the proband (III-1) wants to know whether he has or has not inherited the abnormal gene. Also, he is considering having children and wishes to know his risk status prior to reproducing. DNA markers closely linked to the abnormal HD gene have been identified. After initial counseling, blood is collected from several appropriate affected and unaffected family members for DNA testing. The **DNA analysis** reveals that the proband (III-1) inherited the marker that was segregating with the gene for HD. He is given a 98% risk of having inherited the HD gene (Figure 5-9). In this case, blood from numerous relatives was needed to perform the analysis. Risk assessment was possible because numerous DNA markers were available and because there is a very low risk of genetic heterogeneity in HD. The proband's first cousin (III-2) also requested predictive testing and was found to have markers AA. Since A is not inherited with HD, the proband's cousin was given a very low risk of having inherited HD.

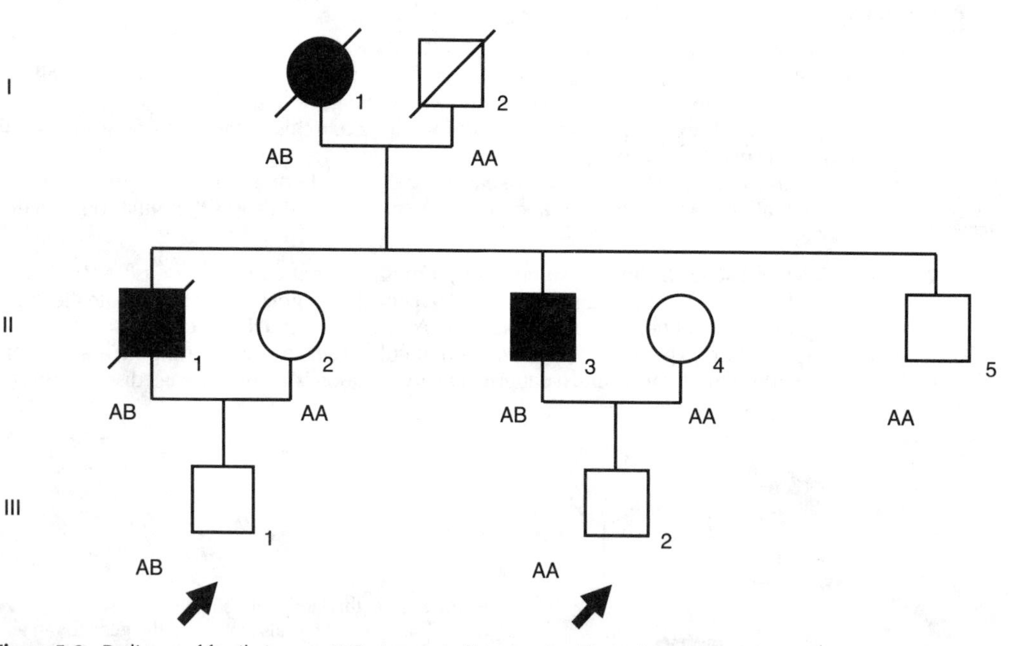

Figure 5-9. Pedigree of family in case 5-C. *Arrows* indicate probands. In this pedigree for Huntington disease (HD), the gene is segregating with the B marker. III-1 has inherited the B marker from his father and has a high risk, while III-2 has inherited the A marker from his father and has a low risk of inheriting HD.

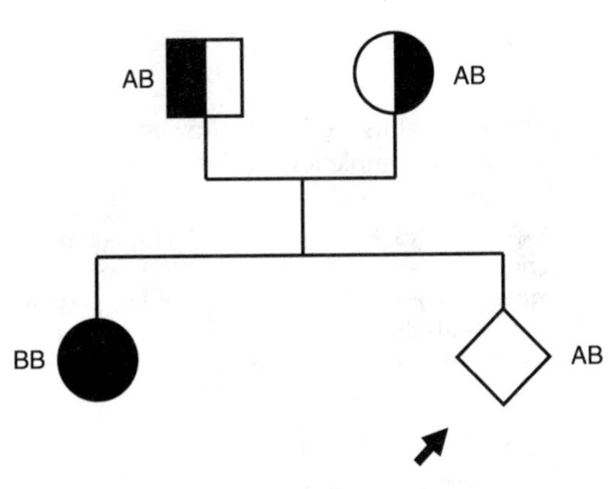

Figure 5-10. Pedigree of family in case 5-D. The *arrow* indicates the proband in this family with phenylketonuria. The mutation is segregating with the B marker. In this instance, the fetus (diamond) has inherited an A marker from one parent and a B marker from the other, and, therefore, is a carrier.

2. **Case 5-D. Indirect DNA diagnosis of phenylketonuria (PKU).** A couple in their mid-twenties has a 5-year-old girl with PKU, an autosomal recessive disorder caused by a defect in the gene for phenylalanine hydroxylase. The affected child is on a special diet and has mild developmental delay. The couple recently discovered that they are expecting another baby and wish to know whether the second child will have PKU. In this case, the gene for phenylalanine hydroxylase has been cloned, but the specific mutation underlying PKU in the 5-year-old is not known. However, variations in the noncoding regions of the PKU gene can be used to determine whether the fetus is likely to be affected by PKU, to be a carrier for PKU, or to have two normal phenylalanine hydroxylase genes. Both the mother and father have AB markers. **DNA analysis** reveals that the affected child inherited a B marker from her mother and a B marker from her father (Figure 5-10). Thus, the mutation must have occurred on a PKU gene that had a B marker in both instances. The mother undergoes chorionic villus sampling at 10 weeks of pregnancy, and analysis of fetal DNA reveals an AB pattern. Thus, the fetus is a carrier for PKU but will not be affected. The accuracy of this prediction is close to 100% because of the very small distance between the DNA variation used to tag the gene for PKU and the actual site of the causative mutation.

IV. DNA BANKING.

Often when the precise mutation responsible for a disease is unknown, DNA from affected and nonaffected relatives is needed to provide DNA analysis and to determine which form of the marker is being inherited with the disease and the normal gene. The most important sources of DNA in such analyses are from persons who are affected by the disease and their parents.

A. **DNA banks** store DNA from crucial relatives in an effort to ensure that no family member will be denied DNA testing due to a lack of available DNA.

B. **DNA can be extracted from any tissue.** However, for convenience, DNA usually is taken from **white blood cells**. In situations that require a large supply of DNA for analysis, white blood cells can be immortalized after exposure to the Epstein-Barr virus. In this way, an unlimited supply of DNA is possible.

STUDY QUESTIONS

Directions: Each of the numbered items or incomplete statements in this section is followed by answers or by completions of the statement. Select the **one** lettered answer or completion that is **best** in each case.

1. All of the following materials are necessary to perform Southern blot analysis for diagnosis of a genetic disease EXCEPT

(A) a membrane for transfer of DNA
(B) DNA from the person being tested
(C) the DNA sequence of the probe being used
(D) a specific restriction enzyme
(E) A DNA probe

2. True statements regarding the use of allele-specific oligonucleotides for diagnosis of genetic disease include all of the following EXCEPT

(A) a precise knowledge of the mutation causing the particular disease is needed
(B) it may be helpful for the diagnosis of sickle cell anemia and α_1-antitrypsin deficiency
(C) the sequence of the whole gene, including flanking regulatory sequences, must be known prior to DNA testing
(D) DNA from only a few family members is needed for diagnosis
(E) this diagnostic method is applicable for only a small number of genetic diseases

3. A woman (II-3) has a nephew (III-1) affected with Duchenne muscular dystrophy (DMD). She wishes to know if she is a carrier for this gene. The pedigree below reveals the family's DNA results using polymorphic DNA markers within the gene.

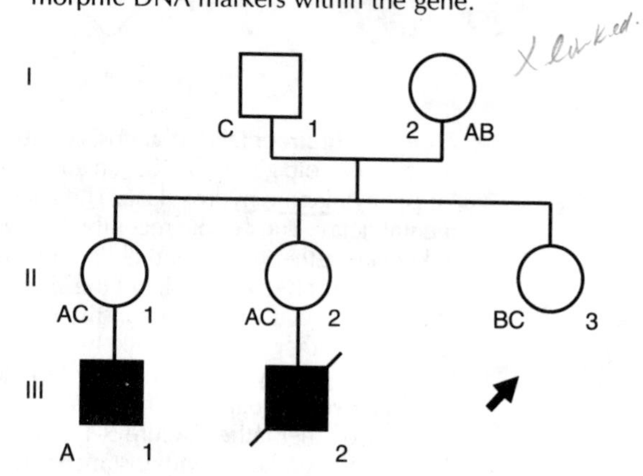

The most likely interpretation of the possibility that II-3 is a carrier for DMD is

(A) very high
(B) mildly elevated
(C) mildly reduced
(D) very low
(E) not able to be determined from this analysis

1-C
2-C
3-D

4. Two parents (I-1 and I-2) of Ashkenazi Jewish descent have had a child (II-1) with Tay-Sachs disease. They are now expecting another child (II-2) and wish to have prenatal diagnosis for this disorder. The pedigree shows the results of DNA analysis.

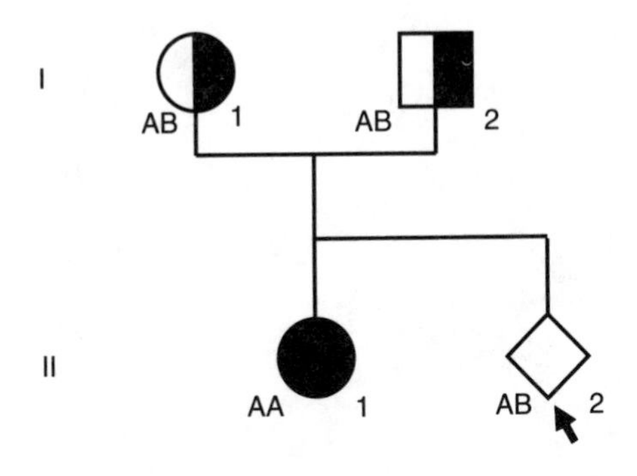

The most likely interpretation of the results for this pregnancy is that the fetus is

(A) affected
(B) completely normal
(C) a carrier
(D) unable to determine from results
(E) unlikely to be the offspring of these parents

4-C

Directions: The group of items in this section consists of lettered options followed by a set of numbered items. For each item, select the **one** lettered option that is most closely associated with it. Each lettered option may be selected once, more than once, or not at all.

Questions 5–9

DNA diagnosis can be performed using linked polymorphic DNA markers for diseases where the gene has not been cloned. If the mutation is known, direct approaches to DNA can be performed. Match the approach best used for DNA diagnosis with the following genetic diseases.

(A) DNA sequence analysis
(B) DNA markers close to the gene
(C) Allele-specific oligonucleotides
(D) DNA markers within the gene
(E) Detection of major rearrangements within the gene

5. α_1-Antitrypsin deficiency
6. Becker muscular dystrophy
7. Hemophilia
8. Duchenne muscular dystrophy
9. α-Thalassemia

5-C
6-E
7-D
8-E
9-E

ANSWERS AND EXPLANATIONS

1. The answer is C *[I A 1].*
To perform Southern blot analysis, DNA is digested with a particular restriction enzyme, electrophoresed in agarose gel, transferred to a membrane and hybridized with a labeled DNA probe. The DNA sequence of the probe being used is not necessary for the analysis.

2. The answer is C *[I D].*
Allele-specific oligonucleotide hybridization is possible if the specific mutation is known allowing synthesis of a mutant and wild type (normal) allele. It is particularly useful when only a small number of mutations cause the disease such as in sickle cell anemia and α_1-antitrypsin deficiency. Only the specific DNA sequence change around the mutation is needed together with DNA sequence information of some flanking DNA. DNA from numerous family relatives is not necessary to assess the presence or absence of the mutation by a particular individual.

3. The answer is D *[II A 3 a; Case 5-A].*
II-3 has a very low risk of being a carrier for Duchenne muscular dystrophy (DMD), which is an X-linked disorder. The grandfather (I-1) must have passed on a C marker to II-1, II-2, and II-3. The grandmother (I-2) has AB markers and must have passed on A markers to II-1 and II-2. The grandson who is affected with DMD (III-1) has an A marker. Therefore, the mutation for DMD must be inherited with the A marker. Grandson III-2 died from DMD. The fact that II-3 inherited the B marker from her mother, instead of the A marker, makes the risk of II-3 being a DMD carrier very low.

4. The answer is C *[II B 2 b (1)].*
In this particular case, it is clear that the defect for Tay-Sachs disease is going along with the A marker. Because the fetus has inherited an A marker from one parent and a B marker from the other parent, it is most likely that the fetus is a carrier.

5–9. The answers are: 5-C *[I D],* **6-E** *[II A 1 b, 3 b];* **7-D** *[III B],* **8-E** *[II A 3 a],* **9-E** *[II A 1 b].*
An α_1-antitrypsin deficiency is most simply diagnosed with the use of allele-specific oligonucleotides for normal and mutant alleles, since α_1-antitrypsin deficiency most commonly arises from a single mutation.

Becker muscular dystrophy (BMD) is allelic to Duchenne muscular dystrophy (DMD). The most common mutations in the dystrophin gene, which causes both diseases, are major gene rearrangements, especially deletions.

Hemophilia is caused by mutations in a very large gene—the factor VIII gene. In this instance, there are a large number of mutations underlying this illness. It is difficult to know for which mutation to look in someone presenting for DNA diagnosis. The simplest approach is to use DNA markers within the gene.

A deletion of the α-globin gene causes α-thalassemia, which is very common in Southeast Asia. Detection of this major gene rearrangement represents the most useful approach for DNA diagnosis.

Population

I. GENES IN POPULATIONS

A. Population genetics focuses on the relative distribution of genes or inherited traits within populations. The factors that change or maintain the frequency of genes or genotypes within the population are studied as well.

B. The Hardy-Weinberg law is used to relate the frequencies of genotypes at a single mendelian locus to the phenotype frequencies in a population.

1. **Definitions**
   a. **Selection** is the process that determines the relative share of genes that individuals of various genotypes contribute to the propagation of a population. Genotypes that are disadvantageous are selected against and contribute relatively fewer genes to the next generation. Genotypes that are advantageous are selected for and contribute a disproportionately greater number of genes to the next generation.
   b. **Migration** is the movement of individuals between populations. Migration can result in exchange of genes between populations and alteration of gene frequencies.
2. **The assumptions** upon which the Hardy-Weinberg law is based are:
   a. **There is no mutation** occurring at the locus.
   b. **There is no selection** for any of the genotypes at the locus.
   c. **Mating is completely random.**
   d. **There is no migration** into or out of the population being considered.
3. These assumptions are **never entirely correct** for any locus.
   a. If they are nearly correct, a locus is said to be in **Hardy-Weinberg equilibrium**.
   b. If the assumptions are seriously violated with respect to a given locus, the Hardy-Weinberg law may not apply.
4. **The Hardy-Weinberg law states:** In a population at equilibrium, for a locus with two alleles, *D* and *d*, having frequencies of p and q, respectively, the genotype frequencies are: $DD = p^2$, $Dd = 2pq$, and $dd = q^2$.
   a. Note that $p + q = 1$ because there are only two alleles at the locus, and the sum of their frequencies (expressed as fractions of the total) must equal 1. This also means that $p = 1 - q$ and that $q = 1 - p$.
   b. Note that $p^2 + 2pq + q^2 = 1$. There are only three genotypes, and the sum of their frequencies (expressed as fractions of the total) must equal 1.
5. **For an autosomal dominant condition,** the frequency p of the mutant gene (*D*) is very small compared to the frequency q of the normal gene (*d*). The frequency of the heterozygote $(Dd) = 2pq$ is consequently much greater than the frequency of the abnormal homozygote $(DD) = p^2$. Thus, the disease frequency for an autosomal dominant condition is approximately equal to the frequency of the heterozygote (2pq).
   a. This relationship ($p^2 << 2pq$) explains why **homozygotes for autosomal dominant diseases are rarely observed**.
      (1) In practice, a **homozygote for an autosomal dominant disease** can be recognized by the fact that **both parents are affected with the disease**.
      (2) **The phenotype of the homozygote is much more severely abnormal than that of the heterozygote** for almost all autosomal dominant diseases in which homozygotes have

actually been observed. This contrasts with the situation in some experimental organisms in which the phenotype is the same in homozygotes and heterozygotes for a dominant trait.

b. If p is very small, q is approximately equal to 1 and the frequency of the disease is approximately equal to the frequency of the mutant gene (2p).

6. Autosomal recessive conditions

a. The Hardy-Weinberg law permits calculation of the disease or carrier frequency from the gene frequency, and vice versa.

(1) **The frequency of the disease** (i.e., of the genotype *dd*) is simply the **square of the gene frequency** (q^2).

(2) Conversely, **the gene frequency** (q) can be calculated as the **square root of the frequency of the disease** (q^2).

(3) The **frequency of the heterozygous carrier** (i.e., of the genotype *Dd*) is equal to twice the product of the component gene frequencies (2pq).

(4) **For rare recessive diseases,** q is very small, and p is very close to equaling 1. Under such circumstances, **the frequency of heterozygous carriers is approximately equal to 2q.**

(5) **The carrier frequency for a recessive disease is always greater than the frequency of the disease itself** because mathematically the value of 2pq must always exceed that of q^2 if $p>q$.

(a) For rare recessives (where $p>>q$), the heterozygote frequency is far greater than the frequency of homozygotes with the disease ($2pq>>q^2$).

(b) Thus, **almost all of the mutant genes in the population for a rare recessive disease are carried in heterozygotes.**

b. **Case 6-A. Sickle cell disease.** Sickle cell disease is an autosomal recessive condition that occurs with a frequency of 1/625 among black Americans. What is the frequency of heterozygous carriers in this population?

The three genotypes in this case are represented as *AA* (homozygous normal), *AS* (carrier), and *SS* (sickle cell disease homozygote). The frequency of the disease genotype, *SS,* is given as 1/625, which is q^2. Thus, $q^2 = 1/625$, $q = 1/25$, and $p = 1 - q = 24/25$ (or approximately = 1).

According to the Hardy-Weinberg law, the frequency of heterozygous carriers for sickle cell disease among black Americans = 2pq = 2(24/25)(1/25) or approximately 2(1/25), which is about 8%.

7. X-linked recessive conditions

a. The Hardy-Weinberg law in its usual form refers only to females who carry two alleles for each X-linked locus. Males only carry one copy of every X-linked gene. For males, the genotype frequencies for *Dy* and *dy* correspond directly to the gene frequencies p and q, respectively (y represents the Y chromosome in the male).

(1) Thus, **the gene frequency (q)** for an X-linked recessive disease is simply **equal to the frequency of that condition among males.**

(2) **The values of p and q are the same in males and females** because it is not known whether a gene will end up in a male or female zygote.

(3) The **frequency of an X-linked recessive disease is much greater in males than in females** because $q>>q^2$ [because q is a fraction < 1; e.g., $1/10 >> (1/10)^2$].

(4) Since X-linked recessive diseases are generally uncommon, the value of q is typically very small compared to p, and $2pq>>q^2$. Thus, **heterozygous carrier females are much more common than homozygous affected females** for X-linked recessive traits.

(5) **If q is very small,** p is approximately equal to 1 and the frequency of female carriers (2pq) is approximately equal to twice the frequency of affected males (2q).

(6) In the population as a whole, **about two-thirds of all mutant X-linked recessive genes are found in carrier females and about one-third in affected males.**

(a) In general, two-thirds of all X-chromosomes and two-thirds of all X-linked genes are found in females; one-third are found in males.

(b) The reason for this distribution is that males and females occur in approximately equal numbers and females have two X-chromosomes while males have only one.

b. **Case 6-B. Hunter syndrome.** Hunter syndrome, an X-linked recessive disease, occurs once in every 100,000 males. What is the frequency of carrier females?

The gene frequency in males (q) is equal to the disease frequency and, therefore, is 1/100,000. The gene frequency in males and females must be the same, so q = 1/100,000 in females as well. The frequency of heterozygous carriers in females is 2pq or approximately 2 q, since p is almost equal to 1. Thus, the frequency of heterozygous carrier females is approximately equal to: $2q = 2(1/100{,}000) = 1/50{,}000$

II. MUTATIONS IN POPULATIONS

A. Genetic load

1. Mutations are continually occurring, and most mutations are detrimental. The genetic load is the **cumulative reduction of reproductive fitness** produced by all detrimental alleles at all loci in all individuals in the population.
2. **The concept of genetic load** depends not only on the mutant alleles that are actually present but also upon the circumstances in which fitness is being determined. **Mutations that are deleterious in one population may be neutral or beneficial in another** population under other circumstances. For example, the sickle cell mutation, which causes a serious disease in the homozygous state, is beneficial to populations living in regions where malaria is endemic because the more common heterozygotes are resistant to malaria.

B. One measure of genetic load related to recessive traits is the increase in death rate that occurs among the offspring of consanguineous parents (see Ch 3 I C 2 c).

1. The number of recessive **lethal equivalents** that, on the average, each person carries can be calculated from this increase. **A lethal equivalent is defined as** a heterozygous gene that would be lethal to homozygotes in inverse proportion to the number of genes involved. For example:
   - a. **One gene,** which, if homozygous, would be lethal to 100% of homozygotes
   - b. **Two genes,** which, if homozygous, would be lethal in half of the homozygotes
   - c. **Three genes,** which, if homozygous, would be lethal in one-third of the homozygotes
2. **Lethal equivalents are calculated from the rate of death** among the children of consanguineous couples, and **the particular gene or genes involved** does not have to be known. It is necessary to use the concept of lethal equivalents to quantitate this death because the number of genetic loci actually involved cannot be determined.
3. On the basis of such studies, it is estimated that **each normal person carries three to five lethal equivalents**. In other words, **every person carries abnormal recessive genes.**
   - a. One implication of this observation is that people who are found to be **carriers of recessive genes** for conditions such as Tay-Sachs disease, sickle cell disease, or cystic fibrosis **do not differ from others because they are carriers. They differ because they know one of the abnormal genes they carry.**
   - b. Another implication is that **any strategy designed to eliminate all abnormal genes in a population can only succeed by eliminating all people in the population.**

N·B →

C. Mutations are rare in the population but common among patients with dominant or X-linked diseases that reduce fertility. For example, consider the autosomal dominant disease **achondroplasia,** the most common kind of short-limbed dwarfism.

1. **New mutations.** About 90% of all cases of achondroplasia arise as new mutations, but the **frequency of new mutations** of the normal gene for achondroplasia is only about **1/100,000**.
2. This apparent contradiction is explained by the fact that **two ratios with the same numerator but very different denominators** are being compared.
   - a. The number of achondroplastic dwarfs with new mutant genes is the numerator in both cases, but in the former instance, the denominator is the **total number of achondroplastic dwarfs** (a small number), while in the latter case, the denominator is the **total number of genes,** normal and abnormal, at the achondroplasia locus in the population (a very large number).
   - b. Thus, if a population of 500,000 people included 11 achondroplastic dwarfs and 10 of these represented new mutations, then 10/11 (91%) of the achondroplastic dwarfs would represent new mutations, but the mutation frequency would be about 1/100,000.
     - **(1)** Each normal person has two normal genes. $500{,}000 - 11 = 499{,}989$ normal people who have 999,978 normal genes.
     - **(2)** 11 dwarfs have 11 normal genes and 11 achondroplasia genes.
     - **(3)** 10 of the achondroplasia genes are new mutants, so the mutation frequency is $10/(999{,}978 + 11 + 11) = 10/1{,}000{,}000 = 1/100{,}000$.

D. A paternal age effect is observed with some **new dominant mutations** in which the **average age of fathers** of affected children is significantly increased.

1. This observation implies that there is an increased risk of diseases due to new dominant mutations among the children of older fathers. The exact magnitude is unknown, **but the risk for all new dominant mutations combined is unlikely to be more than 0.5%–1% among the children of fathers who are over 40 years old.**

2. There appears to be **no** independent **association** of new dominant mutations **with the age of the mother**. (Maternal age does affect the rate of nondisjunction that is responsible for the occurrence of trisomy in offspring. See Chapter 2 II B 3 b.)

E. Mutation–selection equilibrium is the maintenance of constant gene frequencies in a population through replacement by mutation of abnormal alleles lost by death or reproductive failure. If the frequency of a monogenic disease remains constant from generation to generation and if other causes of gene gain or loss (e.g., migration) are negligible, then **all genes lost in one generation by selection must be replaced in the next generation by mutation.**

1. **If an autosomal dominant disease is a genetic lethal** (i.e., if every carrier of the abnormal gene dies or is otherwise incapable of reproduction), **every case must arise as a new mutation.**
   a. In such circumstances, **the frequency of mutation for the abnormal gene can be calculated directly as one-half the frequency of the disease.** Every person represents two genes, and affected patients would be expected to be heterozygotes with one abnormal gene and one normal gene. If the number of people with the disease is D and the total number of people in the population is N, then the disease frequency is D/N, and the mutation frequency is D/2N.
   b. In practice, **diseases that are complete genetic lethals are usually not recognized as autosomal dominant traits** because they do not exhibit the characteristic pattern of transmission in families. This method of calculating mutation rates is therefore applied to just the subset of patients who are known to represent new mutations (i.e., those who are born to unaffected parents).
   c. **Mutation frequencies** have been calculated for several dominant diseases by this method. These frequencies have ranged **from about 10^{-6} to about 10^{-4}**.

2. **For an X-linked recessive disease that is genetically lethal in males, the frequency of new mutations is one-third the frequency of the disease.**
   a. The reason for this is, again, that females have two X chromosomes and males have only one. If the population includes equal numbers of males and females, two-thirds of all X-chromosomes occur in females and one-third in males. Similarly, **two-thirds of all X-linked genes (both normal and abnormal) occur in females and one-third in males.** A gene for an X-linked recessive disease that occurs in a male is expressed as the disease, but the same gene in a female is likely not to be expressed because the female's allelic gene is normal (i.e., she is a heterozygous carrier).
   b. If an X-linked disease is a genetic lethal, all males who carry the abnormal gene at that locus will die or otherwise fail to reproduce. Thus, **the one-third of all abnormal alleles contained in males are lost in every generation.** If the disease frequency does not change, **the abnormal genes that have been lost must be replaced in the next generation by new mutations.** Since genes are distributed randomly at fertilization, one-third of the new mutant genes at the disease locus end up in affected males and two-thirds end up in carrier females.
   c. In a **sporadic case** of an **X-linked recessive disease** that is genetically **lethal in males, there is only a two-thirds chance that the mother is a carrier** because there is a one-third chance that the affected boy represents a new mutation.

III. EFFECTS OF MEDICAL INTERVENTION ON DISEASE GENE FREQUENCIES

A. Some people have expressed concern that treatment of genetic disease would result in an increased frequency of such conditions in the population because abnormal gene carriers who would otherwise die will be permitted to survive and reproduce. In general, **this concern is unfounded.**

1. **Treatment of autosomal recessive diseases** has almost no effect on gene frequencies because most abnormal alleles are carried in unaffected heterozygotes.

2. **Treatment of a dominant or X-linked disease** could increase the disease frequency, but the change would probably be quite small.
   a. **Many dominant diseases do not affect reproduction** because they have late onset or are relatively mild. Thus, carriers may reproduce normally even without treatment.
   b. **Genetic counseling** in families of patients with serious autosomal dominant or X-linked diseases tends to discourage mating and thus decreases the likelihood that such traits will be transmitted to the next generation.

3. Conversely, **genetic counseling has very little effect on reducing the frequency of abnormal genes in a population**.

4. **Prevention and treatment of genetic disease benefits individual patients and families.** The effect on disease frequencies in the population is, in general, negligible.

B. Eugenics can be defined as improvement of the human species by selective breeding. For centuries, selective breeding has been used successfully in plants and domesticated animals. The best specimens were crossed to improve the quality of the stocks.

1. **Positive eugenics** seeks to increase the propagation of "more desirable" types of people.

2. **Negative eugenics** seeks to decrease the propagation of "less desirable" types.

3. **Eugenic policies became quite popular and were implemented** in Europe and North America during the early decades of the twentieth century.
   a. Some of the proponents of eugenics were **eminent scientists,** but their advocacy was often based on unconvincing data or on interpretations or extrapolations that went far beyond the facts.
   b. The popularity of eugenics was partly due to its **resonance with the contemporary social and cultural milieu,** which included broad acceptance of racism, imperialism, and social darwinism.

4. Eugenics was used to provide "scientific" support for **restrictive immigration policy, laws prohibiting marriage** on the basis of race and certain handicaps, and **laws requiring sterilization** of people considered to have serious genetic defects. A perversion of eugenic doctrine was used as a "scientific" basis for the extermination of Jews and other ethnic minorities by the Nazis during World War II.

5. Our current understanding of genetics provides **little scientific support** for eugenic programs, all of which must entail subjugation of the interests of the individual to the interests of future society, and thus some **loss of personal freedom.**

STUDY QUESTIONS

Directions: Each of the numbered items or incomplete statements in this section is followed by answers or by completions of the statement. Select the **one** lettered answer or completion that is **best** in each case.

1. Cystic fibrosis is the most common autosomal recessive disease among white Americans, occurring with a frequency of 1/2500 births. What is the approximate frequency of heterozygous carriers for cystic fibrosis in this population?

(A) 1/25
(B) 2/25
(C) 1/50
(D) 1/2500
(E) 2/2500

2. Sickle cell disease, an autosomal recessive condition that occurs with a frequency of 1/625 among black Americans, produces anemia, pain crises, and an increased risk of certain infections. What is the approximate chance that the clinically normal sister of a person with sickle cell disease will have an affected child if she marries an unrelated black man?

(A) 1/4
(B) 1/25
(C) 1/50
(D) 1/75
(E) 1/625

3. Albinism is an autosomal recessive disease that produces a lack of skin and hair pigmentation. The frequency of the gene for albinism in a population is known to be 1/190. What is the approximate risk for albinism in the child of an albino woman who marries an unrelated and unaffected man with no family history of albinism?

(A) 1/95
(B) 1/190
(C) 1/380
(D) 1/570
(E) 1/760

4. Each of the following observations can be explained by mutation–selection equilibrium EXCEPT

(A) new dominant mutations occur more often among the children of fathers who are over age 40 than among the children of younger fathers
(B) the frequency of mutation for an autosomal dominant disease that is genetically lethal is equal to one-half the frequency of the disease
(C) new mutations are often observed among patients with autosomal dominant diseases that substantially impair reproduction
(D) the chance of a new mutation is 1/3 in a boy who is the first in his family to be affected by an X-linked recessive disease that is genetically lethal in males
(E) the mothers of some heterozygous female carriers of X-linked recessive diseases that are lethal in males are not carriers themselves

5. Duchenne muscular dystrophy is an X-linked recessive disease that produces progressive loss of muscle function. Affected boys usually die by early adulthood; they are unable to reproduce. The frequency of Duchenne muscular dystrophy is about 1/25,000 males. What is the approximate frequency of carrier females for this disease?

(A) $\frac{2}{\sqrt{25{,}000}}$
(B) $\frac{2}{3 \times \sqrt{25{,}000}}$
(C) $\frac{2}{25{,}000}$
(D) $\frac{1}{25{,}000}$
(E) $\frac{2}{3 \times 25{,}000}$

1-A
2-D
3-B
4-A
5-C

6. Which one of the following is a correct statement regarding genetic load?

(A) Effective treatment of autosomal recessive diseases is likely to increase the frequency of such diseases in a few generations
(B) Genetic counseling is likely to reduce the frequency of autosomal dominant diseases substantially
(C) Mutations that are deleterious in one population may be beneficial in others
(D) Lethal equivalents are largely confined to populations with a high rate of consanguinity
(E) Eugenic policies, if conscientiously applied, could eliminate almost all abnormal genes from the population

---

6-C

Directions: The group of items in this section consists of lettered options followed by a set of numbered items. For each item, select the **one** lettered option that is most closely associated with it. Each lettered option may be selected once, more than once, or not at all.

Questions 7–10

Stenosis of the aqueduct of Sylvius is sometimes inherited as an X-linked recessive trait. Affected individuals have severe hydrocephalus and do not reproduce. In each of the following pedigrees, the affected boys have X-linked aqueductal stenosis. Match the recurrence risk for a future son of the woman indicated by the asterisk in each pedigree with the appropriate value.

(A) Essentially 0 (risk of mutation)
(B) 1/6
(C) 1/4
(D) 1/3
(E) 1/2

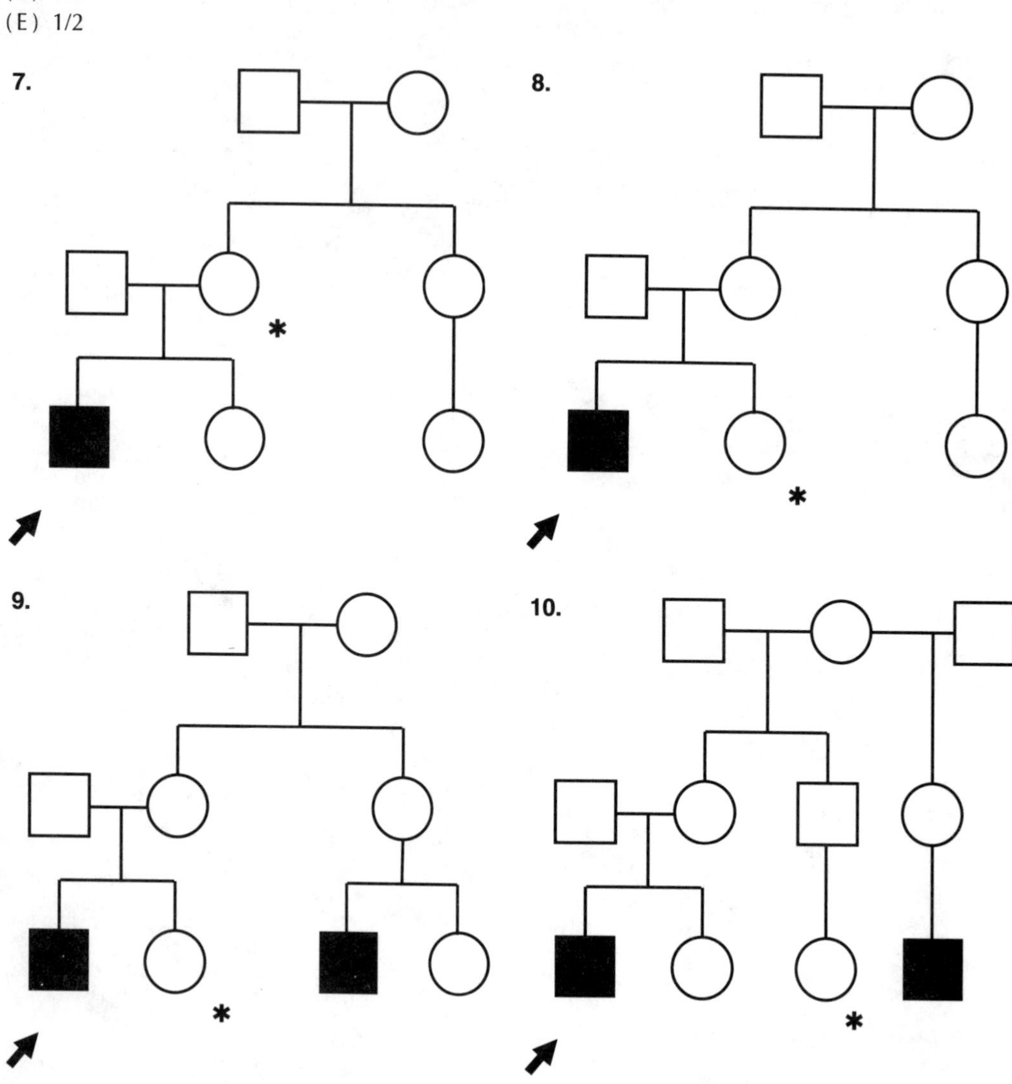

7-D 9-C
8-B 10-A

ANSWERS AND EXPLANATIONS

1. The answer is A *[I B 4].*
According to the Hardy-Weinberg law, the frequency of affected homozygotes for an autosomal recessive disease is q^2. In this case, $q^2 = 1/2500$, and $q = \sqrt{(1/2500)} = 1/50$. The heterozygote frequency is represented by 2pq where $p = 1 - q$. Since this is an uncommon disease, the heterozygote frequency is approximately equal to 2q or $2/50 = 1/25$.

2. The answer is D *[I B 4].*
An unaffected woman whose brother has sickle cell disease has a 2/3 chance of being a carrier of the abnormal gene. The disease occurs among homozygotes with a frequency of $1/625 = q^2$. The frequency of the abnormal gene (q) is $\sqrt{(1/625)} = 1/25$.

According to the Hardy-Weinberg law, the frequency of carriers in the general population is $2pq = 2 \times (24/25) \times (1/25)$ or about 2/25, which is the chance that the woman's husband carries the sickle cell gene. The chance that the woman would have an affected child is determined by multiplying the chance that she is a carrier by the chance that her husband is a carrier multiplied by the chance that the child would be affected if both parents are carriers:

$$\frac{2}{3} \times \frac{2}{25} \times \frac{1}{4} = \frac{4}{300} = \frac{1}{75}$$

3. The answer is B *[I B 6].*
If the frequency of the abnormal gene (q) is 1/190, then the frequency of carriers is approximately $2q = 2/190 = 1/95$. The woman is homozygous for the disease gene, so she will transmit it to her offspring. If her husband is a heterozygous carrier, he will have a 1/2 chance of transmitting the abnormal gene to his children. The chance that this woman would have an affected child is the product of her risk of carrying the gene, her husband's risk of carrying the gene, and the risk that they both would transmit it to a child. This is:

$$1 \times \frac{1}{95} \times \frac{1}{2} = \frac{1}{190}$$

Note that in this case the chance that both parents will transmit the abnormal gene is 1/2 and not the usual 1/4 because the mother MUST transmit the abnormal gene, and the father has a 1/2 chance of transmitting it.

4. The answer is A *[II E 1 a, 2].*
Mutation–selection equilibrium accounts for the maintenance of constant disease (and gene) frequencies in a population by replacement of all genes lost to selection in one generation by new mutations in the next generation. New mutations do occur more often among the children of older fathers, but this appears to be related to the continuing cell division that occurs among spermatogonia throughout life. Selection and gene loss are not a factor in the paternal age effect.

Since every person represents two genes at each locus, there are twice as many genes at each locus as there are people in a population. The frequency of individuals with a disease due to a mutant gene is thus twice the frequency of new mutant genes among all genes at that locus.

For a genetic lethal, all cases of the disease must represent new mutations. A dominant disease that substantially impairs fertility would disappear in a few generations if new mutations did not occur. Among people affected by such diseases, new mutations are common.

According to mutation–selection equilibrium, 1/3 of boys who have sporadic cases of an X-linked recessive disease that is lethal in males represent new mutations. In the other 2/3 of cases, the mothers are carriers. These women represent new mutations themselves in half of the cases because new mutant genes may first appear in either males or females.

5. The answer is C *[I B 7 a (5)].*
For an X-linked recessive disease, the gene frequency (in both males and females) is the same as the frequency of the disease in males. In this case, $q = 1/25{,}000$. The frequency of female heterozygotes is indicated by the term 2pq in the Hardy-Weinberg law. If the disease is rare (i.e., if q is very small), the frequency of carrier females is approximately equal to 2q or 2/25,000 in this case.

6. The answer is C *[II A 2].*
Whether or not a gene exerts a deleterious effect depends on many factors such as environmental conditions, population density, and the "genetic background" (i.e., what other genes are common in the population). A gene that is deleterious under one set of circumstances may be beneficial under others.

Treatment of autosomal recessive diseases has almost no effect on gene frequencies because most abnormal autosomal recessive alleles are carried in clinically normal heterozygotes. Conversely, genetic counseling has little beneficial effect on the overall frequency of dominant diseases because many are due to new mutations, have onset after the usual period of reproduction, exhibit incomplete penetrance, or have variable and often mild expression.

Lethal equivalents (genes that, when homozygous, cause death in some affected individuals) are measured in consanguineous matings but are found in almost everyone. Each normal person carries approximately 3–5 lethal equivalents in a heterozygous state.

Eugenic policies are unlikely to eliminate most deleterious genes because almost every normal person carries a few abnormal genes, because new mutations continually occur and because whether a gene is considered normal or not depends on many variable genetic and nongenetic factors.

7–10. The answers are: 7-D, 8-B, 9-C, 10-A *[I B 7 a].*
X-linked aqueductal stenosis is an X-linked recessive disease that is genetically lethal in males. In the pedigree in question 7, there is just one affected boy. According to mutation–selection equilibrium, there is a 1/3 chance that this case arose as a new mutation and a 2/3 chance that the mother carries the abnormal gene. If she is a carrier, the risk that her son would inherit the abnormal gene and be affected is 1/2. Thus, the overall risk that her son would have X-linked aqueductal stenosis is:

$$\frac{2}{3} \times \frac{1}{2} = \frac{2}{6} = \frac{1}{3}$$

In the pedigree in question 8, the son of the sister of the proband is at risk. Again, the mother's risk of being a carrier of the gene for X- linked aqueductal stenosis is 2/3 because of mutation–selection equilibrium. Her daughter has a 1/2 chance of inheriting the abnormal gene if the mother is a carrier. If the daughter is a carrier, her son has a 1/2 risk of inheriting the abnormal gene from her and being affected with the disease. The overall risk that the daughter's son would have X-linked aqueductal stenosis is thus:

$$\frac{2}{3} \times \frac{1}{2} \times \frac{1}{2} = \frac{2}{12} = \frac{1}{6}$$

The pedigree in question 9 differs from the previous pedigrees because two boys are affected with X-linked aqueductal stenosis, and they are related to each other through their mothers. It is, therefore, very likely that the proband's mother and her sister both carry the abnormal gene. Two separate new mutations are an extremely unlikely explanation for this pedigree. If the proband's mother is a carrier, the proband's sister has a 1/2 chance of being a carrier as well. If the proband's sister is a carrier, her son has a 1/2 chance of inheriting the abnormal gene and being affected with the disease. The overall risk that the proband's sister's son would have X-linked aqueductal stenosis is:

$$\frac{1}{2} \times \frac{1}{2} = \frac{1}{4}$$

In the pedigree in question 10, the mother of the proband is almost certainly a carrier of the abnormal gene because her maternal half-sister also has a boy with X-linked hydrocephalus. In this family, however, the risk to the sons of the mother's brother's daughter is a concern. This woman (the one with the asterisk) cannot have inherited the abnormal gene because her father did not inherit it from his mother; if he did, he would have been affected with the disease. If a woman does not carry the abnormal gene, she cannot transmit it to her son, so the risk of the proband's cousin's son being affected is extremely small.

7
Genetic and Environmental Interaction

J.M. Friedman, Michael R. Hayden, Barbara McGillivray

I. INTRODUCTION. Arguments are sometimes heard about the relative importance of "nature" and "nurture" in determining human traits, but there are few instances in which either normal development or a disease state occurs without interaction of both genetic and nongenetic factors. Even if only genetic factors are considered, it is important to remember that mutant genes do not act in isolation; they are influenced by the action of many other genes. Inborn errors of metabolism are largely determined by genetic mutations, but sometimes can be treated or ameliorated by dietary or other nongenetic manipulations. Pharmacogenetics provides a vivid demonstration of the interaction of genotype and environmental agents. Most common isolated congenital anomalies and most common diseases of middle life (e.g., diabetes mellitus, hypertension) exhibit multifactorial inheritance (i.e., they are due to combination of genetic and nongenetic factors).

II. BIOCHEMICAL BASIS OF DISEASE. Genes either specify the ultimate amino acid sequence of a protein, or they control the rate or timing of synthesis of a specific protein. The immediate consequence of a gene mutation is a change in the quality or quantity of a specific protein. More than 500 single gene biochemical abnormalities are partly or completely understood, with a greater number yet to be classified. Rather than attempting to remember each individually, the physician should be familiar with how various types of mutations influence the final protein and determine the clinical pathology.

A. Relationship of genes and enzymes. Beadle and Tatum (1941) refined the concept of "one gene—one polypeptide" to state the following.

1. **All biochemical processes are genetically controlled.**
2. **Each biochemical pathway is made up of steps,** each of which are under the control of a different enzyme.
3. **Each enzyme is encoded by one or more genes.**
4. **A single gene mutation will result in an alteration in a biochemical reaction;** therefore, all single gene disorders should be the result of an abnormal protein molecule.

B. Protein functions and disorders. Proteins function not only as enzymes but also are involved in metabolism and structure. In each function, mutation can cause specific genetic disorders.

1. **Enzymes** act as catalysts to mediate biochemical reactions and are usually proteins. The enzyme substrates include carbohydrates, organic acids, lipids, and amino acids. Some examples of single gene disorders (see Ch 3 I B, C) involving enzyme defects include the following.
   a. **Galactosemia** (see Ch 3 I C 3 a). The most common enzyme deficiency involves galactose-1-phosphate uridyl transferase. Infants present with hepatomegaly, jaundice, and hypotonia. Death may occur from sepsis. The enzyme deficiency results in accumulation of galactose, which cannot be converted to glucose. Treatment involves elimination of lactose from the diet.
   b. **Phenylketonuria (PKU)** [see Ch 5 III C 2; Case 5-D]. Hepatic phenylalanine hydroxylase converts the essential amino acid phenylalanine to tyrosine. Deficiency of phenylalanine hydroxylase leads to increased levels of both phenylalanine and its metabolites, which results in mental retardation and pale coloring in affected children. Treatment involves severe restriction of oral phenylalanine.

c. Tay-Sachs disease [see Ch 5 II B 2 b (1)]. Hexosaminidase A is involved in complex lipid metabolism by removing sugars from a particular branched long-chain lipid (ceramide), which acts as a surface membrane receptor in the brain. In Tay-Sachs disease, the enzyme abnormality leads to massive accumulation of the lipid in brain cells (a storage disorder). Affected children gradually lose skills (e.g., sitting) and sight. Death occurs at a young age. There is no treatment yet.

d. Mucopolysaccharidosis type I (MPS I)

(1) Deficient levels of α-L-iduronidase result in three overlapping clinical phenotypes, which are all part of MPS I.

(a) **Hurler syndrome (MPS IH)** is the most severe with children having stiff joints, hepatosplenomegaly, corneal clouding, and progressive mental retardation. The children die of respiratory infection or cardiac failure.

(b) **Scheie syndrome (MPS IS)** has some similar features (e.g., joint stiffness, corneal clouding) to MPS IH, but the effects are milder, and the children do not develop mental retardation. The diagnosis may not be made until early adulthood.

(c) **Hurler-Scheie syndrome (MPS IHS)** is an intermediate phenotype with joint involvement, corneal clouding, but little mental retardation.

(2) All three clinical conditions are associated with **abnormal mucopolysaccharide excretion, and low α-L-iduronidase levels**. The most likely explanation of these results is that Hurler and Scheie syndromes result from different alleles of the α-L-iduronidase gene, while the Hurler-Scheie syndrome is a compound.

2. Transport. Membrane transport proteins are involved in the movement of various substances across cell membranes, but transport proteins may also be involved with transport through the body, or transport inside the cell. Deficiencies are usually only clinically significant in homozygotes.

a. β-thalassemia. As a consequence of a number of different mutations affecting the β-globin gene, there is decreased synthesis of β chains. Homozygotes have severe anemia and require repeated blood transfusions. The normal function of the hemoglobin molecule is to transport oxygen from the lungs to the peripheral tissues.

b. Cystic fibrosis (CF) [see Ch 5 II B 3]. Recently, the gene involved in CF has been found to encode for a transmembrane conductance regulator protein involved in the transport of chloride across the membranes of cells in the pancreas, lungs, and sweat glands. It is postulated that the defect involves chloride permeability and leads to chronic obstructive lung disease and pancreatic insufficiency. Treatment is still only symptomatic.

3. Cell structure. One protein class comprises the structure or framework of extracellular molecules (e.g., collagens), the cell membrane and cytoskeleton, and the organelles within the cell. Deficiencies may lead to clinically significant disease in both heterozygotes and homozygotes. Two examples are:

a. Duchenne muscular dystrophy (DMD). Dystrophin is a structural protein thought to be important for the integrity of the muscle cell membrane. Defects in the X-linked dystrophin gene are most often deletions, which lead to either decreased or absent production of dystrophin (see Ch 5 II A 3 a).

b. Osteogenesis imperfectas are a group of disorders involving defects in the production of either type I or type III collagen.

(1) **Type I collagen** is the major structural protein of bone and is a helical structure composed of two procollagen proα_1(I) chains and one procollagen proα_2(I) chain.

(2) Infants with a particularly severe type of osteogenesis have been found to have a **point mutation** involving a substitution of a glycine, which results in a bulkier molecule.

(3) Affected infants are heterozygous for the mutation and usually die from severe fracturing and bone crumpling.

4. Intercellular metabolism. A number of protein products function as hormones, as receptors for hormones or metabolites, or as signal transducers. Disorders may be inherited as autosomal recessive or autosomal dominant. Red–green color blindness is a common X-linked condition involving specific light receptors. More significant disorders include the following.

a. Familial hypercholesterolemia (FH), which is one of the most common disorders in humans (1/500 individuals are heterozygous for the condition). See Chapter 3 II A 1 c and Chapter 5 II A 3 c for more information on FH.

b. Thyroid hormone deficiencies. A number of thyroid hormonogenesis defects exist (e.g., thyroid peroxidase deficiency, abnormal carrier substance, abnormal iodine receptor), all of which are inherited as autosomal recessive traits.

(1) Some thyroid hormone deficiencies, such as **thyroid peroxidase deficiency,** involve an inability to convert accumulated iodide to organically bound iodine. A number of abnormal alleles are now defined that can determine the clinical severity.
(2) The end result is **congenital hypothyroidism,** which presents as an infant with coarse features, delayed growth, and eventual mental retardation.
(3) Treatment involves administering exogenous thyroid hormone.

C. Types of protein alterations. Protein synthesis is the result of transcription of DNA into mRNA in the nucleus of the cell and translation of mRNA into protein in the ribosome. Transcription is, therefore, nucleotide-to-nucleotide, while translation is nucleotide-to-amino acid sequence. The regulation of transcription and translation determines the normal process of development. The process can be broken down into the following steps.

1. **Transcription and RNA splicing** (see Ch 1 II B 1, 2). Alterations in the amount of transcribed RNA will be seen when deletions or mutations involve regulation, splice sites, or introduce a stop codon. This results in an inability for the complete DNA sequence to be read and converted to mRNA. The **thalassemias** result from a decreased or absent amount of the globin product.

2. **Translation** (see Ch 1 II B 3). If a mutation results in a frameshift, transcription may occur, but there will not be translation into a polypeptide.

3. **Secondary and tertiary structure of the polypeptide.** Mutations in the secondary and tertiary structure result in production of a protein; however, abnormal folding, resulting in an unstable molecule, may occur.
   a. Mutations in the collagens may produce abnormal crosslinking, which results in a weak structural protein.
   b. An example of weak structural proteins due to collagen mutations is seen with the stretchy skin and hyperextensibility in certain types of **Ehlers-Danlos syndrome**.

4. **Three-dimensional structure.** The completed protein may be destined to occupy a specific position in the cell structure. Alterations in key amino acid sequences may interfere with this function.
   a. In **methylmalonic aciduria,** the enzyme lacks a leader sequence that prevents entry of the enzyme into the peroxisome.
   b. Infants present with vomiting and failure to thrive.

5. **Association with other proteins.** Individual proteins may combine with other proteins outside the cell before becoming biologically active. If the mutation interferes with the formation of these macromolecules, the function of the protein is disturbed. In some rare types of **thalassemia,** the mutations do not interfere with the formation of the individual globins, but with the interfacing between the globins and other components to form hemoglobin.

D. Biochemical disease. Although many of the mutations responsible for metabolic disease have been characterized, the clinician and the biochemist first note certain features in the patient or in the patient's biochemistry profile. Until the basic defect is clarified, what the **clinician sees** is the **consequence of a variety of mutation types**.

1. **Failure of formation of a product.** If there is a block in a metabolic pathway, the end result will be a reduction or absence of a particular metabolite (e.g., glycogen storage disease leads to hypoglycemia). The basic defect could be with the enzyme, a cofactor, or a receptor.

2. **Accumulation of precursors** or excessive production of substances derived from precursors via alternate pathways. There may be toxic effects from the accumulated substrates or physical changes with the storage (e.g., hepatosplenomegaly in some of the storage disorders).

3. **Overproduction of a normal product** may occur if the defect involves a lack of feedback control over a particular pathway. In **gout,** purine overproduction leads to painful deposition of crystals in joints.

4. **Transport and receptor diseases.** An abnormal receptor or transport protein results in an alteration in the binding or cell membrane transport of other molecules. The end result is a deficiency of normal protein. **Androgen insensitivity syndrome** involves defective binding of androgens, a deficiency at the nuclear level, and can result in sex reversal.

5. **Diagnosis of biochemical disease** involves the identification of accumulated or missing metabolites, the measurement of specific enzyme levels, or the identification of protein variants. See Chapter 5 for discussions of techniques used in DNA diagnosis of biochemical diseases and other genetic disorders.

E. **Genotype–phenotype correlations.** An important question to answer is whether the clinical presentation can be predicted from knowledge of the gene involved. Present knowledge of specific mutations suggests the following.

1. **Allelic mutations may result in different clinical phenotypes.** The mutations in the dystrophin gene may lead to DMD or Becker muscular dystrophy (BMD), depending on the amount of functioning dystrophin. In the thalassemias, there may be enormous variation, depending on the type and location of allelic mutations.

2. **The clinical presentation may not be predictable from knowledge of the specific mutation.** Although the basic defect may be known, the actual development of a disease may not be clear. For example, the changes seen in the fragile X gene (i.e., the increased size of a repeat segment) do not yet explain the speech delay and other predictable findings of fragile X syndrome.

3. **Inheritance patterns are not always recessive.** The majority of enzyme deficiency conditions (e.g., PKU) are manifested only in the homozygote, although the normal heterozygote may have intermediate levels of the involved enzyme. However, many biochemical diseases (e.g., FH, type I osteogenesis imperfecta) involving structural or transport proteins are manifested in the heterozygote; an affected homozygote may be rare.

III. PHARMACOGENETICS

III. PHARMACOGENETICS refers to the influence of genes on the determination of the response to drug therapy. These responses may either take the form of an exaggerated physiological response to a drug, a resistance to drug effects, or an increased frequency of side effects. In some instances, pharmacological agents can also precipitate and trigger the effects of certain genetic diseases. Other factors that might influence the responsiveness to drugs include ethnic background and the presence of specific human histocompatibility antigens or human leukocyte antigen (HLA) groups. In general, certain genes may affect the metabolism of drugs, trigger the clinical presentation of certain genetic diseases, or increase the frequency of certain side effects.

A. **Genes affecting drug metabolism.** Polymorphisms in certain genes can have significant effects on the activities of certain enzymes that are crucial in the metabolism of certain drugs. The following findings re-emphasize the importance of increased awareness of the possibility of differences in drug response and in dose requirements among different patients from various ethnic and racial groups.

1. **Acetylation** is critical for the metabolism of many different drugs. Polymorphisms in the rate of acetylation in different patients can result in different effects of drugs in certain persons.
   a. For example, patients who have a phenotype of **slow acetylation** for drugs such as **isoniazid,** which is an important drug for treating tuberculosis, are more likely to **develop the side effects** of isoniazid therapy (e.g., peripheral neuropathy even in conventional doses).
   b. On the other hand, patients who have **rapid acetylation** of drugs like isoniazid may be **inadequately treated** since the isoniazid levels in the blood will be decreased. Therefore, a patient taking isoniazid for tuberculosis may become more prone to recurrence of the disease.
   c. Genetic polymorphisms of acetylation have been described for many drugs that are commonly used in medicine.

2. **The cytochrome P_{450}-dependent enzyme system** is also involved in the metabolism of many different drugs. Variation in the gene for this enzyme can result in the increased frequency of side effects in certain persons.
   a. About 8%–10% of the **American Caucasian population** and less than 1% of the **Chinese population** are poor metabolizers of drugs that utilize this system.
   b. Drugs that fall into this category include **propranolol,** a β-blocker commonly used in the treatment of hypertension and angina.
   c. The altered activity of this cytochrome P_{450} enzyme results in an **altered sensitivity** to propranolol with increased frequency of side effects at conventional doses in a significant number of American Caucasians.

3. Individuals vary markedly in their responses to **ethanol**. There are significant ethnic differences in the pharmacological affects of alcohol due to the rates that individuals from different backgrounds metabolize ethanol.
   a. Persons of **Asian descent,** including Chinese and Japanese individuals, **metabolize ethanol at a higher rate than Europeans,** with more rapid production of by-products of ethanol and increased clinical effects.
   b. The variation in metabolism of ethanol in different population groups is due to **polymorphism in the enzyme alcohol dehydrogenase**.

B. Pharmacological effects of drugs on certain genetic diseases

1. **Glucose-6-phosphate dehydrogenase (G6PD) deficiency.** Patients who are deficient in this enzyme might develop rapid hemolysis as an adverse reaction to treatment with antimalarial agents. This is seen in people of **Mediterranean** and **African descent**.
   a. People with G6PD deficiency may remain healthy provided their red cells are not subjected to chemical or pharmacological stresses.
   b. The locus controlling G6PD is situated on the X chromosome; therefore, this condition is inherited as an **X-linked recessive trait**.

2. **Malignant hyperthermia** is usually triggered by anesthetic agents such as halothane. Affected people present with muscle spasm and hyperthermia with subsequent acidosis. If untreated, death occurs.
   a. This condition is inherited as an **autosomal dominant trait**.
   b. At the present time, at-risk individuals wanting to know whether they have inherited the gene for this disorder must undergo a **muscle biopsy,** which is assessed for an abnormal contraction in **response to caffeine or other stimulatory agents**.
   c. The discovery of the gene for malignant hyperthermia on chromosome 19 offers the future possibility of DNA-based testing for this disease.

3. **HLA and adverse reactions.** In some instances, the adverse reactions to drugs may be controlled or influenced by genes in the HLA region, which, in turn, influence the immune response. For example, patients receiving gold therapy who have the HLA haplotype DRW3 are more likely to have side effects than individuals who do not have this HLA haplotype.

4. **Succinylcholine sensitivity.** The enzyme cholinesterase hydrolyzes choline esters including acetylcholine and succinylcholine, which is a widely used muscle relaxant. Approximately 1 in 3000 Caucasian individuals is homozygous for an atypical cholinesterase allele, which results in an ability to break down succinylcholine. Such people will have prolonged muscle relaxation and apnea after anesthesia with succinylcholine. The abnormal cholinesterase can be determined through a laboratory test.

C. Clinical concerns in pharmacogenetics

1. It is important to recognize that genes can significantly alter the therapeutic responsiveness to drugs, and in some instances these drugs may precipitate clinical manifestations of genetic diseases. A key to detecting these susceptibilities is a **thorough family history**.
   a. A family history may reveal the presence of a particular adverse reaction to a drug in a first-degree relative, and in those instances, the possibility of an autosomal dominant disorder must be considered.
   b. Appropriate investigations must be done prior to subjecting the person to a particular stress such as an anesthetic.

2. The presence of unusual adverse reactions to pharmacological agents may reflect the presence of a **rare genetic allele** in that particular individual, which predisposes the person to the adverse event.

3. **Certain drugs may be teratogenic** in some individuals due to variations in genes of the mother or the fetus (see Ch 8 II E, F). In other instances, **genetic traits might protect some individuals** from the harmful effects of certain drugs.
   a. The **challenge is to identify patients and their families who are susceptible to adverse drug reactions** by virtue of their genetic constitution.
   b. Remember that drug metabolism, including the process of absorption, distribution, binding to receptors, and degradation, is determined in part by genetic factors.

IV. MULTIFACTORIAL INHERITANCE is responsible for most normal phenotypic differences among individuals as well as for many congenital anomalies and common diseases of adulthood.

A. Definitions and terminology

1. The term **multifactorial** is used in medicine to describe traits that are due to a combination of many genetic and nongenetic factors. If only the genetic factors are considered, the inheritance is called **polygenic**.
   a. This definition of a multifactorial trait implies both a familial nature and an environmental dependence. A child of tall parents is more likely to be tall than is a child of short parents, but poor nutrition (an environmental factor) may compromise the growth of either child.
   b. The genetic predisposition to multifactorial traits is usually inherited from both parents.
   c. The **genes and environmental factors affecting a given multifactorial trait may vary among different individuals**. Black hair may be the result of genes inherited from an Asian genetic background in one individual, of genes inherited from an African genetic background in another person, and of hair dye in a third individual whose genetic background would ordinarily produce brown hair.

2. **Quantitative traits** are those that **can be measured**. Examples include height, weight, and serum cholesterol level.
   a. In general, **quantitative traits exhibit continuous variability**. This means that they can assume an unlimited number of intermediate values between extremes. This contrasts with **discrete** or **qualitative traits,** which are all-or-none phenotypes like Down syndrome or achondroplasia.
   b. **Quantitative phenotypes** are typically inherited as **multifactorial traits** in normal individuals.

B. Normal variation

1. **Most phenotypic variation among normal individuals** is due to multifactorial traits. Examples include the range of skin pigmentation, weight, and intelligence seen among normal people.
2. **Extreme values** may occur as a result of normal variation. A man is unlikely to be 4′10″ tall just on the basis of normal variation, but it is possible.
3. **Children tend to resemble their parents** with respect to normal traits because of the underlying genetic components. Tall parents tend to have tall children. For normal quantitative traits, a child's value tends to resemble the average of his or her parents' values. This average is called the **midparent value**.
4. **Each parent contributes half of the genetic factors** for any multifactorial trait in his or her child, on the average. This means that the children of a parent who exhibits an extreme value for a multifactorial trait tend to have less extreme values themselves, a phenomenon known as **regression to the mean**.
5. Although quantitative traits are continuously variable, **only a limited range of values may occur**. For example, human adults with heights of 5″ or 40′ do not exist.
6. **Quantitative characteristics** that exhibit continuous variation among normal individuals may also exhibit a **major alteration as a result of genetic or environmental factors of large effect**. Thus, achondroplasia or thyroid ablation by radiation may greatly reduce the height of a child in whom average stature would otherwise have been expected.

C. Multifactorial diseases. Many diseases are also inherited as multifactorial traits.

1. **Some multifactorial diseases are defined clinically in terms of quantitative traits.** For example, hypertension is diagnosed on the basis of a patient's blood pressure being above a certain value.
   a. The difference between "normal" and "disease" in such cases is arbitrary. Blood pressure is a continuous variable, and the pathological effects of having "high" blood pressures of 140/110, 120/85, or 125/81 compared to "normal" blood pressures of 120/80 or 110/75 is only one of degree.
   b. Inheritance of such diseases can be thought of in terms of extreme manifestations of a multifactorial quantitative trait (see IV A 2).

2. Other multifactorial diseases differ qualitatively from the normal state. Cleft palate is one example. In such cases, a multifactorial predisposition to the disease is inherited, and the disease either occurs or does not occur, depending on the aggregate strength of the predisposing factors.

   a. The **multifactorial-threshold model** (Figure 7-1) is a formal explanation of how a multifactorial predisposition can produce a trait that is qualitatively distinct from normal.

   b. According to this model, all of an individual's genetic and environmental disease-predisposing factors, when considered together, constitute liability. **Liability follows a normal distribution in the population.**

   c. If an individual's **liability exceeds a certain threshold value, he or she will have the disease**. If the liability does not exceed that threshold, the disease will not occur.

3. Diseases transmitted as multifactorial traits

   a. Most **isolated congenital anomalies** appear to be multifactorial traits. Examples include cleft lip with or without cleft palate, congenital heart disease, and anencephaly.

   b. Many **common diseases of adulthood** also seem to be multifactorial traits. Noninsulin-dependent diabetes mellitus, arteriosclerotic heart disease, and schizophrenia are examples.

4. Clinical characteristics of multifactorial diseases include the following.

   a. The disease tends to be **familial but does not usually follow any monogenic pattern of inheritance**.

   b. Multifactorial disorders tend to **occur more frequently in one sex** than in the other. Examples include:

      (1) Pyloric stenosis, which is more common in males

      (2) Systemic lupus erythematosus (SLE), which is more common in females

   c. The recurrence risk is the same for all relatives who share the same proportion of genes.

      (1) For example, each sibling, child, and parent shares half of his genes with a proband, on the average.

      (2) The recurrence risk in all such first-degree relatives would be expected to be about the same.

   d. The recurrence risk in a family drops off quickly as the relationship to an affected individual becomes more remote. For example, the recurrence risk for cleft lip with or without cleft palate is about 4% for the sibling but only about 0.5% for the first cousin of an affected child.

   e. The recurrence risk in first-degree relatives of an affected individual is usually about equal to the square root of the frequency of the condition in the population.

      (1) This means that the recurrence risk is lower in populations exhibiting a lower incidence of a condition.

      (2) For multifactorial **congenital anomalies** that have population incidences of about 1/1000, the **recurrence risk** in siblings or children of an affected individual is usually **2%–4%**.

      (3) For multifactorial **diseases of adulthood** that have population prevalences in the range of 1%, the **recurrence risk** in siblings and children is usually in the range of **5%–10%**.

   f. Multifactorial diseases are **more common among the children of consanguineous parents**. This is because parents who are related are more likely to share similar disease-predisposing genes.

   g. If one of a pair of **monozygotic twins** has a multifactorial disease, the **chance that the second twin is also affected** (i.e., is concordant) is less than 100% and **usually less than 50%**.

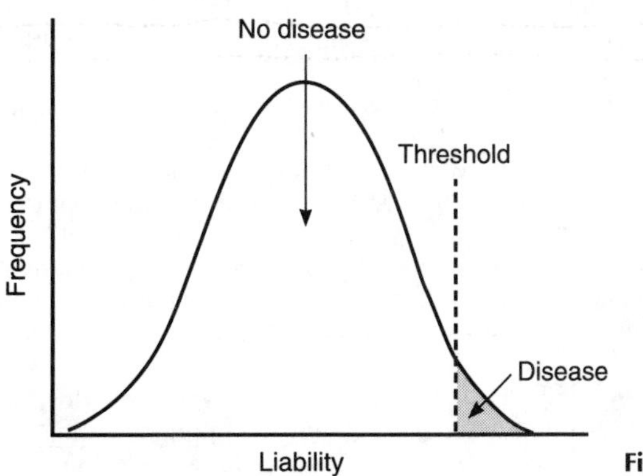

Figure 7-1. The multifactorial-threshold model.

h. The **concordance** for multifactorial conditions **in dizygotic twins is much less** than the concordance in monozygotic twins and usually is similar to the recurrence risk in ordinary siblings.

5. Factors that influence the recurrence risk for multifactorial disorders in a family can include the following.

a. The recurrence risk is higher if more than one close relative is affected.

(1) For example, the risk for cleft lip with or without cleft palate in a child who has one affected sibling is about 4%. If the child has two affected siblings, the risk is about 10%.

(2) This reflects the fact that a couple who have already had more than one affected child are more likely to have more predisposing factors (i.e., more liability) than a couple with just one affected child.

(3) Note that this is quite different from monogenic inheritance in which the risk of recurrence depends on the parental genotypes and is independent of the number of previously affected children.

b. The recurrence risk is higher in the relatives of an affected individual of the less frequently involved sex.

(1) Individuals of the less frequently involved sex must have **more predisposing factors** to manifest the disease so their relatives are at greater risk of being similarly affected.

(2) For example, **pyloric stenosis** occurs five times as often in males as in females. The recurrence risk in the brother of an affected child is 3.8% if the affected child is male, but 9.2% if the affected child is female.

c. The recurrence risk is higher in relatives of more severely affected probands.

(1) Such individuals must have **more predisposing factors** to manifest the disease severely; therefore, their relatives are at greater risk.

(2) For example, the risk of recurrence of **cleft lip with or without cleft palate** is 2.5% in the sibling of a child with unilateral cleft lip but almost 6% in the sibling of a child with bilateral cleft lip and palate.

(3) Note that this is very different from the situation with monogenic and chromosome disorders.

6. Multifactorial inheritance is a diagnosis of exclusion for the following reasons.

a. The **clinical features** of a multifactorial disorder **rarely distinguish** it from similar lesions that have another cause.

(1) For example, **cleft lip and cleft palate,** alone or in combination, usually exhibit multifactorial inheritance.

(2) There are, however, more than 150 other conditions that have cleft lip and/or palate as a feature.

(3) Some of these conditions are due to chromosome abnormalities, others to monogenic disorders, and still others to teratogenic environmental effects.

b. Multifactorial congenital anomalies typically **occur in patients who do not have embryologically unrelated birth defects.**

(1) Cleft lip and club feet are examples of abnormalities that are **embryologically unrelated** to each other.

(2) Cleft lip and cleft palate are embryologically related because morphological alterations that produce one can also produce the other.

(3) A common congenital anomaly such as **spina bifida** in a child who is otherwise completely normal is likely to be multifactorial. Spina bifida in a child who also has congenital heart disease, kidney malformations, and cleft lip is not likely to be multifactorial.

c. The recurrence risks used for multifactorial conditions do not apply to situations in which the condition is not multifactorial.

(1) The recurrence risk for the sibling of a child with congenital heart disease, omphalocele, and club feet cannot be calculated from the recurrence risks estimated for these conditions when they occur as isolated anomalies.

(2) The recurrence risk for the sibling of such a child is determined by the underlying cause of all of the child's anomalies, and this is unlikely to be multifactorial in a child with multiple congenital anomalies.

V. GENETICS OF COMMON ADULT DISORDERS. By age 25, approximately 5% of individuals will suffer from a genetic disease. However, if disorders that have a genetic component in their causation (e.g., heart disease, cancer) are included, more than 50% of the population will suffer from a disorder with a major or minor genetic contribution during their lifetime, many of which are caused by the interaction of genetic and environmental factors. By identifying susceptible individuals, it may be possible to modify known environmental factors and prevent the illness in genetically susceptible persons.

A. Diabetes affects approximately 5% of the population. There are different genetic contributions to insulin-dependent (type 1) and noninsulin-dependent (type 2) for adult-onset diabetes.

1. Type 1 (juvenile-onset, insulin-dependent) diabetes mellitus

a. Genetic background. An **association with HLA-DR3 and HLA-DR4** is documented, and 95% of insulin-dependent diabetics have one or both of these antigens compared with 50% of the general population. The risk to a first-degree relative of a person with insulin-dependent diabetes mellitus is approximately 10%.

(1) Since the subgroups HLA-DR3 and HLA-DR4 can be rapidly identified by molecular analysis, refinement of prediction has now become possible.

(a) The hypothesis is that genes closely linked to these HLA antigens control immune responses, which make a person susceptible to environmental factors such as viruses, which can result in pancreatic damage and subsequent diabetes.

(b) These patients develop antibodies to insulin and islet cells of the pancreas and require insulin therapy.

(2) It is now possible to identify persons at risk for juvenile-onset diabetes by using a combination of molecular, serological, and immunological markers.

b. Environmental aspects. Approximately 50% of monozygous twins are concordant for this disease, indicating that environmental factors—probably viral infections (e.g., rubella)—also play a major role in addition to genetic factors. The greatest risk is to subjects who have antigens HLA-DR3 and HLA-DR4 and a family history of insulin-dependent diabetes mellitus.

2. Type 2 (adult-onset, noninsulin-dependent) diabetes mellitus

a. Genetic aspects. Type 2 diabetes mellitus has a very strong genetic predisposition, and concordance in monozygous twins is close to 100%. The risk to siblings for developing diabetes increases with age. The risk to first-degree relatives of a person with noninsulin-dependent diabetes mellitus is approximately 5%.

b. Environmental aspects. The major factor that unmasks the presentation of diabetes is obesity. One way to prevent onset of type 2 diabetes mellitus is for those individuals who have a first-degree relative with this disorder to maintain optimal body weight. In other words, knowledge of a family history of diabetes can have a significant influence on modification of lifestyle in an effort to prevent the expected outcome.

B. Hemochromatosis

1. Clinical aspects. Hemochromatosis is due to the accumulation of toxic quantities of iron in different organs and has a disease frequency of about 1/300.

a. Iron concentration, which is highest in the liver and pancreas, sometimes reaches 50 to 100 times the normal range.

b. Iron concentration can also increase in the joints, skin, spleen, kidney, stomach, and the thyroid, adrenal, and pituitary glands.

c. As a result, patients present with pain in their joints associated with increased pigmentation in the skin and feelings of weakness and malaise.

d. In time, the iron accumulation affects liver and pancreatic function, causing cirrhosis and diabetes.

e. Hypogonadism, due to involvement of the pituitary with testicular atrophy, and loss of libido also occur.

2. Diagnosis of hemochromatosis is made by a significantly increased transferrin saturation level to between 80% and 100%, which is associated with increased serum ferritin and serum iron.

3. Treatment. The clinical manifestations of this disease are **completely preventable** if recognized early and treated with **regular phlebotomy**. If untreated, hemochromatosis results in premature death due to diabetes, cirrhosis, or cardiac failure.

4. Genetic aspects. About 1/10 Caucasian individuals are carriers of hemochromatosis, which is an autosomal recessive trait. The disease is associated with HLA haplotype A3, and it has been shown that the gene causing this disorder is also closely linked to the HLA region. In some instances, there may be a family history of hemochromatosis in first-degree relatives. The presence of hemochromatosis in a parent and child reflects the high frequency of the carrier state for this disease.

C. Cancer

1. **Genetic contribution to common malignant tumors of adulthood.** Genes play a major role in the development of cancer; there are some genetic syndromes predisposing to malignancy and some tumors that follow a strict mendelian inheritance. The following factors support a genetic contribution to common cancers.
   a. **Involvement of multiple siblings or first-degree relatives.** In some instances, this may appear as an autosomal dominant trait, and the risk may approach 50% for an at-risk offspring.
   b. **Multiple primary tumors.** This has also been found to increase the risk for relatives in the case of colonic tumors.
   c. **Early onset.** The more atypical the age of onset for a particular cancer, the more likely that genes are playing some part in its causation.
2. **Recent advances.** Recently, tumor suppressor genes (in particular, the p53 gene) have been shown to play a key role in the development of human cancer in different tissues.
   a. Mutations in the p53 gene cause a loss of suppressor function and can activate this gene as an **oncogene**.
   b. In the future, it may be possible to **detect change** in the p53 gene in certain tissues **and predict** who is likely to develop certain cancers.
3. **Genetic implications of some common cancers.** The common cancers of adult life include cancers of the breast, colon, lung, prostate gland, and stomach, as well as cancers affecting the central nervous system (CNS).
   a. **Breast cancer.** Genetic studies have convincingly shown a significant increase in the frequency of breast cancer in close relatives of a person with the disease.
      (1) **Risks for breast cancer**
         (a) In the **general population,** the risk of breast cancer is around 7%.
         (b) If a **first-degree relative is affected,** a woman has two to three times the population risk.
         (c) If there are **two first-degree relatives** with bilateral premenopausal onset, a woman's risk might be as high as 50%.
      (2) The **presence of breast cancer in a first-degree relative** should increase the early detection screening for a woman. Breast self-examination should begin earlier, and in some instances, examinations would include more frequent mammograms.
      (3) Many of the hereditary forms of breast cancer have early onset, bilateral occurrence, and are multicentric.
   b. **Gastrointestinal cancer.** The genetic contribution is low for esophageal cancer, slightly higher for gastric cancer, and highest for colon cancer. The presence of colon cancer in a first-degree relative, especially occurring at an early age, should prompt more frequent testing for occult blood in the stool and sigmoidoscopy examinations in first-degree relatives.
   c. **Lung cancers,** especially bronchial carcinomas, were rare before the twentieth century. Most epidemiological factors have focused on environmental factors, such as smoking. While the major cause for lung cancer is tobacco, genetic factors also have an important, although smaller, role in conferring predisposition to this environmental factor. Recently, the first-degree relatives of lung cancer patients have shown a risk that is two to three times higher than the risk for lung cancer in the general population.

D. Coronary artery disease remains the most frequent cause of morbidity and mortality in Western industrialized countries, accounting for more deaths than all forms of cancer combined.

1. The **role of genes in the causation of atherosclerosis** is supported by family studies and concordance rates of 65% for coronary artery disease in monozygous twins.
2. **Genetic hyperlipidemia** presents with cholesterol levels that are markedly elevated and, in some instances, might have been elevated since childhood.
   a. The importance of the recognition of a genetic hyperlipidemia is that it **influences therapy and management**.

b. In general, genetic hyperlipidemias do not respond adequately to environmental modulation by diet, and **drug therapy is often needed** to bring down the cholesterol levels.

c. Genetic hyperlipidemia **should be suspected when there is a positive family history** as defined by the presence of angina, heart attack, or stroke in a first-degree relative before the age of 60.

3. Genes are particularly implicated in **premature atherosclerosis** when a first-degree relative has a history of either angina or heart attack before the age of 60.

a. In this group, the **family history** is the most important indicator that the individual is at increased risk for coronary disease.

b. In addition, **clinical examination** may reveal particular findings, including **tendon xanthomatas** and an **arcus cornealis,** which is indicative of a genetic hyperlipidemia. Genetic hyperlipidemia should also be considered if a patient has a **very elevated low-density lipoprotein (LDL) cholesterol level**.

4. Important genetic disorders causing premature atherosclerosis

a. FH (see Ch 3 II A 1 c; Ch 5 II A 3 c) is inherited as an autosomal dominant trait and is penetrant early in life. The molecular basis for this disorder is due to different mutations in the LDL receptor gene.

b. Familial combined hyperlipidemia (FCH) occurs with a frequency of approximately 1%–2% in the general population. The diagnosis of FCH can be made in the presence of an elevated LDL cholesterol or triglyceride level in the patient and siblings or other first-degree relatives. The pathogenesis of the disorder is unknown.

VI. CONGENITAL ANOMALIES

A. Perspective

1. An **anomaly** is any abnormal deviation from the expected type in structure, form, or function. **Congenital** means present at birth.

2. Congenital anomalies are common causes of morbidity and mortality, especially in childhood.

3. Many congenital anomalies are not apparent at birth, although, by definition, all are present at birth. External malformations, such as polydactyly or cleft lip and cleft palate, are generally noted in the immediate newborn period, but internal or functional anomalies (involving, for example, the heart, kidneys or brain) may not become apparent until later in life.

4. Similar congenital anomalies in different individuals may have different causes. This phenomenon is called **etiologic heterogeneity**.

a. In general, **the cause of a congenital anomaly cannot be determined by its appearance alone**.

b. To determine the cause, one must consider the **pregnancy history** and **family history** of the child as well as any **other congenital anomalies or other abnormalities** that may be present.

c. It is particularly important to determine if a congenital anomaly is **isolated** (i.e., the only anomaly in a child who is otherwise completely normal) or **part of a more generalized pattern of anomalies**.

(1) Isolated congenital anomalies usually (but not always) are of **multifactorial** etiology (see IV C 3).

(2) Multiple congenital anomalies in a child are usually **not** of multifactorial etiology.

B. Types of congenital anomalies

1. Malformations are morphologic defects resulting from **intrinsically abnormal developmental processes**.

a. Malformations may have a variety of **different causes**.

b. A malformed structure exhibits abnormal development **early in embryogenesis**.

c. Examples include polydactyly (too many fingers or toes), oligodactyly (too few fingers or toes), most spina bifida, most cleft palates, and most kinds of congenital heart disease.

2. Disruptions are morphologic defects resulting from **breakdown of, or interference with, an originally normal developmental process**.

a. Disruptions may result from the effects of a teratogenic agent or may have an unknown cause; they are **usually not caused by single gene or chromosome abnormalities**.
b. Disruptions **may occur at any time during gestation**. The manifestations tend to differ with gestational timing, however.
c. Examples include **amniotic band disruptions** (see Figure 7-2) and most cases of **porencephaly** (cystic lesions within the brain).

3. **Deformations** are abnormalities of form or position of a part of the body **caused by nondisruptive mechanical forces**.
   a. Deformations result from **mechanical interference** with the normal growth, function, or positioning of the fetus in utero. Constraining factors that may predispose to deformations include:
      (1) First pregnancy (uterus more resistant to stretch)
      (2) Small maternal size
      (3) Small uterus
      (4) Uterine malformation
      (5) Large uterine fibroids
      (6) Small maternal pelvis
      (7) Early engagement of the fetal head into the mother's pelvis
      (8) Unusual fetal position (e.g., transverse lie)
      (9) Oligohydramnios
      (10) Large fetus
      (11) Multifetal gestation
   b. Deformations are **usually not caused by chromosome or single gene abnormalities** unless the deformations are associated with malformations in the same infant.
   c. Deformations **usually develop in the second half of pregnancy** when the size of the fetus is large in comparison to the size of the uterus.
   d. Because deformations are caused by mechanical factors, they can often be **treated by mechanical means**. For example, positional deformations of the feet can usually be treated successfully by physical therapy or casting.
   e. Examples of deformations include positional abnormalities of the feet and moulding of the head.

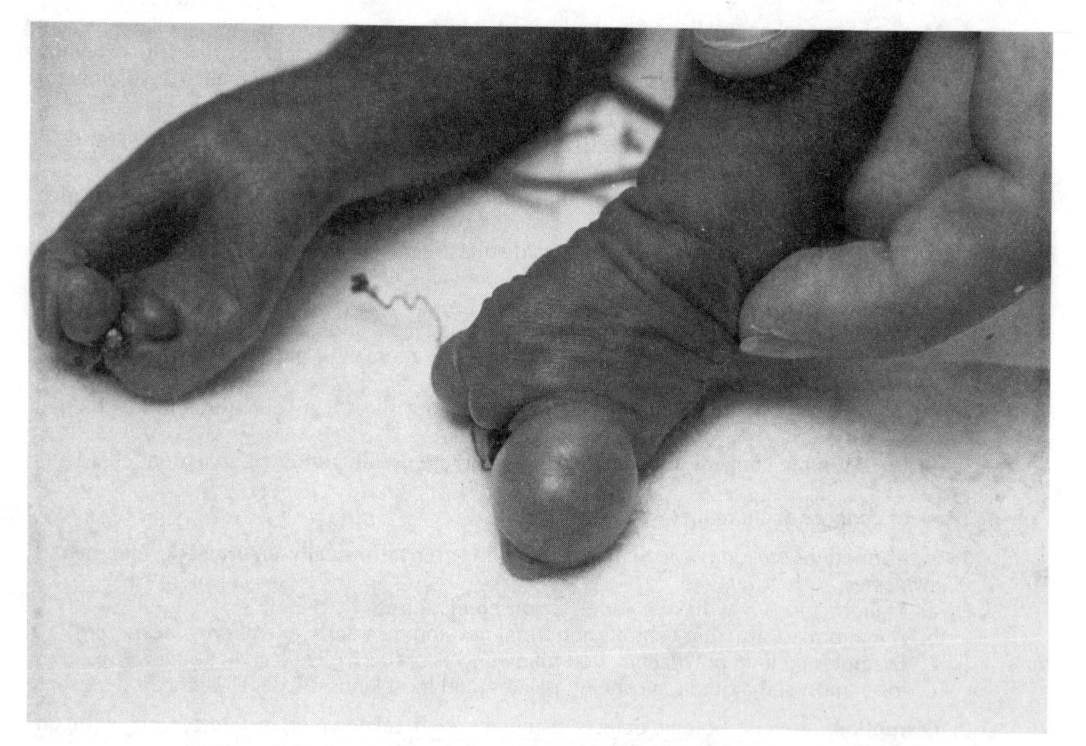

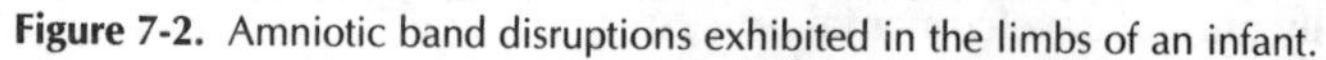

Figure 7-2. Amniotic band disruptions exhibited in the limbs of an infant.

4. Dysplasias are morphologic defects caused by **abnormal organization of cells into tissue.**
   a. Dysplasias are **often due to single abnormal genes.**
   b. Dysplasias usually develop during embryogenesis.
   c. Examples include hemangiomas and thanatophoric dysplasia (a form of lethal short-limbed dwarfism).

C. Patterns of anomalies should be sought in individuals with two or more congenital defects. Such patterns may be of four types:

1. Sequences are patterns of anomalies derived from a single known or presumed structural defect or mechanical factor.
   a. A sequence can be thought of as a **cascade of anomalies,** all of which derive directly or indirectly from a single primary defect.
   b. A sequence may have **several different causes,** just as a malformation may have several different etiologies.
   c. One example is the **Robin sequence,** which consists of broad U-shaped cleft palate, small mandible, and a tendency for the tongue to obstruct the airway.
      (1) The primary error of morphogenesis in this instance is a **small jaw (micrognathia).** This, in turn, causes the oropharynx to be too small to contain the tongue normally while the palate is forming. The tongue remains partially in the nasopharynx and prevents the palatal shelves from approximating so that they can fuse.
      (2) The **broad cleft palate** is a more serious problem in a child than the small jaw, but the cleft palate is a secondary consequence of the micrognathia.
      (3) There are many different causes of Robin sequence since many different factors can cause an embryo to have a small jaw.
   d. Another example is the **oligohydramnios sequence.**
      (1) Features of this sequence include pulmonary hypoplasia (which is often fatal), positional deformities of the hands and feet, and Potter's facies (a compressed appearance of the face and body).
      (2) Oligohydramnios sequences can occur as a **consequence of decreased amniotic fluid** caused by any factor. Commonly encountered causes include amniotic fluid leak and decreased amniotic fluid production due to placental hypoperfusion or fetal anuria.

2. Developmental field defects are patterns of anomalies resulting from the disturbed development of a morphogenic field or a part thereof.
   a. Developmental (or morphogenic) fields can be thought of as regions of the embryo that develop in a related fashion. The derivative structures are not necessarily close spatially in an infant or adult.
   b. Similar developmental field defects in different children may have **different causes.**
   c. An example is **holoprosencephaly,** a serious abnormality of midline brain development that is often associated with eyes that are set too closely together, hypoplasia of the nose, and midline cleft lip and palate. All of these abnormalities can be traced to a single localized defect of the prechordal mesoderm in the third to fourth week of embryonic development.

3. Syndromes are patterns of anomalies, all of which are pathogenically related.
   a. In clinical genetics, the term "syndrome" implies a **similar etiology in all affected individuals.**
   b. The implication of causal specificity does not apply to the word "syndrome" in any other medical context.
   c. Examples include Down syndrome (caused by trisomy for chromosome 21), fetal alcohol syndrome (caused by maternal alcohol abuse during pregnancy), and Marfan syndrome (caused by a dominantly inherited single gene mutation).
   d. Syndrome recognition is one goal of clinical genetic evaluation because it **permits precise genetic counseling.**

4. Associations are patterns of anomalies that occur together more frequently than expected by chance but are not identified as sequences, syndromes, or developmental field defects. In other words, this pattern simply represents a **nonrandom grouping of congenital anomalies** in an individual.
   a. The same group of anomalies may occur in a sequence, a syndrome, or a developmental field defect as well.
   b. A similar association may have **different causes** in different children.

c. Examples of associations include abnormal external ears and renal anomalies; single umbilical artery and cardiac defects; or vertebral, cardiac, and renal anomalies.
d. As more is learned about the pathogenesis of congenital anomalies, some children with certain associations are likely to be reclassified as having a syndrome, a sequence, or a developmental field defect.

D. Minor anomalies are unusual morphologic features that are of **no serious medical or cosmetic consequence**.

1. **Examples** include dermatoglyphic alterations, abnormal scalp hair whorls, alterations in the shape of the auricles, and bifid ribs.
2. Minor anomalies are important because they **may be characteristic of certain patterns of anomalies** and permit their classification as specific syndromes. Down syndrome, fetal alcohol syndrome, and many other syndromes are recognized clinically on the basis of characteristic patterns of minor anomalies rather than the presence of specific major anomalies.
3. Recognition of minor anomalies is also important because the presence of multiple minor anomalies in a child suggests a generalized disorder of early embryogenesis. **The more minor anomalies a child has, the more likely he or she is to have an associated major congenital anomaly.**
4. **Certain minor anomalies tend to be associated with certain malformations.** For example, absence of the occipital hair whorl or ocular hypertelorism tend to be associated with brain malformations.

STUDY QUESTIONS

Directions: Each of the numbered items or incomplete statements in this section is followed by answers or by completions of the statement. Select the **one** lettered answer or completion that is **best** in each case.

1. Hurler and Scheie syndromes are both mucopolysaccharide disorders. Hurler syndrome presents in early childhood with severe bony changes and coarse facies, while Scheie syndrome is much milder and later in onset. If the two disorders are allelic, which one of the following results would be expected?

(A) Both syndromes should show similar levels of enzyme activity
(B) If cells from each patient are cultured together, the increased mucopolysaccharide excretion characterizing both should normalize
(C) Sibships should occasionally demonstrate some children with Hurler syndrome and some with Scheie syndrome
(D) Offspring of one Hurler heterozygote parent and one Scheie heterozygote parent should have an intermediate disorder, not one or the other
(E) Molecular analysis will not demonstrate any differences with the genetic sequence

2. Which one of the following statements about quantitative traits is true? They

(A) are the number of structures of a given type that occur in a person's body, such as the number of fingers, ribs, or kidneys
(B) usually exhibit multifactorial inheritance
(C) are more frequent in monozygotic twins than in dizygotic twins
(D) are continuously variable and may assume any value
(E) usually exhibit more extreme values in children than in either parent

3. All of the following characteristics suggest that a disease under consideration is inherited as a multifactorial condition EXCEPT the

(A) disease occurs more frequently in women than in men
(B) disease occurs more often among the children of an affected patient than among the grandchildren
(C) risk of recurrence of the disease in a child is greater if both parents are affected than if only one parent is affected
(D) incidence of the disease is 4% in the population and 50% among the children of an affected parent

4. All of the following clinical features may indicate an inborn error of metabolism EXCEPT

(A) a child who is the second pregnancy to parents who are known to be inbred through generations
(B) the parents of a child with mental retardation who have also had two early miscarriages as did the maternal grandmother
(C) an infant who was initially bright and alert and is now losing skills and paying little attention to the external environment
(D) a previously well infant with no dysmorphic features who collapses at 2 days of age

---

1-D
2-B
3-D
4-B

5. Succinylcholine is a muscle relaxant widely used during anesthesia. About 1/3000 Caucasians are unable to degrade succinylcholine resulting in prolonged apnea after an operation. This effect is due to impaired activity of the enzyme cholinesterase, which inactivates succinylcholine. Mr. K, aged 40, has previously had difficulty breathing after anesthesia. All of the following statements regarding his case are correct EXCEPT

(A) a biochemical test should be performed to assess his ability to breakdown succinylcholine
(B) the absence of a family history of this problem is likely to exclude this as a genetic problem
(C) closer postoperative support may be necessary
(D) family members should be aware of this problem
(E) knowledge of this history might alter the drugs used in anesthesia

6. A 6-year-old African boy developed anemia after treatment of malaria with primaquine. All of the following suggestions should be recommended for this boy EXCEPT

(A) he is likely to have glucose-6-phosphate deficiency (G6PD)
(B) his sisters should not be treated with primaquine
(C) treatment of malaria should be changed in his case
(D) it is likely that his mother is a carrier for G6PD deficiency
(E) some drugs, including sulfonamide antibiotics, should be used with caution

7. An 18-year-old man has a strong family history of premature atherosclerosis; his father and grandfather each died of a heart attack in their 30s. The young man's father was known to have high cholesterol. All of the following statments regarding the proband are true EXCEPT

(A) he has up to a 50% chance of having high cholesterol levels
(B) dietary therapy is unlikely to bring down his cholesterol levels if they are elevated
(C) his siblings should have their cholesterol measured
(D) treatment for hypercholesterolemia should begin in his 30s
(E) modification of other risk factors, such as smoking, is especially important for this young man

5-B
6-B
7-D

Directions: The group of items in this section consists of lettered options followed by a set of numbered items. For each item, select the **one** lettered option that is most closely associated with it. Each lettered option may be selected once, more than once, or not at all.

Questions 8–12

Match the following clinical findings with the most likely type of alteration in protein function.

(A) Enzyme deficiency
(B) Abnormality of membrane transport protein
(C) Cell structure abnormality
(D) Altered cellular receptor
(E) Abnormal control of growth and differentiation

8. Affected males with Menkes syndrome, an X-linked disorder, have a progressive loss of skills and are found to have increased intracellular levels of copper with decreased serum levels

9. A female infant has multiple episodes of pneumonia, skin infections, and encephalitis; she has a combined immune deficiency; carriers in the family may be identified

10. A young woman presents with joint dislocations, stretchy skin, and easy bruisability; electron microscopy demonstrates abnormal appearing collagen fibrils; both her father and sister have had similar problems

11. Hereditary spherocytosis causes anemia in a woman because of an increased destruction of red blood cells with increased fragility

12. A phenotypically female infant presents with inguinal hernia and at surgery has a testicular structure identified in the hernia; her chromosomes are 46, XY, and she is diagnosed as having androgen insensitivity

---

8-B
9-A
10-C
11-C
12-D

ANSWERS AND EXPLANATIONS

1. The answer is D *[II B 1 d].*
The Hurler-Scheie offspring show the effect of two allelic mutations. Children with both Hurler and Scheie syndromes have been described. Both syndromes are the result of an abnormality in the enzyme α-L-iduronidase; in Hurler syndrome the enzyme is severely deficient, while in Scheie syndrome the enzyme has better residual levels. In classic co-culturing experiments, mixing the two cells did not correct the defect since the same enzyme was involved (in contrast to experiments using Hunter and Hurler syndrome cell lines, where separate enzymes are involved). Both Hurler syndrome and Scheie syndrome are inherited as autosomal recessives, and offspring of Hurler heterozygous parents produce children with Hurler syndrome, while offspring of Scheie heterozygote parents produce children with Scheie syndrome.

2. The answer is B *[IV A 2, B 3–5].*
Quantitative traits are those that can be measured on a continuous scale. They usually exhibit multifactorial inheritance and have a normal population distribution, with average values being most common. Monozygotic twins tend to resemble each other closely with respect to quantitative phenotypes, but such traits occur in everyone. Although quantitative traits are continuously variable, the range of values that occurs is limited by physiologic and anatomic constraints. Because children inherit only half of the genes that determine a quantitative multifactorial trait from each parent, a parent who manifests an extreme value for a trait tends to have children who exhibit less extreme values.

3. The answer is D *[IV C 4].*
Isolated congenital anomalies and common diseases of adulthood often are inherited as multifactorial traits. Such conditions characteristically occur in one sex more frequently than in the other. The incidence of multifactorial diseases is higher among individuals who are more closely related to an affected person than in those who are less closely related because the former are likely to share with the proband more genetic and nongenetic factors that predispose to the disease. Multifactorial diseases typically exhibit a higher recurrence risk in families with more affected members because such families are likely to possess and thus transmit more factors that contribute to the disease. The incidence of multifactorial diseases in the first-degree relatives of an affected patient typically equals approximately the square root of the population incidence.

4. The answer is B *[II A–E].*
Inborn errors of metabolism, whether enzyme or cofactor problems, are usually inherited as autosomal recessives. A clue may be the history of consanguinity. Some enzyme deficiencies lead to a buildup of substrate, which may be toxic or may be stored in various organelles. The process is ongoing, leading to progressive involvement. The initial presentation of some children with metabolic abnormalities is with the onset of feeding immediately after birth. Until that time, the placenta and maternal enzyme function protect the fetus. The history of both a developmentally delayed child and miscarriages is more suggestive of an inherited chromosome rearrangement such as a translocation (see Ch 2 III E–F).

5. The answer is B *[III B 4].*
Serum cholinesterase is the enzyme that breaks down succinylcholine. In some populations, particularly those of Caucasian descent, about 1/3000 individuals has an abnormal cholinesterase enzyme that is unable to degrade succinylcholine, which may result in prolonged apnea after anesthesia. The family history of this particular disorder might not be positive, but it does not exclude this as a genetic problem. If the physician is aware of this particular reaction with succinylcholine, postoperative support may be necessary to deal with the prolonged apnea. Family members should certainly be aware of this particular problem and should inform the anesthesiologist of this prior to surgery.

6. The answer is B *[III B 1].*
Glucose-6-phosphate dehydrogenase (G6PD) deficiency is a very common X-linked enzyme disorder that affects approximately 5% of black males in the United States. Because G6PD deficiency is an X-linked recessive disorder, an affected male's sisters would be carriers but could be treated with primaquine and other drugs that may unmask the deficiency in males. For this particular boy, treatment of malaria should be changed to another drug that might not precipitate hemolysis leading to anemia. It is likely, of course, that this boy's mother is a carrier for this disorder and that she has passed it to her son.

7. The answer is D *[V D 1–4].*
Treatment of high cholesterol should strongly take into account the family history. In this instance, with the father having a high cholesterol level and a history of a heart attack in his mid 30s, it is most likely that there is a genetic hyperlipidemia in this family. In this instance, dietary therapy is usually insufficient to bring down the levels and in view of the significant family history of very marked premature atherosclerosis, treatment should begin before the age of 30 in an effort to reduce the likelihood of this particular outcome. Of course, in this particular man, as in other patients with hyperlipidemia, modification of other risk factors is essential to try to reduce the risk for later atherosclerotic events.

8–12. The answers are: 8-B, 9-A, 10-C, 11-C, 12-D *[II C 3 b, D 1, 4].*
Menkes syndrome involves an abnormal copper transport protein, such that a differential is seen with copper both intra- and extracellularly. The condition leads to severe cerebral degeneration and death in infancy.

One type of severe combined immune deficiency is secondary to adenosine deaminase deficiency (an enzyme involved in white blood cell function). In many enzyme deficiencies, heterozygotes are identified by intermediate enzyme levels. The inheritance is often autosomal recessive.

Both the woman with the connective tissue problem (a subtype of Ehlers-Danlos syndrome) and the woman with hereditary spherocytosis have an abnormal structural protein (with spherocytosis, the spectrin forming the red cell membrane skeleton is abnormal). Structural protein abnormalities are often autosomal dominant in inheritance.

The child with androgen insensitivity (see Ch 10 III) has an abnormal androgen receptor, such that androgen is not presented to the nucleus. The XY child consequently develops as a phenotypic female.

8 Teratogenesis and Mutagenesis

J. M. Friedman

I. INTRODUCTION. Both teratogens and mutagens can cause alterations in the structure and functioning of the body, but the mechanisms differ. Teratogens cause damage by altering embryonic or fetal development directly. Mutagens cause changes within the genetic material that may lead to inherited disease if the germ cells are affected or to cancer if somatic cells are involved.

Epidemiology is the study of the frequency and distribution of disease within groups of people and of factors that affect this frequency and distribution. Epidemiologic studies are an important means of evaluating potential teratogens and mutagens in human populations.

II. TERATOGENESIS

A. A teratogen is an agent that can produce a permanent alteration of structure or function in an organism after exposure during embryonic or fetal life. Teratogens include environmental factors, medications, drugs of abuse, and occupational chemicals.

B. Clinical teratology is concerned with the following:

1. The relationship between the **anomalies in a child** and teratogenic exposure
2. The **risk of anomalies** for a child of a woman who has been exposed to a teratogen
3. The **risks to a pregnant woman** of treatment or exposure to a given agent

C. Principles of clinical teratology

1. **The circumstances and nature of exposure** are as important as the nature of the agent itself in determining teratogenic risk. For example:
   a. **Thalidomide** is not teratogenic if the exposure is just holding a pill in one's hand.
   b. However, any agent that is given to a pregnant woman in a manner and dose that is toxic to her may present a danger to her fetus. This is true of even "benign" agents such as table salt.
2. **An embryo is not a little adult.**
   a. The likelihood that an agent will produce a teratogenic effect in an embryo or fetus on the basis of its pharmacologic action in an adult or even in a child cannot be predicted.
   b. The biochemical processes involved in the development of an organ or structure may have no relationship to the processes involved in maintaining the function of that organ or structure once it has formed.
3. **Teratogens act at vulnerable periods of embryogenesis and fetal development.**
   a. In general, the embryo is **most sensitive** to damage **between 2 and 10 weeks after conception** (4 to 12 weeks after the beginning of the last menstrual period). During this time, most structures and organs are differentiating and forming. Each structure has its own period of greatest sensitivity within this time.
   b. The **first 2 weeks after conception** is generally considered to be a period that is **resistant to the induction of malformations** by teratogens. At this point, the embryo consists of few cells, and damage is usually either repaired completely or results in death of the embryo.
   c. **By 10 weeks** after conception, most structures in the embryo have been formed, so **malformations are unlikely to be produced by subsequent exposures**.

(1) However, disruptions, disturbances of growth, or altered tissue maturation may be caused by teratogenic exposures throughout pregnancy.
(2) Such effects are most likely to be manifested as fetal growth retardation and abnormal central nervous system (CNS) function.

4. **Individual differences in susceptibility to teratogens exist.**
   a. In **experimental animals,** it is clear that genetic differences can greatly alter the susceptibility to teratogenic damage.
   b. In **humans,** it is likely that both maternal and fetal susceptibility are important in determining the likelihood of teratogenic effects.

5. **Combinations of exposures to teratogenic agents may have effects different from those resulting when exposures occur individually.**
   a. In some cases, the risk inherent in exposure to combinations of teratogens may be substantially greater than the sum of risks of individual exposures.
   b. In other cases, exposure to certain agents may decrease the risk usually associated with exposure to another agent.

6. **Teratogenic exposures tend to produce characteristic patterns of multiple anomalies** rather than single defects.
   a. The recognition of most human teratogens has resulted from clinical identification of an unusual pattern of congenital anomalies among the children of women who had similar exposures during pregnancy.
   b. Examples include the embryopathies due to rubella, alcohol, and isotretinoin.

D. **Teratogenic factors** are thought to be responsible for about **10% of all congenital anomalies**. These factors fall into several groups.

1. **Maternal metabolic imbalance** includes alterations of the maternal internal milieu that are teratogenic. These factors are environmental with respect to the embryo or fetus but are intrinsic to the mother.
   a. The children of women with **insulin-dependent diabetes mellitus** have a risk of congenital anomalies that is two to three times greater than that of the general population.
      (1) The **most common malformations** among infants of diabetic mothers are **congenital heart disease** (occurring in 2%–3%) and **neural tube defects** (occurring in 1%–2%). Some other congenital anomalies, such as **caudal dysplasia** and **proximal femoral hypoplasia,** are rare, but they occur much more often in the infants of diabetic mothers than in other infants.
      (2) **Many of these congenital anomalies can be detected prenatally** by high-resolution ultrasound examination, fetal echocardiography, or measurement of the α-fetoprotein concentration in maternal blood or amniotic fluid. **Prenatal diagnosis should be offered to all pregnant women with insulin-dependent diabetes mellitus.**
      (3) **Malformations develop early in gestation.** Improved control of maternal diabetes later in pregnancy cannot influence the frequency of malformations in infants of diabetic mothers. Thus, **pregnancies in women with insulin-dependent diabetes should be planned** so that excellent control can be maintained from the moment of conception.
   b. **The children of women with phenylketonuria (PKU),** particularly if untreated during pregnancy, are almost certain to be born with mental retardation, microcephaly, and other anomalies. Such children usually do not have PKU themselves but are damaged by developing in a uterine environment of hyperphenylalaninemia.
   c. **Children of women with certain endocrinopathies** such as androgen-secreting tumors are at high risk of being born with abnormalities resulting from such pathologic hormone exposure.
   d. **Pregnant women with systemic lupus erythematosus (SLE)** have an increased risk for fetal cardiac conduction defects, which may be transient or permanent and which may cause fetal death.

2. **Infectious agents** can involve the embryo or fetus transplacentally.
   a. **Syphilis** can cause congenital infections that present at birth or in early infancy. Features may include stillbirth, various cutaneous lesions, rhinitis, hepatosplenomegaly, meningoencephalitis, and osteochondritis.
   b. **Congenital toxoplasmosis** (*Toxoplasma gondii*) may be asymptomatic or present with a variety of abnormalities. Severely affected infants may exhibit chorioretinitis, hydrocephaly or microcephaly, intracranial calcification, and mental retardation.

c. Rubella (German measles) embryopathy produces fetal growth retardation, hepatosplenomegaly, purpura, jaundice, microcephaly, cataracts, deafness, congenital heart disease, and mental retardation.

d. Congenital cytomegalovirus (CMV) infection may produce fetal growth retardation, hepatosplenomegaly, hemolytic anemia, purpura, jaundice, intracranial calcification, and microcephaly.

e. Varicella (chickenpox) infection in utero is rare but may be devastating. Manifestations include cutaneous lesions or scarring and disruptions such as limb reduction defects or cortical atrophy.

f. Congenital infection with **human immunodeficiency virus (HIV)** results in the development of AIDS and death in early childhood. Microcephaly and growth failure may also occur.

g. Parvovirus infection of the fetus may produce severe anemia, hydrops, and death.

3. Ionizing radiation causes DNA damage and can injure the developing embryo.

a. In utero exposure to large doses of ionizing radiation such as those that are used for cancer radiotherapy can produce microcephaly and mental retardation. Diagnostic radiographs during pregnancy are rarely associated with radiation doses that pose a substantial risk to the embryo or fetus but should be avoided if possible.

b. Radioactive iodine is concentrated in the fetal thyroid gland after the thirteenth week of pregnancy and may produce cretinism.

4. Environmental agents and occupational chemicals are often of concern to pregnant women.

a. Only two such chemicals are **known teratogenic agents** in humans.

(1) Maternal poisoning during pregnancy by food contaminated with **methyl mercury** has resulted in damage to the fetal CNS. Ataxia, weakness, and features of cerebral palsy may be produced.

(2) Maternal ingestion of large amounts of **polychlorinated biphenyls (PCBs)** during pregnancy produces fetal growth retardation, diffuse hyperpigmentation, and natal teeth.

b. Suspected environmental teratogens. The following agents are suspected of being teratogenic in humans.

(1) Hyperthermia, regardless of cause, that produces sustained elevation of maternal body temperature to levels substantially above normal (e.g., 40° C). This is usually due to fever but may occur with extreme use of saunas or hot tubs.

(2) Lead

c. Unlikely teratogens. It is unlikely that maternal use of a video display terminal (VDT) during pregnancy poses any measurable risk to the fetus.

5. Drugs of abuse not only damage the health of the mother but also interfere with the development of the embryo or fetus.

a. Established human teratogens

(1) Alcohol

(a) Classic fetal alcohol syndrome occurs among the children of women with chronic, severe alcoholism during pregnancy.

(i) Mothers of infants with fetal alcohol syndrome usually drink more, and often much more, than **3 ounces of absolute alcohol daily**—the equivalent of about six beers, six glasses of wine, or six mixed drinks every day.

(ii) Features of fetal alcohol syndrome include growth deficiency, mental retardation, behavioral disturbances, and typical facial appearance (Figure 8-1). The facies are characterized by short palpebral fissures, hypoplastic midface, long and flat philtrum, and narrow upper lip vermilion. Congenital heart disease and structural brain anomalies are common.

(b) The risks of maternal **binge drinking** during pregnancy have not been clearly defined but may be substantial.

(c) Smaller amounts of alcohol during pregnancy have been associated with less severe disturbances of growth, intellectual performance, and behavior among offspring. **No safe level of maternal drinking during pregnancy has been established.**

(2) Cocaine. Maternal use of cocaine during pregnancy has been associated with placental abruption and the occurrence of vascular disruptions such as encephaloclastic lesions in the fetus.

(3) Maternal abuse of **organic solvents** such as toluene by inhalation is suspected of causing fetal brain damage.

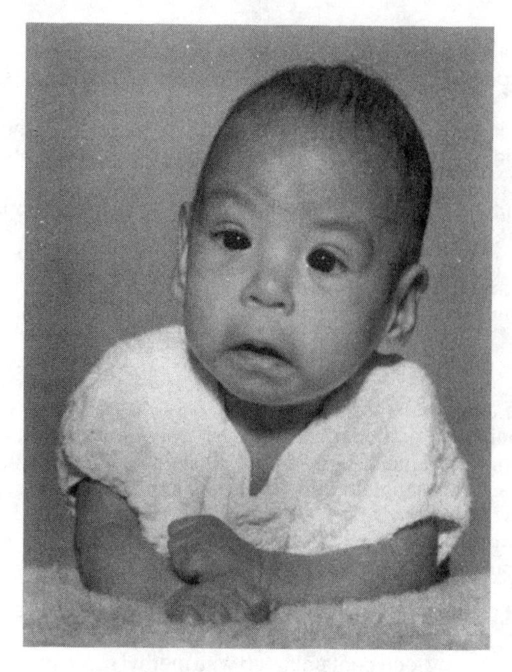

FIGURE 8-1. Infant with fetal alcohol syndrome—typical facies. (Photograph courtesy of T. Kellerman.)

b. Recreational drugs **not proven to be teratogenic** in humans with usual exposures include:
- **(1) Marijuana**
- **(2) Heroin**
- **(3) LSD**

6. Medications

a. Maternal treatment with some medications is known to cause embryonic or fetal damage in humans.

(1) **Thalidomide** exposure in the first trimester of gestation may produce limb reduction defects, facial malformations, and other congenital anomalies.

(2) **Aminopterin and other cytotoxic drugs** kill rapidly growing cells in the fetus and cause growth deficiency and a variety of other anomalies.

(3) An increased rate of congenital anomalies is observed among the children of epileptic women treated with anticonvulsant medications during pregnancy.

(a) Whether this increased risk of congenital abnormalities is due to the mother's epilepsy, a teratogenic effect of the medications, or some combination thereof is uncertain. In general, however, **therapy with multiple anticonvulsant drugs is associated with a higher risk than therapy with a single anticonvulsant**.

(b) Drugs that have been implicated as having teratogenic effects include **phenytoin, trimethadione, paramethadione, carbamazepine,** and **valproic acid**.

(i) Maternal treatment with **valproic acid** during pregnancy has also been associated with an increased risk of **neural tube defects** among offspring.

(ii) **Prenatal diagnosis** of neural tube defects should be made available to women who have been treated with valproic acid during early pregnancy.

(4) Maternal use of **androgenic hormones** during pregnancy can produce virilization of the external genitalia of female infants.

(5) Daughters of women treated with **diethylstilbestrol (DES)** during early pregnancy often exhibit abnormalities of the vaginal epithelium and cervix and have a greatly increased risk of clear cell adenocarcinoma of the vagina or cervix.

(6) Maternal treatment with **lithium** during early pregnancy has been associated with an increased frequency of certain kinds of congenital heart disease in offspring.

(7) CNS anomalies, ocular malformations, midface hypoplasia, and radiographic evidence of epiphyseal stippling may occur among the children of women treated with **warfarin** during pregnancy.

(8) **Isotretinoin** and **etretinate** are vitamin A congeners that can cause craniofacial, brain, cardiac, and other serious fetal malformations.

(9) Maternal treatment with **angiotensin converting enzyme (ACE) inhibitors** such as **captopril** and **enalapril** late in pregnancy has been associated with oligohydramnios, fetal

renal failure, and fetal or neonatal death. ACE inhibitors are not known to pose a substantial teratogenic risk in early pregnancy.

b. Maternal treatment with the following medications is **suspected, but not proven, to be teratogenic**.

(1) **Penicillamine**
(2) **Quinine**
(3) **Vitamin A** in very large doses (> 25,000 IU/day)

c. Some medications that are commonly used in pregnancy have not been shown to be teratogenic in conventional human doses.

(1) **Metronidazole**
(2) **Aspirin**
(3) **Contraceptives**
  (a) Oral contraceptives
  (b) Vaginal spermicides
(4) **Bendectin**

III. MUTAGENESIS

A. Definition and overview.
A **mutagen** is an agent that can alter the DNA or chromosomes.

1. While teratogens act only during embryonic or fetal development (see II A), **mutagens may act at any time of life**. Thus, mutations may occur in the gamete, zygote, embryo, fetus, child, or adult.

2. Teratogens affect the development of a tissue, organ, or structure. In contrast, **a mutation always affects a single cell**.

a. If this single cell is a germ cell, the mutation may be transmitted to subsequent generations.

b. If a single cell in a very early embryo sustains a mutation, many tissues of the embryo (including the germ cells) may be affected as embryogenesis progresses.

c. If a single cell in an embryo, fetus, child, or adult sustains a mutation, only cells derived from the mutated cell will carry the mutation. Most cells in the individual will not contain the mutation.

B. Clinical importance of environmental mutagens

1. **Radiation and thousands of chemicals** can produce mutations in bacteria, fruit flies, mice, rats, or cultured human and animal cells. Examples of such experimental mutagens include urethane, cyclophosphamide, and dioxin.

2. **Somatic cell mutations** play an important part in the development of human neoplasia (see Ch 9 III C 4 b), but it is usually not possible to attribute these mutations to exposure to a specific mutagenic agent.

3. **No chemical agent has been proven to produce gametic mutations in humans.** It is likely, however, that mutagens do affect human germ cells but that available studies are just unable to show this effect. There are several explanations for this.

a. **Most potential mutations are repaired to the normal state** by efficient DNA correction mechanisms that exist within human cells.

b. **Most mutations are recessive** and will be expressed only when they occur in homozygotes. Thus, many generations may pass before any effect is seen.

c. **The rate of disease due to new gametic mutations in the general population is probably substantially less than 1%.** Many different conditions can be produced by such changes. It is very difficult to carry out studies in humans that are large enough to show a measurably increased rate of disease due to new mutations under these circumstances.

4. From a practical point of view, concerned patients can be told that there is no evidence that **remote preconceptional exposure** of either parent to any mutagenic environmental agent substantially increases the risk of congenital anomalies among offspring.

a. Even though the risk of inducing birth defects in future children by exposure of the parents to mutagens is small in the individual case, **the risk for the population as a whole may be substantial**.

b. There may be no completely safe level of exposure to radiation or to a mutagenic chemical. Thus, **absolute safety is usually not possible**.

IV. EPIDEMIOLOGY OF CONGENITAL ANOMALIES

A. Definitions and overview

1. **Epidemiology** is the study of relationships between factors that determine the frequency and distribution of disease within or between populations.
2. **Birth defects epidemiology** seeks to demonstrate associations between a particular congenital anomaly or congenital anomalies in general and various potential etiologic factors.
3. **Etiologic heterogeneity**
   a. Clinically indistinguishable **congenital anomalies** in different patients may have **different causes**. For example, cleft palate may be caused by a chromosome abnormality in one patient, by a teratogenic exposure in another, and by a single-gene mutation in a third patient. In most patients, however, cleft palate results from a multifactorial predisposition. It is sometimes possible to determine the cause of the congenital anomalies in an individual patient by dysmorphic evaluation (see Ch 7 IV C 5–6).
   b. In epidemiologic studies, **it is important to consider etiologic heterogeneity when trying to determine whether a particular exposure is likely to be associated causally with a given congenital anomaly**.

B. Types of studies. Studies of associations between possible causative factors and congenital anomalies may be of several types. Most often, such investigations deal with exposure of the mother to putative teratogens early in pregnancy.

1. **Case reports** are simply descriptions of patients with a given congenital anomaly or pattern of anomalies whose mothers were exposed to certain agents during gestation.
   a. Case reports are valuable in raising suspicion regarding the possible teratogenic effect of an agent but **cannot be used to prove that the association is causal**.
   b. **Most associations observed in individual case reports are spurious** (i.e., the associations are due to chance or some noncausal relationship).
2. **A clinical series** consists of cases that are similar with respect to exposures or similar anomalies.
   a. If a **recurrent pattern of congenital anomalies** is observed among infants born to women with similar exposures during pregnancy, a causal relationship may be suggested, particularly if the pattern of anomalies or the exposure or both are rare.
   b. Almost all human teratogens have been recognized initially in clinical series. Thus, the **most effective means of recognizing human teratogens is the alert clinician** who notices a similar unusual pattern of congenital anomalies among the children of women who have had a similar unusual exposure during pregnancy.
   c. Because of the biases inherent in collecting cases for a clinical series, this approach usually **cannot be used to estimate the magnitude of risk** associated with exposure to an agent.
3. **Cohort studies** compare the frequency of congenital anomalies among the children of women exposed during pregnancy to the agent in question to the frequency among children of women who were not exposed.
   a. Cohort studies **can be used to estimate the risk and statistical significance** of an observed association.
   b. Cohort studies are often subject to serious **biases of ascertainment and recall** and may be difficult to interpret because of **confounding factors**. Appropriate experimental design and analysis are essential to proper interpretation of cohort studies.
   c. Cohort studies of congenital anomalies are usually **difficult** and quite **expensive** to perform.
4. **Case–control studies** compare the frequency of maternal exposure during pregnancy to a given agent among children with and without congenital anomalies.
   a. Case–control studies **can be used to estimate the risk and statistical significance** of an observed association.
   b. Like cohort studies, case–control studies are often subject to serious **biases of ascertainment and recall** and may be difficult to interpret because of **confounding factors**. Appropriate experimental design and analysis are essential to proper interpretation of case–control studies.

5. **Animal experiments** are essential in determining mechanisms involved in teratogenic effects. Such studies may also be useful in identifying agents that should be suspected of having teratogenic potential in humans, but **animal studies alone cannot prove that an exposure is teratogenic in humans**.
   a. Experimental animals **differ from humans** in physiology, metabolism, and embryonic development.
   b. **Exposures** used experimentally often differ in **dose and route** from those encountered in humans.
   c. Humans may exhibit a wide range of **individual susceptibilities** to teratogenic effects, but these variabilities are usually controlled in experimental animals by employing genetically inbred strains.
   d. Animal experiments deal with **populations**. In humans, the physician is generally concerned about the risks for an **individual patient**.

C. Interpretation of studies

1. **Types of associations**
   a. An association between the occurrence of a congenital anomaly in a child and a prenatal exposure in the mother may be due to **chance**. This is most likely if the exposure or the anomaly or both are relatively common or poorly defined. Even rare coincidences are expected to occur occasionally in a country as large as the United States.
   b. An association between a congenital anomaly in a child and a prenatal exposure may occur because **both the anomaly and the exposure are related to another variable**. For example, maternal cigarette smoking during pregnancy might be associated with small head size in infants because women who smoke are more likely to drink alcohol and maternal drinking is associated with small head size in infants.
   c. An association is said to be **causal** if the exposure produced the congenital anomaly. The following factors support the interpretation that an association is causal:
      (1) Consistency of the association in studies of different design and in different populations
      (2) Dependence of the effect on dose and time of exposure
      (3) Existence of a similar effect in an experimental animal model
      (4) Biologic plausibility

2. **Cohort studies and case–control studies can be evaluated statistically.** Four parameters are often used to assess the outcome of such studies.
   a. **Statistical significance** is an expression of how likely the association is to have occurred due to chance alone. For example, if an association is statistically significant at a level of $p < 0.05$, the probability is less than 0.05 or 1/20 that it would have occurred by chance alone.
   b. **Absolute risk** is the rate of occurrence of a condition among individuals who have had a given exposure. If 20% of infants born to women who took isotretinoin in the first trimester of pregnancy are found to have congenital anomalies in a study, then the absolute risk is 20%.
   c. **Relative risk** is the ratio of the rate of a condition among those exposed to the rate among those not exposed. A relative risk of 3.0, for example, implies that an anomaly occurs three times as frequently among the offspring of pregnancies with a given exposure as among the offspring of pregnancies in which that exposure did not occur.
      (1) A very high relative risk may be associated with a small absolute risk if the defect under consideration is rare.
      (2) For example, an exposure that produces a relative risk of 10 would only increase the frequency of a congenital anomaly that normally occurs with a rate of 1/100,000 to 1/10,000.
   d. **Attributable risk** is the proportion of cases of a condition in exposed individuals that can be attributed to the exposure. **Population attributable risk** is the proportion of cases of the condition in the population as a whole that results from the exposure.
      (1) The population attributable risk provides an estimate of the **amount by which the rate of a condition might be reduced by prevention of an exposure**.
      (2) For example, in a study of childhood deafness, the population attributable risk for maternal rubella is found to be 10%. One could, therefore, estimate that the overall frequency of childhood deafness would be reduced by 10% if rubella infection of pregnant women could be entirely eliminated.

STUDY QUESTIONS

Directions: Each of the numbered items or incomplete statements in this section is followed by answers or by completions of the statement. Select the **one** lettered answer or completion that is **best** in each case.

1. Maternal abuse of which one of the following substances is known to be teratogenic in humans?

(A) Marijuana
(B) LSD
(C) Heroin
(D) Alcohol
(E) Coffee

2. Which one of the following statements about mutagens is correct? Mutagens are

(A) agents that can produce a permanent alteration of structure or function in an organism after exposure during embryonic or fetal life
(B) common causes of congenital anomalies in humans
(C) responsible for causing most autosomal recessive diseases
(D) responsible for causing most autosomal dominant diseases
(E) capable of affecting both somatic cells and germ cells in humans

Questions 3–5

A study is done on low birth weight infants born in a district of Finland over a 12-month period. The mothers of some of these infants drank soft drinks containing a new artificial sweetener called Nomoflab during pregnancy; the mothers of the rest of the infants did not drink such beverages. The findings of the study are as follows:

	Birth Weight of Infant	
Mothers	**> 2500 g**	**< 2500 g**
Did not consume Nomoflab	2609	123 (4.5%)
Consumed Nomoflab	1915	192 (9.1%)

Statistical significance: $P < 0.001$
Absolute risk = 9.1% among infants of women who drank beverages containing Nomoflab during pregnancy
Relative risk = 2.0
Population attributable risk = 31%

3. Which of the following interpretations is consistent with these data if none of the women had used beverages containing Nomoflab during pregnancy?

(A) The frequency of low birth weight infants would have been more than 1000 times lower
(B) 9.1% of the infants would not have had low birth weights
(C) Only half as many infants would have had low birth weights
(D) The overall frequency of low birth weight infants would have been 31% lower
(E) None of the above

4. Which one of the following statements correctly represents the association observed between maternal use of Nomoflab during pregnancy and low birth weight in the infant?

(A) It is likely to be due to chance alone
(B) It could not have been due to chance alone
(C) It is about two times more likely to be causal than due to chance alone
(D) It is about nine times stronger than expected by chance alone
(E) None of the above

5. The risk of birth weight below 2500 g among infants of women who used Nomoflab during pregnancy in this study is which one of the following?

(A) More than 1000 times as great as the risk among infants of women who did not drink soft drinks containing this sweetener during pregnancy
(B) About nine times as great as the risk among infants of women who did not drink soft drinks containing this sweetener during pregnancy
(C) About two times as great as the risk among infants of women who did not drink soft drinks containing this sweetener during pregnancy
(D) About 31% as great as the risk among infants of women who did not drink soft drinks containing this sweetener during pregnancy
(E) None of the above

1-D
2-E
3-D
4-E
5-C

6. Five pregnant patients call your office in one hectic day with concerns about possible teratogenic effects of exposures they have had. Which one of the following patients has had an exposure that is most likely to have harmed her fetus? A woman who

(A) attempted suicide by taking 500 aspirin tablets 8 weeks into her pregnancy; was in a coma with salicylate toxicity for 2 days and required treatment with hemodialysis; and has now fully recovered and is 10 weeks pregnant
(B) continued taking her oral contraceptive pills for 2 months after having become pregnant and is currently 12 weeks pregnant
(C) works in the office of a nuclear power plant that has operated normally for the past 12 years while maintaining ambient radiation levels within government safety levels, and is currently 16 weeks pregnant
(D) injects rats with isotretinoin two to three times a week and has done so for the past 18 months; always wears gloves and a gown when injecting the animals or otherwise handling them; and is currently 8 weeks pregnant
(E) has been hospitalized repeatedly for chronic alcoholism; has not had a drink for the past 10 months; and is currently 12 weeks pregnant

7. Five young women who are each planning to become pregnant a few months from now call you for advice because each is concerned about something that she fears may increase her risk for having an abnormal infant. Which of the following women is at increased risk for having an infant with congenital anomalies? A woman who

(A) is an airline flight attendant and is concerned about cosmic ray exposure that occurs during flight
(B) works for an insurance company entering data into a video display terminal for 8 hours every day
(C) currently has rubella
(D) is just completing a course of treatment with isotretinoin
(E) has insulin-dependent diabetes mellitus that has been well-controlled for the past 5 years

8. A substantial teratogenic risk is associated with treatment of a woman in the first trimester of pregnancy with each of the following drugs EXCEPT

(A) lithium
(B) captopril
(C) etretinate
(D) valproic acid
(E) warfarin

6-A
7-E
8-B

Directions: The group of items in this section consists of lettered options followed by a set of numbered items. For each item, select the **one** lettered option that is most closely associated with it. Each lettered option may be selected once, more than once, or not at all.

Questions 9–11

Match each statement with the kind of teratology study it most closely describes.

(A) Case report
(B) Clinical series
(C) Double-blind controlled trial
(D) Cohort study
(E) Case–control study

9. Compares the frequency of maternal exposure to an agent during pregnancy between two groups of children, of which one has congenital anomalies and the other does not

10. Is particularly valuable in establishing a characteristic pattern of congenital anomalies produced by maternal exposure to a given teratogen during pregnancy

11. May be useful in raising suspicion of a teratogenic effect, but associations observed are usually not causal

---

9-E
10-B
11-A

ANSWERS AND EXPLANATIONS

1. The answer is D *[II D 5 a (1), b (1)–(3)].*
Alcohol abuse is the most frequently encountered human teratogenic exposure. Chronic maternal alcohol abuse during pregnancy can cause the fetal alcohol syndrome. Maternal use of marijuana, heroin, LSD, or coffee during pregnancy is not thought to increase the risk of malformations among offspring.

2. The answer is E *[II D 3; III A, B 1–3].*
Ionizing radiation and various chemicals have been shown to cause mutations in human somatic cells, and it is likely that such mutagens affect human germ cells as well, although this has not yet been formally proven. Germ line mutations have been induced in various experimental animals with mutagenic agents.

Teratogens, not mutagens, produce permanent alterations of structure or function after exposure during embryonic or fetal life. New mutations, whether spontaneous or caused by environmental agents, are uncommon causes of congenital anomalies in humans. Most induced mutations are recessive, but most dominant and recessive diseases result from altered genes inherited as such from heterozygous carrier parents.

3–5. The answers are: 3-D, 4-E, 5-C *[IV B 3, C 2].*
In a cohort study such as this one, the population attributable risk provides an estimate of the amount by which the rate of a condition would be reduced if the exposure did not occur. In this instance, the population attributable risk is 31%, so the data are consistent with a 31% reduction in the frequency of low birth weight if none of the women drank beverages containing Nomoflab during pregnancy.

The statistical significance of an association is an expression of how likely the association is to have occurred by chance alone. In this study, the association is statistically significant at a value of $P < 0.001$, indicating that the probability is less than 1/1000 for the association to have occurred just by chance. Although this is a low probability, chance occurrence is not impossible.

The relative risk is an expression of how much more frequent an anomaly occurs among the exposed group than among those not exposed. The observed relative risk of 2.0 indicates that low birth weight is about twice as common among the infants of women who drank Nomoflab-containing soft drinks during pregnancy as among the infants of women who did not.

The observed absolute risk of 9.1% simply indicates that 9.1% of infants of women who drank beverages containing Nomoflab during pregnancy were of low birth weight. This figure includes infants who would have been of low birth weight even if their mothers had not drank such beverages. This background rate is 4.5% in the study.

6. The answer is A *[II D 3, 5 a (1), 6 a (8), c].*
Although maternal use of aspirin in usual therapeutic doses during pregnancy is not thought to be associated with a measurable teratogenic risk, the situation is much different in the woman who attempted suicide. Since the mother suffered serious toxic effects of her aspirin overdose, there is a danger that the fetus may also have suffered toxic (or teratogenic) effects as well. As a rule, any agent that is taken by a pregnant woman in a manner and dose that is toxic to her may present a danger to her fetus. Oral contraceptives in usual doses do not appear to present a serious teratogenic risk, even if the exposure occurred during the critical period of embryogenesis. It is unlikely that the woman who works in a nuclear power plant or the woman who works in the laboratory was exposed to a sufficient amount of the agent to raise concern, despite the fact that both radiation and isotretinoin are well-recognized human teratogens and the "exposures" occurred during the critical period of embryogenesis. Fetal alcohol syndrome is due to maternal ingestion of large amounts of alcohol during pregnancy, and the chronic alcoholic did not drink during the current pregnancy so the fetus is not at risk for alcohol embryopathy.

7. The answer is E *[II C 1, 3, D 1 a, 2 c, 3, 4 c, 6 a, (8)].*
Children of women with insulin-dependent diabetes mellitus have a two- to threefold increased risk of congenital anomalies. Good control of diabetes in pregnancy from the time of conception is important, but even this does not reduce the teratogenic risk completely. Although ionizing radiation is a well-recognized human teratogen, it is unlikely that either the flight attendant or the woman who works with the video display unit will encounter enough radiation in her job to be of concern. The woman who now has rubella should have recovered completely from her infection and be immune to the disease when she becomes pregnant in a few months. Similarly, the woman who is completing her course of treatment with isotretinoin now should not be at risk for teratogenic effects related to this drug in a few months, assuming that she does not take the medication again at that time.

8. The answer is B *[II D 6].*
Maternal treatment with captopril and other angiotensin converting enzyme (ACE) inhibitors late in pregnancy has been associated with fetal anuria, oligohydramnios, or death. Maternal treatment with captopril in the first trimester is not thought to pose a substantial risk of damage to the embryo or fetus. Lithium, etretinate, valproic acid, and warfarin are all considered to be human teratogens. Maternal treatment with any of these agents in usual therapeutic doses during the first trimester of pregnancy poses a teratogenic risk to the embryo or fetus.

9–11. The answers are: 9-E, 10-B, 11-A *[IV B 1–4].*
Case–control studies compare the frequency of exposure to a putative teratogen during pregnancy in two groups of mothers—those whose children have a congenital anomaly and those whose children do not. Case–control studies can be used to calculate the statistical significance and to estimate the relative risk of an observed association. Such calculations can also be made from cohort studies in which the frequency of congenital anomalies is determined among the children of women who were exposed to a given agent during pregnancy and among the children of women who were not so exposed.

Most human teratogens have been identified by recognizing a recurrent and characteristic pattern of congenital anomalies among a series of children whose mothers were exposed to the same agent during pregnancy. Clinical series are thus an extremely important method of detecting teratogenic effects in humans, even though such studies usually do not permit calculation of statistical significance or estimation of risks.

Case studies are sometimes valuable in raising suspicion of a teratogenic effect, but more often the observation of a congenital anomaly in an infant whose mother was exposed to a particular agent during pregnancy is just coincidental. This is especially likely if the exposure or congenital anomaly or both are relatively common.

9
Genetic Paradigms in Human Disease

J. M. Friedman and Barbara McGillivray

I. INTRODUCTION. Genetics is relevant to the nature of medicine in many ways besides the diagnosis and management of genetic disease in families. This chapter provides four examples of such applications: genetic markers, cancer, embryonic development, and behavior.

II. GENETIC MARKERS are used to determine the relationship of individuals to each other and the genetic similarity of blood or tissue to an individual (see Ch 4 I).

A. Twins

1. From a genetic perspective, there are **two kinds of twins**.
   a. **Monozygotic (MZ; identical) twins** have all of their genes in common. MZ twins are derived from a single zygote that produces two separate embryos.
   b. **Dizygotic (DZ; fraternal) twins** have half of their genes in common, on average. DZ twins develop from two separate zygotes and bear the same genetic relationship to each other as siblings do.
2. **Genetic markers can be used to determine zygosity.**
   a. **MZ twins** have identical alleles at all loci tested.
   b. **DZ twins** differ from each other at a substantial number of loci.
   c. Genetic markers can be used in a similar fashion to determine zygosity within sets of triplets or higher **multiple births**.
   d. Genetic markers rarely need to be used to distinguish **twins of unlike sex** who are almost always DZ twins.

B. Paternity testing

1. Paternity testing requires **samples from the child, the mother, and the putative father**.
2. In general, **paternity can be excluded, but not proven, with certainty,** using genetic markers. Two kinds of exclusions are possible.
   a. **Paternity is excluded if neither of the putative father's two alleles at a locus is present in the child.** For example, if the putative father is homozygous for a genetic marker that the child does not have, paternity is excluded.
   b. **Paternity is excluded if the putative father lacks an allele that is present in the child but not in the mother.** For example, if a child has a marker that neither the mother nor the putative father has, paternity is excluded.
3. The more markers tested and the less common the alleles seen, the more likely it is that a wrongly accused man can be excluded as the father of a child.
4. A **probability of paternity** can be calculated based on the number of markers tested and the frequency of the alleles for which the child is informative. With DNA fingerprinting, which employs extremely polymorphic markers [see Ch 4 I B 1 a (2) (b)], probabilities of paternity that are very high can usually be obtained with the real father.

C. Other forensic uses of genetic markers include:

1. **Determining the origin of blood, semen, or tissue specimens** obtained at crime scenes. Small amounts of DNA or protein can be used to test for specific genetic markers characterizing a victim or suspect.

Table 9-1. MN Blood Group

Phenotype	Genotype
M	*MM*
N	*NN*
MN	*MN*

2. **Identifying missing persons.** Genetic markers can be compared with the person and his or her parents or other family members to establish true biological relationships.

D. **Blood typing** is important in preventing transfusion reactions. **Blood groups** reflect cell surface antigens coded by genes that are usually inherited in a simple mendelian fashion. These antigens are most often identified by serological tests.

1. **The MN blood group** consists of two alleles that are codominantly expressed (Table 9-1). Therefore, the genotype can be determined directly from the phenotype. The MN blood group is not usually important for blood transfusions or determining mother–fetus compatability.
2. **The ABO blood group** is the most important group in terms of blood transfusions.
   a. The ABO blood group consists of **three major alleles**.
      (1) The **A allele is expressed codominantly**. People who have an A allele express A antigen on the surface of their red blood cells. People who have no A allele spontaneously make antibody to A antigen.
      (2) The **B allele is expressed codominantly**. People who have a B allele express B antigen on the surface of their red blood cells. People who have no B allele spontaneously make antibody to B antigen.
      (3) The **O allele is recessive to the A and B alleles**. The O allele does not produce a corresponding cell surface antigen.
   b. The genotypes and phenotypes of the ABO blood group are shown in Table 9-2. The antigens and antibodies that occur with the various blood types appear in Table 9-3.
   c. **Rules of blood transfusion**
      (1) **Blood cells expressing an antigen should be transfused only into persons who also have that antigen.** For example, AB blood may be given to individuals of type AB, but not to individuals of type A, B, or O.
      (2) **Blood cells lacking an antigen may be transfused into persons who have that antigen.** For example, O blood may be given to individuals of type A, B, or AB as well as to those of type O.
3. **The Rh blood group** contains an allele (usually called *D*) that produces a major cell surface antigen. Individuals whose cells carry this major antigen are referred to as being **Rh$^+$**. Individuals whose cells lack the major Rh antigen are said to be **Rh$^-$**. The phenotypes and genotypes for the major Rh antigen are in Table 9-4.
   a. **Rh$^-$ individuals are capable of making an antibody to the major Rh antigen but do not do so unless they have been exposed to Rh$^+$ blood.** Sensitization of an Rh$^-$ person usually occurs because of pregnancy with an Rh$^+$ fetus or transfusion of Rh$^+$ blood.
   b. **Hemolytic disease of the newborn (erythroblastosis fetalis)** is caused by destruction of fetal red blood cells by maternal antibody. **Rh incompatibility is the most common cause of severe hemolytic disease of the newborn.**
      (1) The "setup" for Rh hemolytic disease of the newborn occurs when the **mother is Rh$^-$ and the father is Rh$^+$**.

Table 9-2. ABO Blood Group

Phenotype	Genotype
O	*OO*
A	*AO* or *AA*
B	*BO* or *BB*
AB	*AB*

Table 9-3. Antigens and Antibodies of the ABO Blood Group

Blood Type	Antigens on Red Blood Cells	Antibody in Serum
O	Neither A nor B	Both anti-A and anti-B
A	A	Anti-B
B	B	Anti-A
AB	Both A and B	Neither anti-A nor anti-B

(2) If the father's genotype is *DD,* all of the mother's pregnancies by him are at risk. If the father's genotype is *Dd,* only half of the mother's pregnancies by him are at risk, on average.

(3) Rh hemolytic disease of the newborn **usually does not occur in the mother's first pregnancy** because sensitization is required before she begins to make anti-Rh antibody. Once hemolytic disease of the newborn does occur, however, it tends to become worse in each subsequent affected pregnancy.

(4) **Rh hemolytic disease of the newborn is usually preventable** by avoiding sensitization of Rh^- girls and women. This can be done by avoiding transfusion with Rh^+ blood and by treating Rh^- pregnant women with anti-Rh antibody (RhoGAM) injections.

c. The Rh blood group is complex. In addition to the major (*D*) antigen, there are other minor Rh antigens (usually called *C* and *E*) that can occasionally be of clinical significance.

E. Tissue typing is necessary for successful organ or tissue transplantation.

1. **The major histocompatibility complex (MHC)** is the most important genetic region in determining tissue types for transplantation. In humans, the MHC is called the HLA (human leukocyte antigen) complex.
   a. The HLA complex is composed of **several closely linked genetic loci** on chromosome 6.
      (1) **Class I HLA loci** include ***HLA-A, HLA-B,*** and ***HLA-C***. These loci function in recognition of foreign antigens by cytotoxic T lymphocytes. Class I antigens are particularly important in graft rejection.
      (2) **Class II HLA loci** include ***HLA-DR, HLA-DP,*** and ***HLA-DQ***. They are involved in recognition of foreign antigens by helper T lymphocytes.
      (3) **Class III HLA loci** code for a few of the components of the **complement** cascade. These genes include *C2, C4S, C4F,* and *BF*.
   b. Most class I and class II HLA loci are **highly polymorphic**.
   c. Because the loci of the MHC are closely linked, all of the alleles in the complex are usually transmitted as a unit to the next generation. This unit, called a **haplotype,** comprises the set of alleles contained in the HLA complex of one chromosome 6.
   d. Most HLA genes produce cell surface antigens that can be detected by serological or other methods. These HLA **antigens are transmitted as simple codominant traits**.

2. **A general rule in organ transplantation** is: The greater the similarity between HLA types of the donor and recipient, the better the chance for avoiding rejection.
   a. **MZ twins are ideal organ donors** because they are completely identical with respect to their HLA types.
   b. **A parent and child are usually haploidentical,** which means they have in common one HLA haplotype that is identical by descent.
   c. **Siblings usually have a 25% chance of being HLA identical,** a 50% chance of being haploidentical, and a 25% chance of being completely different with respect to the HLA complex.

Table 9-4. Rh Blood Group

Phenotype	Genotype
Rh^+	*DD* or *Dd*
Rh^-	*dd*

III. CANCER AND GENETICS

A. General nature of neoplasia

1. A **neoplasm** is an abnormal tissue that grows beyond normal cellular control mechanisms. Neoplasms may involve almost any tissue in the body and may be either **benign** or **malignant,** depending on whether or not they possess the potential to spread to other parts of the body, that is, to **metastasize**.
2. **Neoplasia** is the pathological process that results in the development of neoplasms. It appears to be a **multistep process** that varies for patients with histologically similar tumors as well as for patients with different kinds of tumors.

B. Malignancy as a phenotype

1. **Most neoplasms are of multifactorial etiology** in the sense that both inherited and noninherited genetic factors are involved in their pathogenesis. In some instances, strong environmental predispositions to neoplasia exist; often these act through somatic genetic mechanisms.
2. **A small percentage (5% or less) of patients with neoplasms have a strong predisposition that is inherited as a simple mendelian trait.**
   a. The tumors in these patients do not differ in type or histology from tumors that occur in most patients with neoplasms, but, in general, patients with monogenic predispositions to neoplasia are characterized by the following features.
      (1) The neoplasms have an **earlier age of onset**.
      (2) The neoplasms tend to be **bilateral and/or multifocal**.
      (3) Only **one or a few specific types of neoplasms** occur in each mendelian condition. There are many such conditions, however, and as a group, they involve many different types of tumors.
      (4) In some conditions, there are **associated phenotypic abnormalities** that are characteristic of the specific mendelian trait.
   b. In individuals with mendelian conditions, the risk of developing a malignancy may be very high (75%–100%), but most cells of the affected types do not develop into tumors. Thus, **a predisposition to develop neoplasia is inherited, not the neoplasia itself**. Even in patients with strong genetic predispositions, other events are necessary to cause the cells to become neoplastic.
   c. **Mendelian conditions that produce strong predispositions to neoplasia** are of several types:
      (1) **Syndromes of multiple benign or malignant neoplasms**
         (a) Affected individuals develop increasing numbers of characteristic types of tumors throughout life.
         (b) These conditions are inherited as **autosomal dominant** traits.
         (c) Examples include:
            (i) Neurofibromatosis
            (ii) Multiple endocrine neoplasia
            (iii) Multiple polyposis of the colon
            (iv) Li-Fraumeni syndrome
      (2) **Abnormalities of DNA or chromosomal repair**
         (a) Affected individuals exhibit abnormal DNA repair or increased frequencies of chromosomal breakage.
         (b) These syndromes are inherited as **autosomal recessive** traits.
         (c) Examples include:
            (i) Xeroderma pigmentosum
            (ii) Fanconi pancytopenia syndrome
            (iii) Ataxia telangiectasia
      (3) **Immunodeficiency syndromes**
         (a) Affected patients have congenital abnormalities of immunological function.
         (b) These syndromes are inherited as **recessive traits,** either autosomal or X-linked.
         (c) Examples include:
            (i) X-linked agammaglobulinemia
            (ii) Wiscott-Aldrich syndrome
      (4) **Some monogenic predispositions to neoplasia that do not fall into these categories.** Examples include tylosis (keratosis palmaris et plantaris), which is associated with esophageal cancer, and α_1-antitrypsin deficiency, in which hepatocellular carcinoma may occur.

3. Chromosome abnormalities and neoplasia

a. Some **constitutional chromosome abnormalities** are associated with an increased frequency of certain kinds of malignancy.

(1) Leukemia occurs in about 1% of patients with **Down syndrome**.

(2) Retinoblastoma develops regularly in children with constitutional **deletions of chromosome 13q14**.

(3) Wilms' tumor of the kidney often occurs in children with constitutional **deletions of chromosome 11p13**.

b. Acquired chromosome abnormalities arise in most malignant neoplasms. In these instances, the patient does not usually have a constitutional chromosome abnormality, but karyotypic changes develop during the formation of the neoplasm.

(1) Often, a neoplasm exhibits cytogenetic changes that are characteristic of a specific tumor type. For example, a t(8;14) occurs in most cases of Burkitt's lymphoma and a t(9;22) is found in most cases of chronic myelogenous leukemia (CML). The der(22) t(9;22) of CML is sometimes called the Philadelphia (Ph) chromosome.

(2) Some chromosome abnormalities are associated with differences in the malignant behavior of a neoplasm. In acute myelocytic leukemia, for example, the presence of t(8;21) is a good prognostic sign while the presence of t(9;22) is a poor prognostic sign.

(3) Frequently, neoplasms exhibit multiple chromosome abnormalities. Some of these cytogenetic alterations may be characteristic of the specific tumor type, others may be changes that occur in many different kinds of neoplasms, and still others may be more or less unique to an individual patient.

(4) Cell lines exhibiting different karyotype abnormalities are often seen within a single neoplasm. Sometimes, almost every cell has a karyotype that varies somewhat from others studied from the same tumor.

(5) In general, however, **all of the cell lines are clonally related,** which means it is possible to trace the cytogenetic evolution of the lines back to a single abnormal progenitor.

(6) As the neoplasm progresses (i.e., becomes more malignant), **the karyotype tends to become more abnormal**.

4. Most malignant neoplasms appear to have a clonal origin. This means that all of the neoplastic cells derive from a single abnormal progenitor. In contrast, normal tissues develop from cells that are not clonally related after the first few postzygotic divisions.

a. Cells that do not exhibit neoplastic behavior but are developmentally related to a neoplasm may share in the neoplasm's clonal origin. This indicates that an **initial somatic mutation** led to clonal proliferation of a cell line but **additional events** were necessary to produce neoplasia.

b. The clonal origin of malignancies is consistent with a multistep neoplastic process because multiple independent events are much more likely to occur sequentially in a single cell lineage than simultaneously in many different cells.

C. Malignancy as a genotype. Two classes of genes that are involved in the neoplastic process have been studied in detail: **oncogenes** and **tumor suppressor genes**.

1. Oncogenes are normal genes that when altered, inappropriately expressed, or overexpressed can lead to neoplasia.

a. Oncogenes are widely **conserved in evolution**.

b. The oncogenes that are part of the normal genome of vertebrate cells are called **proto-oncogenes** or **c-oncogenes** to indicate that in their normal state and under normal physiological controls, these genes do not cause tumors.

c. Proto-oncogenes are generally thought to have important **functions in cell growth and differentiation**. Examples include the following.

(1) The proto-oncogene ***c-src*** codes for a **cytoplasmic protein kinase** that affects phosphorylation of certain amino acids in target proteins and thereby influences the activity of the target proteins.

(2) Some proto-oncogene products appear to be **growth factors and growth factor receptors** involved in the development of certain specific cell types.

(a) The proto-oncogene ***c-erbB*** codes for the epidermal growth factor receptor, which has protein tyrosine kinase activity. Therefore, *c-erbB* could also be considered one of the proto-oncogenes that codes for a cytoplasmic protein kinase.

(b) ***Int-2*** is a proto-oncogene related to the fibroblast growth factor gene.

(3) The product of the proto-oncogene ***c-jun*** is a transcription factor that regulates gene expression.

(4) The proto-oncogene ***c-ras*** closely resembles a signal transduction mediator gene. Mediators of signal transduction are involved in transmission of messages across the plasma membrane.

d. The **oncogenic potential** of a proto-oncogene can be **activated** in a variety of ways:

(1) **Point mutation**

(a) The ***c-ras*** proto-oncogene provides the best studied example of activation by point mutation. **Transfection** of certain cell lines with DNA from tumor cells containing a **mutated *ras* oncogene transforms** the phenotype of the cell lines to a phenotype that is characteristic of neoplastic cells.

(b) The mutations that activate *ras* are highly specific, usually involving one of only two amino acids within the *ras* protein.

(2) **Other structural alterations of the oncogene product,** an example of which is the characteristic **t(9;22) chromosomal rearrangement associated with CML** [see III B 3 b (1)]

(a) The translocation moves the *c-abl* proto-oncogene from chromosome 9 into a region called *bcr* on chromosome 22.

(b) The oncogene becomes included in the *bcr* transcription unit producing a fusion protein that has a structure substantially different from that of the normal *c-abl* protein.

(3) **Gene amplification**

(a) Gene amplification increases the number of copies of the normal proto-oncogene within a cell.

(b) Unique chromosomal structures called **double minutes** and **homogeneously staining regions** may be formed in neoplastic cells that contain many copies of an oncogene.

(4) **Placing the proto-oncogene under the control of a more active promoter** (e.g., by insertion of a retrovirus promoter near or within a proto-oncogene sequence)

(5) **Placing the proto-oncogene under the control of an enhancer sequence**

(a) This appears to be a mechanism associated with the **chromosome translocations characteristic of some malignancies.**

(b) An example is **Burkitt's lymphoma,** in which *c-myc* on chromosome 8 is translocated into proximity with an immunoglobulin locus, most often locus IgH on chromosome 14, which codes for the heavy chain of immunoglobulin.

e. In in vitro systems, activation of **more than one oncogene** is usually necessary to change a normal cell line into one that exhibits fully neoplastic properties. This observation is in accord with the **multistep nature of the neoplasia**.

f. An activated oncogene exhibits a dominant phenotype at the cellular level. One copy of an activated oncogene is sufficient to produce its oncogenic effect, even in the presence of a normal, nonactivated proto-oncogene of the same type elsewhere in the cell. **Tumorigenesis results from a gain of function.**

g. In experimental animals, the oncogenic potential of most **tumorigenic viruses** has been shown to be due to their inclusion of an **activated oncogene**.

(1) Such oncogenes appear to have **originated in the genome of a vertebrate host** and to have been picked up by the virus during a transducing event.

(2) The structure of the viral version of the oncogene, called a **v-oncogene,** is altered from that of the host, reflecting the mechanism of origin and activation of the v-oncogene.

2. Tumor suppressor genes are normal genes that function to prevent the development of neoplasia.

a. Tumor suppressor genes are widely **conserved in evolution**.

b. Tumor suppressor genes are defined by the effect of their absence. Tumors tend to develop in cells in which *both* normal alleles of a tumor suppressor gene have been lost or inactivated.

(1) Tumor suppressor genes have a **recessive effect at the cell level**. One normal allele is sufficient to prevent the neoplastic effect, even if the second allele has been inactivated or lost. **The development of neoplasia results from a loss of function.**

(2) Loss or inactivation of a normal tumor suppressor gene may be **constitutional** (i.e., affecting every cell of the body) or **acquired somatically** in a single clone of cells.

c. The ***Rb* gene** that prevents the development of retinoblastoma is the best studied example of a tumor suppressor gene. **Retinoblastoma** provides a model for the action of tumor suppressor genes.

(1) Retinoblastoma is a **malignant tumor of the retina,** which develops in early childhood and is usually fatal if not treated early.

(a) The incidence of retinoblastoma is about 1 per 20,000 infants.

(b) Most retinoblastomas occur sporadically, but about 10% of affected children have a parent who also had retinoblastoma. This form of the disease, called **hereditary retinoblastoma,** is inherited as an autosomal dominant trait.

(c) Tumors in hereditary retinoblastoma are often **bilateral** or **multifocal** and exhibit an **earlier age of onset** than sporadic retinoblastomas.

(2) Retinoblastoma develops when **both alleles of the *Rb* tumor suppressor gene are lost or inactivated in a retinoblast**. Loss or inactivation of the *Rb* gene may be **constitutional** or acquired as a **somatic mutation**.

(a) In **hereditary retinoblastoma,** a chromosome containing an allele of the *Rb* gene that has been deleted or inactivated by mutation is transmitted in **typical autosomal dominant fashion**. All of the cells of the body of a person who has inherited this abnormality begin with only one normal copy of the *Rb* gene.

(i) A retinoblastoma develops in a person who has inherited this mutation only if the **second *Rb* allele is also inactivated or lost somatically**.

(ii) Note that although the abnormal Rb phenotype (i.e., loss of tumor suppression) is **recessive at the cellular level,** the predisposition to hereditary retinoblastoma is transmitted **dominantly at the level of the whole organism**.

(b) The *Rb* gene lies in band q14 of chromosome 13. Children with multiple congenital anomalies due to cytogenetically apparent **constitutional deletions of chromosome 13q14** have lost one normal *Rb* allele in every cell of their bodies. Development of retinoblastoma occurs in children with such deletions **if the other *Rb* allele is inactivated or lost somatically** in a retinoblast.

(c) In most people with sporadic unilateral retinoblastoma, **loss or inactivation of both Rb alleles occurs somatically** in the clone of cells that gives rise to the tumor.

(3) **Loss or inactivation of a normal *Rb* allele** may occur by a variety of mechanisms.

(a) **Mutation producing absent or altered transcripts** is a common mechanism of inactivation of the *Rb* allele transmitted in hereditary retinoblastoma. This is a less common mechanism of *Rb* allele inactivation in somatic cells.

(b) **Deletions of the *Rb* locus** account for most other cases of hereditary retinoblastoma. Most of these deletions are submicroscopic; that is, they are not apparent on conventional cytogenetic studies.

(c) **Loss of a whole chromosome 13** is a common somatic event associated with loss of a normal *Rb* allele from a cell. In some cases, duplication of the other chromosome 13, which carries a deleted or inactivated *Rb* allele, is seen with loss of the normal chromosome 13.

(d) **Mitotic crossing over** between a chromosome with a normal *Rb* allele and one with an inactive *Rb* allele can result in homozygosity for the inactivated state in some of the daughter cells.

d. **Knudson's hypothesis** states that the development of tumors such as retinoblastoma requires **two separate mutations**.

(1) In hereditary retinoblastoma, the first mutation is inherited and the second acquired somatically.

(2) In most sporadic retinoblastomas, both mutations occur somatically.

(3) This explains why, in comparison to sporadic retinoblastomas, hereditary retinoblastomas usually exhibit earlier age of onset and bilateral or multifocal occurrence. The occurrence of a second mutation in a cell that already carries a first (inherited) mutation is much more likely than two independent somatic mutations in a single cell lineage.

(4) Molecular studies have proven that Knudson's hypothesis is correct for the *Rb* gene in retinoblastoma.

e. In general, **each tumor suppressor gene locus is involved in controlling the development of several different kinds of tumors**. For example, the *Rb* locus is involved in the development of at least some cases of osteosarcoma and small cell carcinoma of the lung.

f. A general method of identifying chromosome regions containing tumor suppressor genes is by observing a **reduction to homozygosity** of markers in certain chromosomal regions within tumor cells. Reduction to homozygosity implies loss of one chromosome or a part thereof with elimination of one allele of a tumor suppressor gene.

(1) For example, if a patient is heterozygous for a series of markers on chromosome 17p in his normal tissues but is found to show only one allele at each of these loci in tissue from a neoplasm, reduction to homozygosity is said to have occurred.

(2) Reduction to homozygosity may occur by a variety of mechanisms [see III C 2 c (3)].

(3) This phenomenon **accounts for some of the characteristic chromosome abnormalities associated with specific neoplasms.**

(4) Observing reduction to homozygosity implies that **the retained chromosome homologue carries a tumor suppressor gene that has been lost or inactivated** either somatically or by an inherited mutation.

3. **Other genes are also involved in the development of neoplasia.**

a. The various **monogenic conditions that predispose to tumor development** suggest the nature of some of these additional genes (see III B 2 c), including those that control repair of DNA damage and immunological function.

b. **Some genes appear to act both as oncogenes and as tumor suppressor genes, depending on the circumstances.**

(1) Gene ***p53*** **acts as a tumor suppressor gene in normal cells** and is lost or inactivated during the development of many different neoplasms. Tumor development is associated with loss of **function.**

(2) **Transfection of normal cell lines with certain kinds of mutant *p53* DNA** can cause the cells to acquire neoplastic characteristics. This transfection assay has been used to define oncogenes [see III C 1 d (1) (a)] and implies that tumor development is associated with a **gain of function.**

(3) **Mutations of *p53*** have been shown to occur in **Li-Fraumeni syndrome,** a dominantly inherited condition that strongly predisposes to several kinds of malignancy.

4. **Genes involved in the development of neoplasms in rare mendelian tumor predisposition syndromes often are also involved in common tumors in the general population.**

a. Examples include the *Rb* gene and the *p53* gene, as discussed above.

b. **Mutation is a critical component of the neoplastic process.** Mendelian tumor predisposition syndromes often involve constitutional mutations of genes that more frequently affect neoplastic development through somatic mutation or acquired chromosomal abnormalities.

c. **Development of neoplasia is a multistep process.**

(1) Although the specific steps may differ in different patients or different tumors, often there is **considerable overlap of the genes involved.**

(2) **Environmental factors** that predispose a person to the development of neoplasia may act by promoting mutation or alteration of expression of cellular genes, thereby disrupting their normal control over division and differentiation.

(3) Thus, **environmental factors, inherited genetic factors, and acquired somatic mutations and chromosome alterations all may interact in the development of neoplasia.**

(4) Prevention may involve intervention at any one or more of these steps.

IV. DEVELOPMENTAL GENETICS

A. **Embryonic development** is controlled by the genome. The zygote contains all of the information necessary to produce an entire organism from a single cell.

1. Although the structural changes involved in embryogenesis have been well described for many years, **very little is known about the molecular processes** that control human embryonic development.

a. Some insight has been gained recently regarding molecular and genetic control of embryogenesis in other organisms, especially *Drosophila* and *Caenorhabditis elegans,* but it is uncertain how analogous these systems are to the human.

b. Molecular and genetic analysis of embryogenesis in the mouse, the **best investigated mammal,** is less advanced.

c. Thus, current understanding of the genetic and molecular nature of **human embryogenesis** is **largely based on analogy.**

2. The **major events of early mammalian embryogenesis** include:

- **a. Cell lineage specification,** which is the allocation of cells into lineages that have more restricted developmental potential
- **b. Segmentation,** which is the establishment of periodic structures along the length of the embryo
- **c. Regional specialization** in both segmental structures (e.g., vertebrae) and nonsegmental structures (e.g., the gastrointestinal tract)

3. Molecular mechanisms involved in these processes are largely unknown.

- **a. Maternal gene products laid down in the cytoplasm of the ovum do not appear to be of major importance** in early mammalian embryogenesis. In invertebrates, maternal gene products are responsible for initial determinative events in the embryo.
- **b. Predetermined clonal lineages do not appear to be of critical importance** in mammalian embryogenesis. In invertebrates, differentiation often occurs within specific lineages that are clonal derivatives of individual progenitor cells. Once the lineage is established, the subsequent differentiation of many generations of cell progeny is largely or entirely determined.
- **c.** In mammals, **cell fate is usually determined incrementally by a series of interactions with neighboring cells and the extracellular environment**. Although the developmental potential of an initially differentiated cell population is restricted, many alternate final fates remain possible. The ultimate differentiated state of a cell derived from this initial population depends on the combined effect of many subsequent determinative events.
- **d.** Once differentiation occurs, it is largely irreversible, implying that **fixed patterns of transcription, translation, and processing are established**.
- **e. Two cellular control mechanisms** are important in mammalian embryogenesis:
  - **(1)** The **establishment of morphogen gradients** that induce qualitatively different responses of cells to different morphogen concentrations along the gradient
  - **(2)** The **establishment of regulatory cascades,** in which activation or inactivation of one gene affects the function of other genes, which in turn affects the function of other genes and so forth.

4. A few **genes and gene products** have been recognized that seem to be **important in mammalian embryogenesis**. Most of these genes were identified on the basis of homology to genes involved in development in *Drosophila* or other organisms.

- **a.** ***Hox* genes** occur in clusters and are expressed along the embryonic central nervous system in overlapping domains extending in an anterior to posterior direction.
  - **(1)** The **order of expression of the *Hox* genes** along the developing neural axis is the same as the 3′ to 5′ order of the genes within the cluster.
  - **(2)** A very similar correlation between gene order and spacial order of expression occurs for homologous genes in *Drosophila,* suggesting that the **mechanism of anterior–posterior specification has been maintained through evolution.**
  - **(3)** ***Hox* genes** contain a conserved DNA sequence called a **homeobox** that codes for a **DNA-binding domain** in the corresponding proteins.
- **b.** Some **proto-oncogenes** (see III C 1 c) appear to have an important role in embryonic development. For example, the *int-1* proto-oncogene is expressed in a highly specific manner in the developing central nervous system of the mouse. This gene is homologous to one involved in segmentation in *Drosophila*.
- **c.** The **genes for growth factors and growth factor receptors** are implicated in induction of embryonic differentiation in some vertebrate systems.
- **d. Genes coding for signal transducers** such as retinoic acid also appear to be involved in differentiation in some vertebrate tissues.

5. Elucidation of additional genes and regulatory mechanisms that are critical to mammalian embryogenesis will require a multifaceted approach. **Promising strategies** include:

- **a. Identification and study of mammalian genes** that are **homologous to genes** with established roles in embryonic development **in other organisms**
- **b. Isolation of the genes responsible for established developmental phenotypes** by chromosome localization and candidate gene approaches
- **c. Production of developmental mutations** by insertion of recoverable genetic elements such as transposons into critical genes
- **d. Production of transgenic and chimeric animals** in which the structure or regulation of developmental genes has been altered

e. **Targeted ablation experiments** in which the effects of destroying cells that express a specific gene are tested
f. **Characterization of genes expressed in pluripotent or differentiating cells** such as embryonic stem cell lines
g. **Studies of temporal and spacial patterns of gene expression**

B. Sexual differentiation. Elucidating the underlying mechanisms responsible for abnormal sexual differentiation in humans has not only aided in the management of patients but has also allowed a gradual unfolding and understanding of a basic developmental pathway; that of sexual differentiation. In fact, the development of the internal and external genitalia is one of the better understood processes of early human development.

1. **Indifferent stage.** Although the chromosomal basis of sex is determined at the moment of conception, the internal structures are indifferent until 6 weeks gestation. These structures consist of the following.
   a. **The gonad** develops at 5 weeks gestation from coelomic epithelium and mesenchyme. The epithelium sends projections (primary sex cords) into the mesenchyme. These will become either seminiferous tubules or primary ovarian follicles.
   b. **Primordial germ cells** migrate into the mesenchyme from the yolk sac by 6 weeks gestation and will become either spermatogonia or ova.
   c. **Wolffian (mesonephric ducts)** appear at 30 days gestation and are paired structures.
   d. **Müllerian (paramesonephric ducts)** appear at 40–48 days gestation and are also paired structures. The two pairs of ducts form the internal genitalia of the fetus. At 6 weeks gestation, the structures are indifferent and are present in both male and female fetuses.

2. **The role of sex chromosomes.** The Y chromosome plays a crucial role in sex development; the embryo inheriting a Y chromosome develops as a male, while the embryo lacking a Y chromosome develops as a female.
   a. **Sex determining genes on the Y chromosome** induce testicular development by an unknown mechanism. It is postulated that the Y-linked genes initiate a cascade effect of both X-linked and autosomal genes to promote the indifferent gonad to become a testis. The candidate gene, *SRY* (for sex-determining Y), lies on the short arm of the Y chromosome very near the pseudoautosomal border and codes for a highly conserved protein.
   b. In the **absence of SRY,** the gonad develops as an ovary, but two X chromosomes are necessary for maintenance of the ovary.

3. **Development of the early testis.** At 6 weeks gestation, under the influence of the Y chromosome, the testicular cords evolve from the sex cords and several cell types differentiate (Figure 9-1).
   a. **Sertoli cells** in the testes of the fetus produce müllerian-inhibiting substance, which acts locally and specifically to suppress the müllerian ducts. Suppression of the müllerian ducts inhibits development of the uterus and fallopian tubes. Müllerian-inhibiting substance is present prior to testosterone.
   b. **Leydig cells** begin to produce testosterone by 7 weeks gestation, probably under the influence of placental human chorionic gonadotropin (hCG). Testosterone influences the development of the wolffian duct and other sensitive structures outside the genital tract.
   c. The **germ cells** differentiate to spermatogonia, which are the stem cells of sperm production.
   d. **Ovarian development** occurs without the Y chromosome and is later than testicular development. At 50 days gestation, secondary sex cords form from the epithelium and surround the germ cells.

4. **Development of male internal and external genitalia**
   a. **Internal genitalia.** Testosterone from Leydig cells complexes with an androgen receptor and stimulates the wolffian duct to become the male internal genitalia, which consists of the epididymis, the vas deferens, and seminal vesicles. Internally, the process results in the ejaculatory duct system for sperm and is complete by 14 weeks gestation.
   b. **External genitalia.** Similar to the internal structures, the external structures are initially indifferent and consist of the urogenital tubercle, the urogenital swellings, and the urogenital folds. Male external genitalia are complete by 14 weeks gestation.
      (1) The **urogenital tubercle** will become the glans penis, which should be tubular with fusion inferiorly.

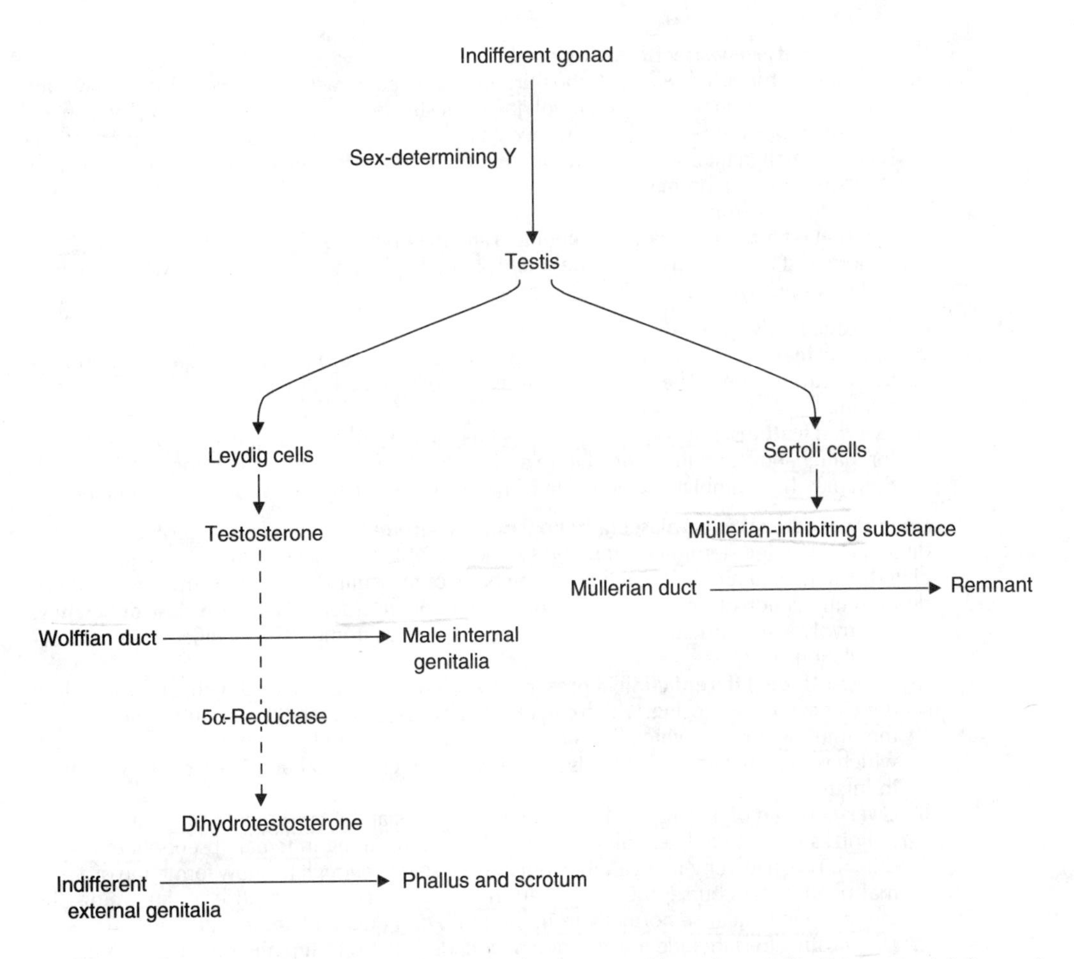

Figure 9-1. Differentiation of cell types involved in the development of the male genitalia. This process is influenced by the Y chromosome and begins at 6 weeks gestation.

(2) The **urogenital swellings** also fuse to become the scrotum under the influence of dihydrotestosterone (DHT).

(3) The **urogenital folds** become the shaft of the penis under the influence of DHT.

5. Testosterone and DHT

a. Testosterone is responsible for development of male internal structures by local diffusion and uptake by a cytosol receptor. At puberty, testosterone causes the changes in body hair, musculature, voice deepening, and penile growth.

b. DHT is formed from testosterone by the enzyme 5α-reductase, which is present in androgen target cells. The external genitalia bind most efficiently to DHT.

6. Development of female internal and external genitalia

a. Internal genitalia. In the absence of a testis, the müllerian duct structures continue to develop and become the fallopian tubes, the uterus, and the upper portion of the vagina. Without testosterone from a testis, the wolffian duct regresses.

b. External genitalia. Externally, the indifferent genitalia show no midline fusion in the absence of androgens. The genital tubercle becomes the clitoris, the urogenital folds become the labia minora, and the urogenital swellings become the labia majora. The presence of ovaries is not required for either internal or external female genitalia formation. The female structures are complete by 14–16 weeks gestation.

C. Abnormalities of sexual differentiation. Differentiation of the sexually indifferent embryo into a sexually differentiated embryo is secondary to a series of events, each of which is concerned with a number of genes. By noting the internal and external features of syndromes involving genital ambiguity or sex reversal, it is often possible to localize the nature of the problem.

1. **Pure gonadal dysgenesis with sex reversal**
   a. **Internal features.** A heterogeneous group, XY females with sex reversal may have an X-linked condition or a deletion involving the testis-determining region on Yp. The gonads are streak, or rudimentary (as in Turner syndrome because of the single X chromosome), but have malignant potential. Without müllerian-inhibiting substance, the müllerian duct structures develop normally into the uterus and tubes. The wolffian duct structures regress without testosterone.
   b. **External features.** Externally, the genitalia are unambiguously female. Affected females are of normal stature but may have the shield-shaped chest and neck webbing, which are typical of Turner syndrome girls.
2. **Mixed gonadal dysgenesis**
   a. **Internal features.** This condition is a variant of Turner syndrome, with mosaicism involving the Y chromosome. There are two cell lines; 46,XY and 45,X with the proportions determining the clinical phenotype.
   b. **External features.** Phenotypic males may have hypospadias, dysgenetic gonads with a risk for malignancy, and the short stature and neck webbing of Turner syndrome. Other children may have ambiguous genitalia or have a classical Turner syndrome presentation.
3. **Congenital adrenal hyperplasia (adrenogenital syndrome)** includes five autosomal recessive disorders involving steroid hormone biosynthesis (Table 9-5). Several disorders can be associated with genital ambiguity, either on the basis of masculinization of a female fetus or undermasculinization of a male fetus. The most common disorder is **21-hydroxylase deficiency,** which involves the structural gene for the adrenal cytochrome P_{450} specific for steroid 21-hydroxylation.
   a. At least **three different clinical presentations** have been associated with 21-hydroxylase deficiency: the salt-losing, which can result in hypovolemia, shock, and subsequent death; the simple-virilizing, which leads to overgrowth and masculinization; and the late-onset, which is extremely rare and leads to virilization in childhood or adolescence rather than in infancy.
   b. **Oversecretion of adrenal androgens,** which begins at the third month of gestation, masculinizes the external genitalia of female fetuses, resulting in female pseudohermaphroditism. The involvement is variable, but may be severe enough to allow formation of a normal phallus. Of course, the gonads are normal ovaries, and because of the timing, the internal genitalia arise normally from the müllerian duct and are female.
   c. **Diagnostic clues** include the absence of gonads in the scrotum, the presence of a uterus, and elevated 17-ketosteroids (17-KS).
4. **Androgen insensitivity.** The spectrum of androgen insensitivity (also called androgen-resistant syndromes) may include genital ambiguity. The abnormality is with the androgen receptor, whose function it is to bind androgen, including both testosterone and DHT.
   a. **Complete androgen insensitivity** is associated with a normal male karyotype and unambiguous female external genitalia, while **incomplete androgen insensitivity** is associated with varying degrees of undermasculinization, including glandular hypospadias.
   b. Because of the variability of expression, the **maternal family history must be carefully evaluated** both for classical expression and the less severely affected males. About one-third of cases are sporadic, but the remainder have affected maternally related relatives.
   c. The **diagnosis of androgen insensitivity** is made by documenting normal male gonads, and

Table 9-5. Types of Congenital Adrenal Hyperplasia

Form	Genitalia	Androgens	Other
21-OH deficiency	Virilized ♀	↑↑	± Salt-losing, most common
11-OH deficiency	Virilized ♀	↑↑	Hypertension
Cholesterol desmolase	Ambiguous ♂	N/A	Salt-wasting
3-ß-OH steroid	Ambiguous ♂	↑↑	Lethal
17-OH deficiency	Ambiguous ♂	↓	Hypertension
17, 20-lyase	Ambiguous ♂	↓	Isolated defect

Note: ↑ = increase; ↓ = decrease; ± = with or without; ♂ = male; ♀ = female.

by specific androgen receptor studies documenting functional absence of the receptor, diminished amounts of normal receptor, or qualitatively abnormal receptor.

d. The **gene for the androgen receptor,** which is located on the long arm of the X chromosome (Xq13), has now been characterized and shows similarities with other steroid receptors (e.g., progesterone, glucocorticoid, and mineralocorticoid receptors).

5. 5α-Reductase deficiency is also known as pseudovaginal perineoscrotal hypospadias, since the infant usually presents with a severely hypospadic penis with the urethra opening on the perineum.

a. Clinical background. Because of an absent or abnormal 5α-reductase enzyme, **testosterone cannot be converted to DHT**. The receptors of the external structures respond most efficiently to DHT and, therefore, are predominantly female when only testosterone is present. The internal genitalia will be normal male.

b. Diagnosis is made by evaluating the production of DHT from testosterone in cultured genital skin fibroblasts.

c. Genetic background. This syndrome is inherited as an autosomal recessive.

d. Treatment. If recognized early, infants can now be treated specifically with DHT, and a good response is usually expected.

D. Clinical presentation. A mental checklist is helpful when examining the external genitalia of a newborn. Genitalia with an indeterminate appearance (large appearing clitoris, severely hypospadic penile structure, partial scrotal fusion, undescended testes) should prompt further investigation (Figure 9-2).

1. Term males. Most term male infants will have descent of testes at least into the upper scrotum at the time of birth. Bilateral cryptorchidism may be associated with hypospadias, with abnormalities of the urinary tract, or with the masculinized female infant. The phallus should measure 2.5 cm in length with the urethral opening at the tip. The scrotum has a midline raphe and is below the penis in orientation.

2. Term females. The female infant's labia majora may not completely cover the labia minora, especially in the preterm infant. The clitoral length should not exceed 1 cm, and there should not be fusion of the labia (at times, some degree of posterior fusion is seen). The vaginal orifice should be visible and is often identified by the presence of a whitish mucus.

3. Evaluation of the infant with ambiguous genitalia. Affected infants may have life-threatening conditions and need assessment quickly with a thoughtful ordering of investigations.

a. After suspicion that an infant's genital appearance is unusual or ambiguous, the infant should be carefully examined for evidence of other malformations.

b. The adequacy of the urinary tract must be confirmed, documenting voiding and using ultrasound or cytoscopy as appropriate.

c. Since a metabolic imbalance may be another consequence, electrolytes and blood glucose should be evaluated on an urgent basis. Additional blood can be drawn for chromosomal studies and determination of 17-ketosteroids.

d. The presence or absence of a uterus may help differentiate the type of abnormality.

e. While the evaluation of the infant is underway, the pregnancy and family history can be reviewed. Important points include the use of any virilizing medications (e.g., danazol), maternal virilization during the pregnancy, previous neonatal deaths, consanguinity, or family history of similarly affected children.

V. GENETICS AND BEHAVIORAL DISORDERS.

Recent advances in the area of genetics and behavior suggest a significant link between genetic factors and major psychiatric illnesses such as schizophrenia and mood disorders. The genetic influence was suggested in the early 1900s by family studies showing an increased incidence of schizophrenia among close relatives of affected patients. Historically, genetic research relevant to psychiatric illness has been limited by the reliance on behavioral or phenotypic traits. Until recently, the genotype could only be inferred.

A. Study methods in behavioral genetics. Four basic methods (Table 9-6)—family, twin, adoption, and prospective studies—are used to assess the genetic component of psychiatric conditions to derive risk figures (empiric figures) both for use in counseling and to determine the possible mode of inheritance of the disorders.

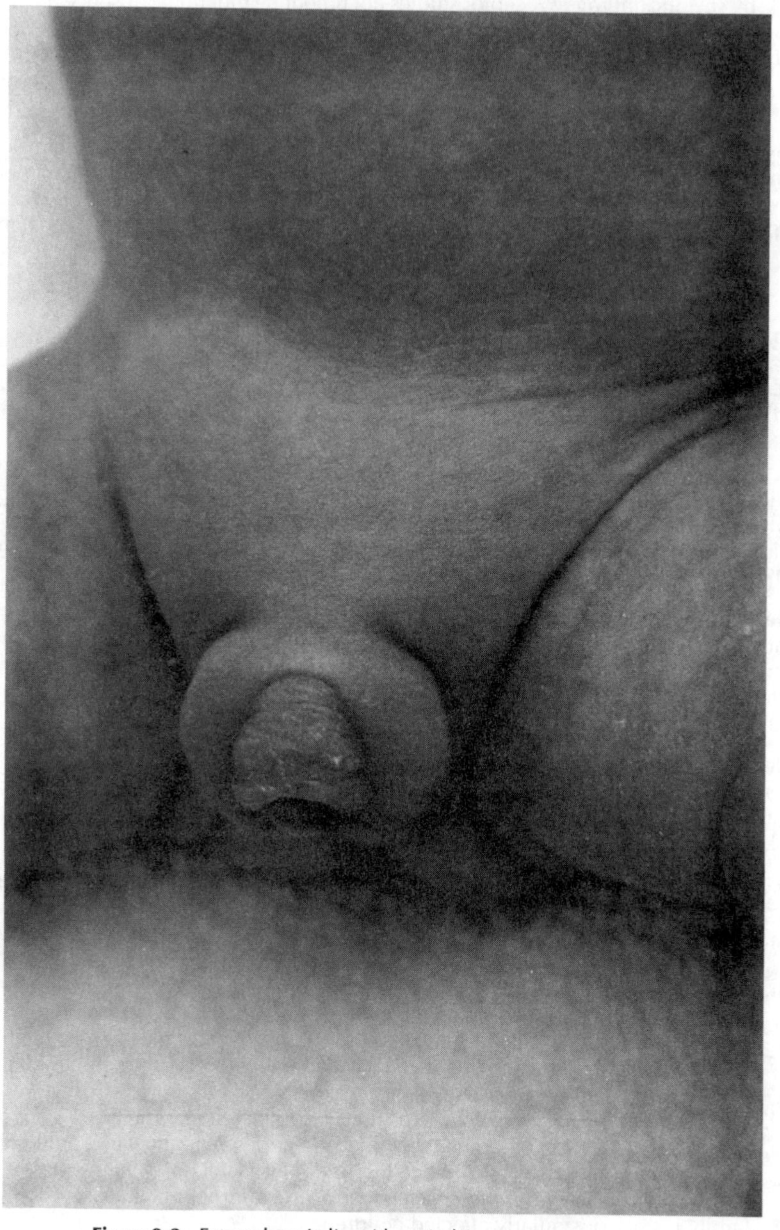

Figure 9-2. External genitalia with an indeterminate appearance.

1. **Family studies** compare the prevalence of a certain psychiatric disorder among relatives of an affected individual to the prevalence in a normal or control population. These studies involve either taking a family history from an affected or unaffected family relative or interviewing individual family members. The pedigree obtained can then be analyzed to compare single gene and multigene models of inheritance.
   a. **Advantages.** This method allows direct contact with the family and should allow maximum information to be gathered regarding the family. Specifically, the following information should be gathered from family studies.
      (1) **Precise diagnosis.** The investigator is able to confirm a diagnosis with direct examination and is more likely to pick up less severely affected family members with careful history taking and individual examination.

Table 9-6. Study Methods in Behavioral Genetics

Method	Features
Family studies	
Pedigree analysis	Fast, inexpensive; underestimates morbidity
Interviews	Direct contact best; blind interview required
Twin studies	Allow heritability studies; need accurate zygosity data; may have self-selection
Adoption studies	Separate genetic from environmental factors; allow long-term prospective studies

(2) **Separation of multiple factors.** If two psychiatric conditions are coexisting, pedigree analysis may allow separation of the conditions to two separate family branches. Families may be excluded to allow cleaner analysis.

(3) **Availability of family for other study methods.** Direct interaction with a family also allows molecular linkage methods to be used in conjunction with pedigree analysis.

b. **Limitations.** The most significant drawback of family studies is their inability to differentiate genetic from environmental influences. Also, unless a specific biologic or genetic marker is available, the genotype usually cannot be identified with certainty. Other problems encountered in family studies include the following.

(1) **Imprecise or incorrect diagnosis.** Although strict diagnostic criteria exist for common psychiatric disorders (i.e., the DSM-III-R of the American Psychiatric Association), an inexperienced interviewer may interpret findings incorrectly. A highly structured interview setting is best for less experienced interviewers.

(2) **Biased data**

(a) A family history (pedigree) obtained by telephone or from only one individual may exclude mildly affected relatives, thus underestimating psychiatric morbidity. Formal family interviews ideally are conducted in person.

(b) The diagnosis of the proband should not be known to the interviewer. Blind interviews are especially important when several diagnostic categories are being considered.

2. **Twin studies** compare the concordance rates for a certain psychiatric disorder in MZ versus DZ twins.

a. **Advantages.** Generally, MZ twins share both genetic and environmental factors, while DZ twins may differ (as with all siblings) because of both genetic and environmental influences. Differences between MZ and DZ twins reflect genetic variation for the observed trait. Twin studies are most useful as a preliminary test of hypotheses regarding genetic or environmental factors.

b. **Limitations.** Although they provide information about the degree of heritability, twin studies do not reveal anything about the genes involved in causing a disorder or the mode of transmission. Other problems with twin studies may result from the following.

(1) **Inaccurate zygosity data.** Only recently have twin studies used accurate methods of typing (i.e., placental membrane or blood grouping data).

(2) **Biased sampling**

(a) **Small samples.** Twin births account for about 1% of total births, and the disorder being studied may also be uncommon. To obtain adequate numbers of twins with a specific disorder, twin registries or clubs are useful.

(b) **Self selection.** Samples may be biased in favor of concordance (i.e., similarly affected twins are more likely to come to attention). The tendency to participate in a study may be related to personality traits shared by the twins, with dissimilar traits more likely to exclude twins from a study.

(c) **Varied sources of twins.** Twins may be drawn from different sources in different studies. This may result in mildly affected patients being compared to those who are more seriously ill.

3. **Adoption studies** compare the prevalence of a psychiatric disorder among the biologic relatives of affected individuals to the prevalence among the adopted relatives. These studies provide a unique opportunity to separate genetic from long-term environmental influences on the development of an individual.

4. **Prospective studies of at-risk individuals** involve long-term comparison of disease prevalence in persons at increased risk and in low-risk controls. These studies allow identification of predisposing environmental factors.

B. Schizophrenia

1. **Definition.** Schizophrenia is a psychosis involving loss of contact with the environment, disintegration of the personality, and disordered or inappropriate thought. The **DSM-III-R criteria** for the diagnosis include:
   a. Psychotic symptoms lasting at least 1 week
   b. Functioning below the highest level previously achieved
   c. Duration of at least 6 months, which may include the prodromal or residual phase of the disturbance

2. **Prevalence.** Published prevalence studies are difficult to compare due to differences among diagnostic standards, life expectancies, and course of illness. The lifetime prevalence of schizophrenia ranges from 0.03% in the Amish to 2.3% in the Swedish. In the United States, prevalence studies average 0.5% among the general population.

3. **Evidence for genetic contribution**
   a. **Family studies.** Recent studies using strict diagnostic criteria and control groups indicate that the risk for schizophrenia among first-degree relatives of probands is increased about 4% over controls.
   b. **Twin studies.** Concordance rates for schizophrenia generally are higher among MZ twins (15%–85%) than DZ twins (2%–10%). The discordance among MZ twins may suggest that environmental factors contribute to schizophrenia.
   c. **Adoption studies.** A large ongoing study among Danish adoptees has identified schizophrenia in 8.7% of close biologic relatives of schizophrenic probands compared to 1.9% of adoptive relatives.

4. **Inheritance of schizophrenia.** Although significant evidence indicates a genetic role in the development of schizophrenia, the exact mode of transmission is unknown. Because direct genotype studies usually are not available, other methods of studying the genetics of schizophrenia are used to define inherited factors segregating with the schizophrenia and to delineate subtypes concordant in sibling pairs.
   a. **Physical and laboratory markers**
      (1) **Monoamine oxidase (MAO) activity.** Platelet MAO activity is inherited and tends to be decreased in schizophrenic patients. It is not clear whether catechol or indoleamine metabolism is important.
      (2) **Dopamine receptor concentration.** Dopamine D2 receptors are increased on both positron and gamma (γ) emission tomography of the brain of schizophrenic patients.
      (3) **Brain morphology.** Lateral ventricular enlargement is seen in some schizophrenics, but the significance is unclear.
      (4) **Neurophysiological markers.** Abnormal smooth pursuit eye movements are seen in 51%–85% of schizophrenics and about 50% of their siblings but in only about 8% of normal controls. The trait is considered to be autosomal dominant with expression as either abnormal tracking or schizophrenia.
   b. **Linkage data**
      (1) Since the first description of a family with mental retardation, dysmorphic features, and schizophrenia segregating with a **partial trisomy of the long arm of chromosome 5 (5q)**, linkage to this cytogenetic region has been established in some large kindreds in which schizophrenia appears to segregate as a dominant trait.
      (2) Other families do not show linkage to this chromosome region. This demonstrates **genetic heterogeneity** even in families with an apparently dominant mode of transmission.

C. Mood disorders

1. **Definition.** Mood disorders are conditions characterized by two forms of pervasive pathologic changes in mood.
   a. **Bipolar disorder** (formerly referred to as manic–depressive illness) is marked by the presence of one or more manic or hypomanic episodes, usually with a history of depressive episodes. Manic episodes may be severe enough to impair functioning.
   b. **Unipolar disorder** (also referred to as depressive disorder) is marked by the presence of one or more major depressive episodes without a history of manic or hypomanic episodes. The DSM-III-R defines the major depressive episodes as periods of depression with a variety of symptoms lasting at least 2 weeks.

2. **Prevalence.** An estimated 0.4%–1.2% of adults have bipolar disorder. Prevalence figures for unipolar disorder vary widely; recent reports indicate that 4.5%–9.3% of women and 2.3%–3.2% of men have the disorder.

3. **Evidence for genetic contribution**
   a. **Family studies.** Hereditary factors have been suspected in the causation of mood disorders since early in this century and are now known to play a major role.
      (1) Generally, family studies show the risk for bipolar disorder among first-degree relatives of affected patients to be higher (19.2%) than the risk for unipolar disorder (9.7%).
      (2) If age of onset is considered in bipolar disorder, the risk is 20% for first-degree relatives of a patient affected before age 40 and 11% with onset after age 40.
   b. **Twin studies** show a striking difference in concordance rates for mood disorders in MZ twins (70%) versus DZ twins (13%). Concordance rates in MZ twins are high regardless of whether the twins are reared apart or not.
   c. **Adoption studies** confirm the importance of genetic factors in bipolar disorder, with a more modest component seen in unipolar disorder.

4. **Inheritance of mood disorders.** Evidence strongly suggests that genetic factors participate in the etiology of mood disorders. A single mode of inheritance has not been documented.
   a. **Possible physical and laboratory markers**
      (1) **Response to drug therapy.** Bipolar disorder shows a better response to **lithium,** whereas unipolar disorder shows a better response to the **tricyclic antidepressants.** These findings suggest involvement with different **neurotransmitters.**
      (2) **Catecholamine levels** have been shown to be increased with mania and decreased with depression. Other brain peptides have similar associations. Since **tyrosine hydroxylase** is the rate-limiting enzyme in catecholamine synthesis, the tyrosine hydroxylase gene is an obvious candidate for further study.
   b. **Linkage data.** Genetic marker studies have examined the relationship between mood disorders and red blood cell types, HLA types, and chromosome variants, to name a few.
      (1) **Blood group O** is increased in bipolar patients compared with unipolar patients or controls in some studies, but reports are conflicting.
      (2) Linkage studies have involved **X-linked** and **autosomal markers**. Linkage between both **deuteranopia** (color blindness) and the **Xg blood group** have been suggested, but such results must be reinterpreted in light of recent mapping data that separate the two loci. Heterogeneity is a possible explanation.
      (3) A large Old Order Amish pedigree has shown a **dominant transmission pattern,** with strong evidence of linkage with the ***H-ras* locus on the short arm of chromosome 11 (11p)**. The tyrosine hydroxylase gene is also on 11p. These results probably reflect heterogeneous causes of the mood disorders and indicate the need for further study.

D. Other disorders

1. **Anxiety disorders**
   a. **Definition.** Anxiety disorders are characterized by anxiety and avoidance behavior; these illnesses do allow perception of reality. Included in this group are anxiety disorders, phobias and obsessive–compulsive disorders. Of these, panic disorder appears to be the most familial in nature.
   b. **Prevalence.** Recent studies show anxiety disorders to be the most frequent psychiatric disorder in the general public, with phobias the most common. Patients who seek treatment are most likely to have panic disorders.
   c. **Evidence for genetic contribution** comes from both family and twin studies. Twin studies

of those with depression secondary to anxiety disorders show a concordance rate of 50% in MZ twins, but only 15% concordance in DZ twins.

2. Personality disorders

a. Definition. Personality disorders are chronic patterns of maladaptive and inflexible behavior that significantly impair function or cause subjective distress. Specific disorders that show genetic influence include schizotypal, antisocial, histrionic, and obsessive–compulsive personality disorders.

b. Prevalence. The prevalence of personality disorders is difficult to ascertain. Anxiety disorders are most frequently found in the general population, while panic disorders are the most frequent in those seeking treatment. Phobias may exist in 5% of the general population.

c. Evidence for genetic contribution

(1) Family studies indicate an increased risk for schizotypal personality disorder among close relatives of schizophrenics. The risk for antisocial personality disorder is greatly increased (5–10 times) in first-degree relatives of affected persons; a lesser risk exists for histrionic and obsessive–compulsive personality disorders.

(2) Twin studies show that concordance rates for personality disorders are generally much higher in MZ twins than in DZ twins.

3. Chromosome disorders associated with behavioral disorders. Combined psychiatric and chromosome disorders are rare. However, some generalizations can be made about sex chromosome aneuploidies.

a. Klinefelter syndrome (Ch 2 II B 2 d). As children, XXY males tend to be immature, with poor judgement and learning problems. The incidence of XXY among schizophrenic males is 0.55% compared to the expected 0.1% in the general population.

b. Trisomy X (Ch 2 II B 2 f). The incidence of XXX females also is higher among schizophrenics than in the general population.

c. XYY syndrome (Ch 2 II B 2 e). Early reports that XYY males were more likely to be aggressive and to commit violent crimes have been dismissed, but affected males do tend to be more immature and impulsive.

d. X monosomy and other Turner variants (Ch 2 II C, D 2 b; III C). Affected individuals have space-form perceptual defects and are immature, although overall IQ is normal. No increase in psychiatric disorders such as schizophrenia has been found.

4. Single gene defects associated with behavioral disorders

a. Fragile X syndrome (Ch 2 IV A 1 a)

(1) Affected males not only have overall developmental delay but also have a disproportionate delay in speech and later abnormal speech patterns. As young children, they exhibit autistic features; true autism is less frequent. As adults, their behavior often is described as friendly and cooperative.

(2) Females are less severely affected, on average, and may have learning disabilities.

b. Phenylketonuria (Ch 5, case 5-D). As well as being mentally retarded, untreated children are irritable and unhappy. These traits disappear with treatment.

c. Homocystinuria. Affected adults are more likely to have unusual personalities with psychoses associated with a schizophrenia-like picture.

d. Adult metachromatic leukodystrophy presents as schizophrenia.

e. Lesch-Nyhan syndrome produces a compulsion to self-mutilation.

STUDY QUESTIONS

Directions: Each of the numbered items or incomplete statements in this section is followed by answers or by completions of the statement. Select the **one** lettered answer or completion that is **best** in each case.

1. A patient with chronic renal failure has five people who are willing to serve as donors for a renal transplant for her: her husband, identical twin, half-brother, sister, and son. Based on the likelihood of HLA identity, the correct arrangement of individuals in order of **decreasing** suitability as a renal transplant donor is

(A) twin, husband, sister, son, half-brother
(B) husband, twin, sister, son, half-brother
(C) twin, son, sister, husband, half-brother
(D) twin, sister, half-brother, son, husband
(E) twin, sister, son, half-brother, husband

2. A young woman whose sister developed a bipolar disorder at age 23 asks her physician if she is at risk for the disorder. No other family members are known to be affected. This woman's risk for developing bipolar disorder is which one of the following percentages?

(A) 1%
(B) 5%
(C) 10%
(D) 15%
(E) 20%

3. Tumors that occur in patients with mendelian conditions that predispose to neoplasia usually exhibit all of the following features EXCEPT

(A) the tissues affected are rarely involved by neoplasia in patients without the mendelian disease
(B) the age of onset is earlier than similar tumors in other patients
(C) the tumors in affected relatives are all of the same type or of a restricted range of types
(D) the tumors are bilateral, multifocal, or both

4. Genes that are important in mammalian embryogenesis include members of all of the following classes EXCEPT

(A) proto-oncogenes
(B) tumor suppressor genes
(C) growth factor genes
(D) *Hox* genes
(E) signal transduction genes

5. Androgen insensitivity could be suspected in all of the following situations EXCEPT

(A) an infant with ambiguous genitalia and no evidence of a uterus on ultrasound
(B) a phenotypic female infant
(C) the child with chromosomal mosaicism (46,XY/45,X) and female genitalia
(D) a male infant with hypospadias

6. A 21-hydroxylase deficiency (congenital adrenal hyperplasia) may be associated with all of the following EXCEPT

(A) phenotypically normal male infants
(B) nondevelopment of müllerian duct structures in affected females
(C) variably masculinized female infants
(D) severe electrolyte abnormalities
(E) variable clinical expression

7. Each of the following mechanisms appears to be of major importance in early mammalian embryogenesis EXCEPT

(A) interactions of cells with neighboring cells and extracellular matrix
(B) morphogen gradients that induce different responses, depending on local concentration
(C) regulatory cascades in which activation or inactivation of one gene affects the function of others
(D) maternal gene products laid down in the ovum that trigger early determinative events
(E) establishment of fixed patterns of transcription

1-E
2-E
3-A
4-B
5-C
6-B
7-D

8. All of the following statements about schizophrenia are true EXCEPT

(A) linkage to an autosomal locus has been proposed
(B) nearly 2% of the population is affected
(C) it is a psychosis characterized by disordered or inappropriate thought
(D) increased concordance rates in MZ twins indicate that genetic factors are important
(E) abnormal smooth pursuit eye movements are positively correlated

Directions: Each group of items in this section consists of lettered options followed by a set of numbered items. For each item, select the **one** lettered option that is most closely associated with it. Each lettered option may be selected once, more than once, or not at all.

Questions 9–13

For each patient listed below, select the blood type that would be appropriate for a transfusion.

(A) A Rh^- or O Rh^- blood
(B) B Rh^+ or O Rh^+ blood
(C) O Rh^-, A Rh^-, B Rh^-, AB Rh^-, O Rh^+, A Rh^+, B Rh^+, or AB Rh^+ blood
(D) AB Rh^+ blood
(E) None of the above

9. A patient who is B Rh^-
10. A patient who is A Rh^+
11. A patient who is AB Rh^-
12. A patient who is O Rh^-
13. A patient who is AB Rh^+

Questions 14–17

Match each statement with the class of gene that it describes.

(A) Oncogenes or proto-oncogenes
(B) Tumor suppressor genes
(C) Both
(D) Neither

14. May be activated by gene amplification, translocation, or mutation
15. Loss of function promotes tumor development
16. Present in the normal cells of normal people
17. Exhibits a dominant effect at the cellular level

8-B	11-A	14-A	17-A
9-E	12-E	15-B	
10-A	13-C	16-C	

ANSWERS AND EXPLANATIONS

1. The answer is E *[II E 2].*
The identical twin must be HLA identical to the patient, so she is the most suitable donor. The sister is the next suitable donor because she has a one in four chance of being HLA identical and a one in two chance of being haploidentical to the patient. The son has almost no chance of being HLA identical to his mother but must be haploidentical to her. The patient's half-brother also has almost no chance of being HLA identical; he has only a one in two chance of being haploidentical. It is unlikely that the patient's husband shares any HLA haplotypes with her.

2. The answer is E *[V C 2, 3 a].*
Although bipolar disorder affects roughly 0.4%–1.2% of the general adult population, the recurrence risk for first-degree relatives is much higher. The risk is highest (20%) when the onset of illness in the proband is prior to age 40. This may reflect the severity of the disorder.

3. The answer is A *[III B 2 a].*
Tumors that develop in patients with mendelian syndromes that predispose to neoplasia are characterized by an earlier than usual age at onset, by multifocal or bilateral occurrence, and by involvement of only one or a few specific tissues. There are many different mendelian disorders that predispose to neoplasia, and among them are syndromes associated with the occurrence of many different kinds of tumors, including those that are most common in the general population.

4. The answer is B *[IV A 4].*
Tumor suppressor genes are not known to be important in mammalian embryogenesis. Several proto-oncogenes appear to be involved in embryogenesis. Growth factors and signal transducers such as retinoic acid can induce differentiation. *Hox* genes appear to be involved in embryonic pattern formation.

5. The answer is C *[IV C 4].*
Androgen insensitivity may be complete or incomplete where the androgen receptor for both testosterone or dihydrotestosterone is completely absent, present but nonfunctional, or dysfunctional with poor activity. This leads to a variety of clinical presentations from an undermasculinized male with hypospadias to a phenotypic female. Because the gonad has developed normally to form a testis and both testosterone and müllerian duct-inhibiting substance function normally, the müllerian duct structures never develop, and a uterus is not seen. The problem is not with the chromosomal makeup of the child; one would expect 46,XY in these infants.

6. The answer is B *[IV C 3].*
A 21-hydroxylase deficiency is associated with underproduction of cortisol, excessive androgens, and sometimes a defect in aldosterone production. This leads to variable degrees of masculinization of chromosomally female fetuses, with the most severe degree being male external genitalia with absent testes. Male infants, although phenotypically normal, may have an increase in pigmentation of the scrotum or a larger phallus, but this is generally not noticed. Diminished aldosterone production leads to salt wasting and severe electrolyte abnormalities. In female infants, the increased androgens are from the adrenal glands and occur after development of the internal genitalia from the müllerian duct. Therefore, such females will have normal development of the ovaries, uterus, and fallopian tubes and are potentially fertile females.

7. The answer is D *[IV A 3].*
Neither maternal effect genes (which lay down products initiating differentiation in the cytoplasm of the egg) nor cell differentiation through specific lineal antecedents appears to be crucial to mammalian embryogenesis. Cell–cell interactions, cell–environment interactions, morphogen gradients, regulatory cascades, and the establishment of fixed patterns of transcription are all critical to embryonic development in many species, including mammals.

8. The answer is B *[V B 2].*
Schizophrenia is seen in about 0.5% of the general population and has an increased recurrence risk in first-degree relatives. The DSM-III-R defines schizophrenia as a psychosis characterized by disordered thought that lasts longer than 6 months. In some families, the disorder is linked with altered smooth pursuit eye movements, suggesting that these families have a dominant disorder involving the alteration of eye movements, schizophrenia, or both. One family study has shown schizophrenia in individuals having partial duplication of the long arm of chromosome 5 (5q), which suggests linkage to this area. Twin studies indicate a higher concordance rate for schizophrenia in monozygotic (MZ) twins than in dizygotic (DZ) twins. However, even MZ twins do not show 100 concordance, which suggests that genetic factors alone do not cause schizophrenia.

9–13. The answers are: 9-E, 10-A, 11-A, 12-E, 13-C *[II D 2 c, 3].*
A patient who is B Rh$^-$ should not be given blood containing either the A or Rh$^+$ antigen. This eliminates the blood in (A), (B), (C), and (D). A patient who is A Rh$^+$ should not be given blood containing the B antigen, but may be given either Rh$^+$ or Rh$^-$ blood. The blood in (A) is, therefore, satisfactory. Type O blood does not display a surface antigen that is recognized by individuals who lack it. A patient who is AB Rh$^-$ may be given type O, A, B, or AB blood but should not be given Rh$^+$ blood. Only the blood in (A) is satisfactory. A patient who is O Rh$^-$ can only be given O Rh$^-$ blood because he or she will form antibodies against any blood with the A, B, or Rh$^+$ antigen. A patient who is AB Rh$^+$ may receive a transfusion with blood of any ABO or Rh type. He or she will not form antibodies to any of these antigens because all are present on his or her own red blood cells. Responses (A), (B), (C), and (D) are all correct, but (C) is the best answer because an AB Rh$^+$ patient has no antibodies to the ABO or Rh antigens.

14–17. The answers are: 14-A, 15-B, 16-C, 17-A *[III C 1, 2].*
Proto-oncogenes can be activated by gene amplification, translocation, mutation, or other means. Once activated, they promote neoplastic development even in the presence of a normal allele of the same proto-oncogene in the same cell. Thus, oncogenes function dominantly at the cellular level.

Tumor suppressor genes must be lost, not activated, to predispose to neoplasia. This loss may occur by mutation that results in failure of transcription, translation, or function of the product; deletion; mitotic crossover; or by loss of the entire chromosome that carries the tumor suppressor gene. Both alleles of a tumor suppressor gene must be lost to promote tumor development; the presence of just one normal allele prevents this effect. Thus, tumor suppressor genes act as recessives at the cellular level.

Both proto-oncogenes and tumor suppressor genes are present in normal cells in normal individuals. Both classes of genes are thought to be important in physiological control of cell differentiation and proliferation.

10
Clinical Genetics: Part I

J.M. Friedman and Barbara McGillivray

I. APPROACH TO CLINICAL GENETICS. Assessing an infant with malformations or dysmorphic features requires knowledge of the normal condition and an appreciation for the alteration of clinical presentations with maturation. To provide accurate information regarding the natural history and the recurrence risks, **a precise diagnosis is needed**.

A. Initial referrals for genetic assessment are usually made by the family physician or a specialty physician. However, concerned families may self-refer or the referral may come from a variety of paraprofessionals (e.g., a public health nurse) involved in the care of an infant with problems.

1. An **abnormal fetus** is usually an urgent referral, and it has the disadvantage of having limited accessibility except with detailed ultrasound or fetal karyotyping. Components of the findings that may concern the parents include:
   - **a.** The natural history of a particular ultrasound finding (e.g., if the problem is lethal)
   - **b.** The extent of malformations present in the fetus
   - **c.** Whether mental retardation may be a component of the findings

2. In a **stillborn infant,** the most serious problem may be the condition of the fetus and the placenta (i.e., the state of maceration, whether it is fixed in formalin, whether the placenta and membranes are available).
   - **a.** The autopsy information may determine whether the geneticist can make a diagnosis.
   - **b.** If an autopsy is incomplete (the fetus may have been examined only externally), counseling is constrained by the available information.

3. A **dysmorphic infant or child** also may be an urgent referral if the anomalies are life-threatening.
   - **a.** The **purpose of the assessment** may be to determine the course of treatment (aggressive therapy versus warmth and comfort) and only later may involve counseling the family. The infant may be sent to a hospital far from the parents upon birth, and the parents may not be able to get to the infant for several days, if at all, to meet with the physician caring for the child.
   - **b.** Details of the family history as well as visual comparison to the family may be lacking.

4. A **dysmorphic adult or an adult with a suspected genetic abnormality** may be referred for advice on reproductive options (including sterilization). Evaluation of dysmorphic features may be limited by:
   - **a.** Lack of knowledge of the appearance of certain syndromes in the adult population
   - **b.** Lack of information about earlier investigations
   - **c.** Death of family members who would have been able to provide details about family history

B. Gathering information. Each situation requires the analysis of a number of pieces of information to formulate a hypothesis regarding the diagnosis and the most likely recurrence risk. Families may be reluctant to divulge details of family members with problems or may not be aware of affected members of earlier generations. The stigma attached to being affected with a dementing disorder may be difficult to overcome at the time of counseling. Establishing a feeling of trust and confidentiality is crucial. All data must be recorded as completely and accurately as possible.

1. **Constructing the pedigree.** Even with the advent of sophisticated molecular analysis, the starting point of information gathering is still the pedigree or the charting of the family history. A

pedigree is a graphic method of representing the generations of a family, with various symbols used for relationships or for particular clinical findings (Figure 10-1).

a. **Steps to taking pedigrees** include the following.
   (1) **Give simple instructions** to the patient as to the information expected and why the information is being sought.
   (2) **Use clearly defined symbols** with a key for interpretation. These symbols may be used to denote different conditions found within the family (see Figure 10-1).
   (3) **Identify the informant, the historian, and the date of the interview** so that the pedigree remains meaningful to others at a later date and can be amended as necessary. The pedigree is usually taken from the person requesting the counseling, but use of other family members (e.g., parents) and documentation (e.g., church records) may be appropriate as well.
   (4) **Ask not only about those alive and affected in a family** but also about miscarriages, stillbirths, children relinquished for adoption, and deceased individuals. Such details are essential to understanding conditions that are lethal or associated with reproductive losses.
   (5) **Ask for details** such as place of birth, size of towns, and the presence of consanguinity to help define recessive conditions.

b. **Analysis of the pedigree** is only possible if both affected and unaffected individuals are shown. Patterns of inheritance may then be clear and differential expression in one sex may be noted. Careful analysis of the pedigree may establish the mode of inheritance.

2. **Reviewing past records.** With the modern problem of limited record storage space, crucial medical records may be destroyed after a number of years. For this reason, obtaining details of past medical information may be a challenge.
   a. After obtaining written permission from the individual regarding his or her medical records (or the closest surviving relative), documentation may be sought from hospital medical record departments, physicians' offices (more likely to be kept for extended periods), coroners' records, or other treating institutions.
   b. Often, medical information is available on biological relatives of an adopted person from the appropriate provincial or state facility.
   c. Types of documentation that are most useful include:
      (1) **Autopsy findings**
      (2) **Reports of surgical procedures**
      (3) **Discharge summaries,** which should review all of the pertinent investigations that a person has undergone
      (4) **Laboratory investigations,** such as chromosome studies
      (5) **Family photographs,** which are very helpful when trying to assess physical characteristics or judge whether another family member was affected or not

3. **Clinical assessment** may involve multiple family members. Important findings should be documented by clinical photographs or videotaping if movement is a distinguishing feature.
   a. **Visual assessment.** Before touching the patient, observe the overall presentation (i.e., check if the individual is behaving in an age-appropriate manner; check if the individual has peculiar movements, facial expressions, or limitations in movement). All examinations should be done with the patient unclothed, although this may be done in a progressive fashion.
   b. **Measurement.** Standard physical parameters should be documented (e.g., head circumference, height, weight), as well as any other measurements appropriate to the problems seen. All measurements can then be compared to published standards, and corrected for age, sex, and racial origin, if possible. Family comparisons may be valid (e.g., many normal family members may have macrocephaly in the situation where you are assessing a child with a large head and other features).
   c. **Extended family**. If possible, other affected or potentially affected family members should be assessed. If a child is suspected of having tuberous sclerosis, the family cannot be adequately counseled regarding recurrence risks unless the parents have been examined.

C. **Counseling.** Certain actions can be taken during the counseling session. (See IV for more detail on counseling.)

1. **Counsel the parents together.** The parents of an affected infant or child need to hear the information together to provide support for each other, to fill in gaps in information, and to learn the information at the same time.

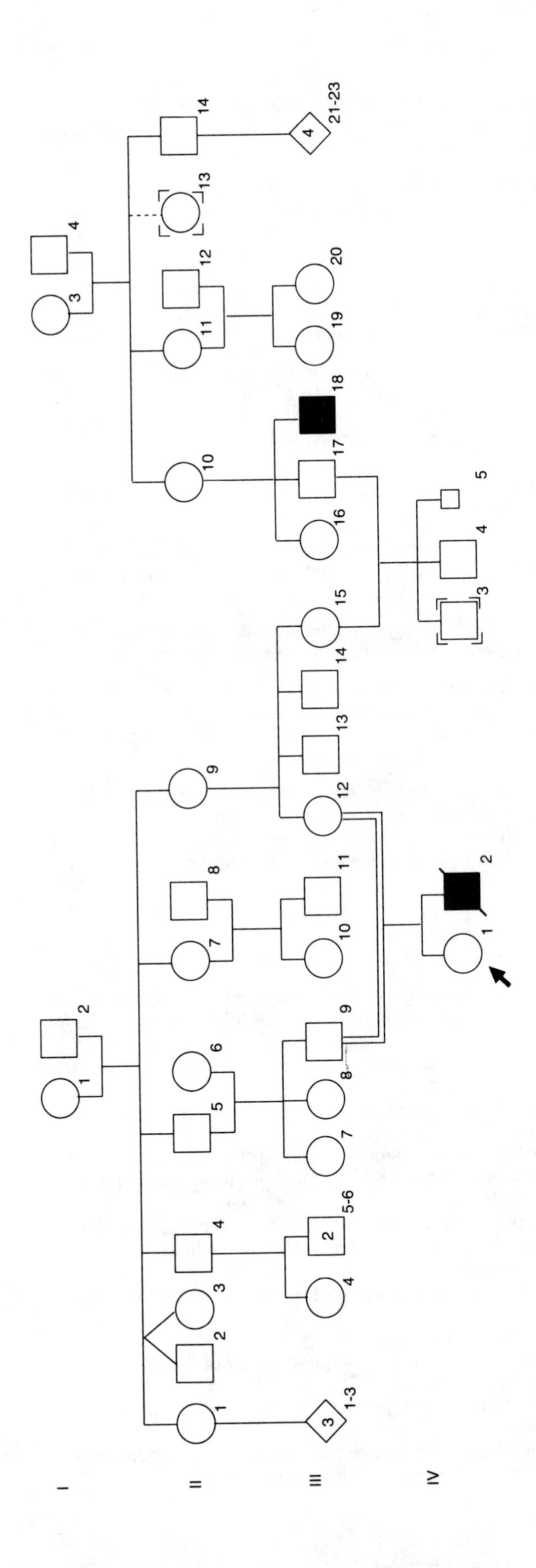

Figure 10-1. Sample pedigree showing a variety of symbols.

2. **Remove distractions.** For the counseling session itself, having the children playing apart from the parents or finding a quiet space in the hospital ward ensures that the parents can be free from distractions and are able to concentrate on the session.

3. **Be prepared to repeat.** Receiving a serious diagnosis is a shock to most families. Realizing that little information may be remembered after the initial diagnosis allows the counselor to repeat the information at a later session, to review at the end of the session, or to document the discussion in a letter to the family.

4. **Use visual aids.** Useful counseling handbooks are available to illustrate modes of inheritance and karyotypes, but drawing or otherwise simply illustrating the counseling points may allow the family to grasp the information. Encouraging questions as the session goes along is helpful.

5. **Ascertain what the family needs.** The actual needs of the family may differ from the perceived needs. The family may wish to know long-term outcome and may not be concerned about the recurrence risk. By directly addressing the family's concerns, the counselor establishes trust and empowers the family to ask questions that might otherwise be thought inappropriate.

D. Follow-up. Many counseling sessions are a one-time occurrence, but some problems should be reassessed. The reasons for doing so include the following.

1. **Lack of a diagnosis.** The child with multiple anomalies may not be diagnosed on the first visit, but with evolution of the features in the child and the information in the literature, a diagnosis may become apparent later.

2. **Counseling other family members.** Information regarding the immediate family may only be discussed with specific permission. Parents, siblings, or other relatives may be seen.

3. **New diagnostic techniques.** Future developments may improve counseling to a family. For example, when the molecular methods become available for the diagnosis of Duchenne muscular dystrophy (DMD), families could be recalled to reassess their risks in light of information gained using the new methods.

4. **Natural history.** Even with an established diagnosis, evolving health concerns may need to be assessed. To provide natural history details for families, children with specific syndromes should be carefully followed to establish the long-term information. Unsuspected long-term complications, further details of developmental achievements, or reproductive risks may be noted.

E. Specific situations

1. **Abnormal stillbirth**

   a. **Gathering information.** Details of the pedigree should include other affected infants, stillbirths, early miscarriages, and a history of consanguinity. To establish the etiology, the following should be performed:

      (1) **Pregnancy history,** including previous pregnancy outcomes, medications used, exposures, and details of fetal growth and movement

      (2) **Complete autopsy,** including examination of the membranes and placenta, radiologic studies of the fetus, and photographs

      (3) **Laboratory investigations** of the fetus and parents, including chromosome studies and other appropriate investigations (e.g., thalassemia evaluations when the fetus has hydrops to rule out α-thalassemia)

   b. **Counseling the family.** Usually, counseling following a stillbirth is an emotionally charged session, as the parents are grieving, have feelings of guilt, and have feelings of anxiety regarding the outcome of future pregnancies. Returning to the hospital is stressful, and parents may view the counseling session as a necessary hurdle before planning another pregnancy.

      (1) Review the literature to help establish a diagnosis.

      (2) Review the findings with the parents, who may request a copy of the autopsy.

      (3) Explain the diagnosis and recurrence risk.

      (4) Discuss reproductive options in subsequent pregnancies, including whether prenatal diagnosis would be available.

      (5) Assess grief response and offer further supportive counseling as appropriate. Provide appropriate literature regarding grief, pregnancy loss, and so on.

2. Developmental delay

a. Gathering information. Diagnosing the child with developmental delay is often difficult, since there are a number of prenatal and postnatal causes to be considered. The family may come to the session frustrated about the time taken to recognize that their child was, in fact, delayed and about the limitations of service available to the family. Alternatively, the geneticist may be the individual first confirming the clinical suspicion of delay. The initial assessment and counseling session will be enhanced with:

(1) A **careful review of the pregnancy,** with particular attention paid to drug, chemical, or infectious disease exposures. Severe bleeding may suggest loss of a twin or a placental abruption, both of which could be associated with fetal compromise. Significant prematurity implies a number of risk factors for developmental delay.

(2) **Previous pregnancy history.** Recurrent losses or a similarly affected child may suggest a chromosomal or single gene etiology.

(3) **Family history.** Other affected individuals, pregnancy losses, and consanguinity or incest are details that help to outline a cause for the delay.

(4) **Results of all investigations** (e.g., developmental assessments, hearing and visual testing, and laboratory testing) **or surgical procedures** should be reviewed.

(5) **Physical examination** to observe growth parameters (e.g., to rule out microcephaly), dysmorphic features, and abnormal movements. The physical appearance should be documented with photographs, especially if the child is likely to be seen over a period of time to establish a diagnosis.

(6) **Further investigations.** Developmental delay in an infant may prompt evaluation of possible intrauterine infection (e.g., organism-specific IgG and IgM levels), metabolic abnormalities, or a chromosome abnormality. An older child, especially if male, may undergo fragile X studies. If all children of a couple are affected, either maternal phenylketonuria (PKU) or mitochondrial inheritance should be considered.

b. Counseling. The information discussed during the counseling session varies, depending on whether or not a diagnosis can be made. If no diagnosis is made, the family may be seen again, but empiric recurrence risks may be valuable in the interim. If a specific diagnosis is established, the geneticist needs to consider the emotional impact on the family, as well as the implications for the future. With a specific diagnosis, the parents will require appropriate medical information (e.g., natural history, possible health concerns), recurrence risks for themselves and other family members, and information regarding lay support groups. Contact with another family who has a child with the same disorder may be helpful. The availability of specific prenatal diagnosis should be covered.

II. CALCULATING GENETIC RISKS

A. Inferring genotypes from phenotypes

1. **The most frequent error made in clinical genetics is counseling on the basis of the wrong diagnosis.** A correct and precise diagnosis is essential for accurate genetic counseling.

2. **The possibility of genetic heterogeneity**—the production of the same phenotype by more than one abnormal genotype—must always be considered. For example, retinitis pigmentosa may be transmitted as an autosomal recessive, autosomal dominant, or X-linked recessive trait in various families.

3. When confronted with several different phenotypic abnormalities in a patient or in a family, the possibility of **pleiotropy**—the production of multiple phenotypic abnormalities by a single gene or gene pair—must always be considered. Often the physical abnormalities are not obviously related as is the case in:

   a. **Nail-patella syndrome** (nail dysplasia, absent patellae, nephropathy)

   b. **Myotonic dystrophy** (weakness and atrophy of distal and facial muscles, myotonia, cataracts, cardiac conduction defects, frontal balding, endocrine abnormalities)

4. When assessing a family with a genetic disease or congenital anomaly, it is important to determine who is affected and who is not. This is information about the **phenotype**. To calculate genetic risks, however, it is necessary to know each individual's **genotype**. When determining genotypes within a family, **assume the most likely case** unless there is good reason to do otherwise.

   a. **Individuals with autosomal dominant diseases can generally be assumed to be heterozygotes** rather than homozygotes.

(1) Homozygotes for dominant diseases occur much less frequently than heterozygotes (see Ch 6 I B 4 b).

(2) A homozygote for an autosomal dominant disease should be recognized by the fact that both parents are affected with the disease.

(3) The phenotype of the homozygote for most autosomal dominant diseases is much more severely abnormal than the phenotype of the heterozygote.

b. Parents of a child with an autosomal recessive disease generally can be assumed to be heterozygous carriers of the abnormal gene.

(1) A parent who is homozygous for an autosomal recessive disease would be expected to exhibit the abnormal trait.

(2) The population frequency of heterozygous carriers of autosomal recessive diseases is much greater than the frequency of new mutations for the corresponding genes (see Ch 6 I B 3 e).

c. One of the parents of a carrier of an autosomal recessive disease generally can also be assumed to be a carrier. Although it is possible that both parents of a heterozygote are themselves carriers, this is unlikely unless there is also a homozygote among their children or there is consanguinity.

N.B

d. In a family exhibiting an X-linked recessive disease, a woman can be assumed to be a carrier if she has more than one affected son or if she has an affected son and other male relatives who are related through their mothers. If a woman is the mother of only one boy with a serious X-linked recessive disease and there is **no history of the condition in the family,** she **cannot be a carrier** (see Ch 6 II C 2).

B. Linkage. Within an individual family, linkage can be used to estimate the probabilities of alternative possible genotypes in individual members (see Ch 4 II F 1). For example, consider the following pedigree for an autosomal dominant, adult-onset, degenerative neurologic disease (Figure 10-2). The proband is a child who is too young to manifest symptoms of the disease.

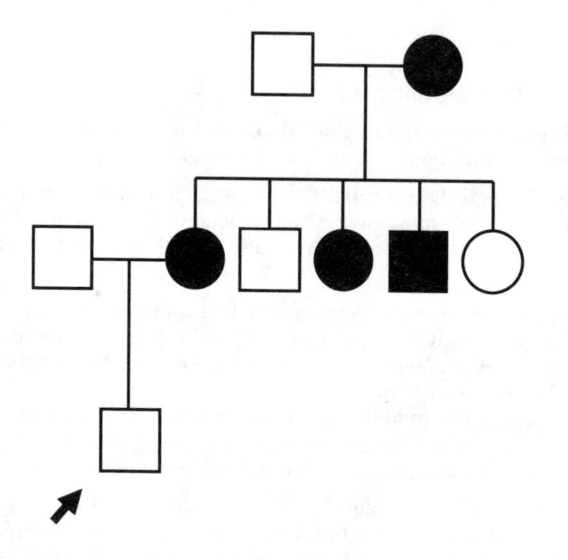

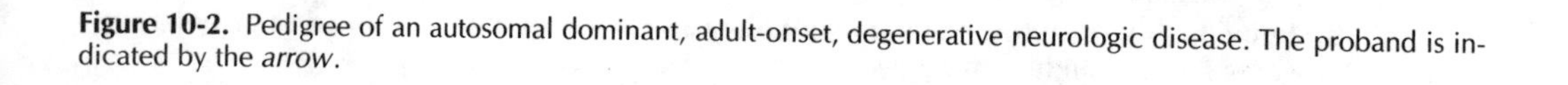

Figure 10-2. Pedigree of an autosomal dominant, adult-onset, degenerative neurologic disease. The proband is indicated by the *arrow*.

1. On the basis of the information provided in the pedigree, the chance that the child has inherited the disease gene is 1/2. If, however, the disease gene locus (*D*) is known to be linked to the locus for a genetic marker (*M*) such as an anonymous DNA sequence, the family can be typed for the marker to determine the genotype of the fetus at the disease locus. In this family, the typing produces the following results (Figure 10-3):

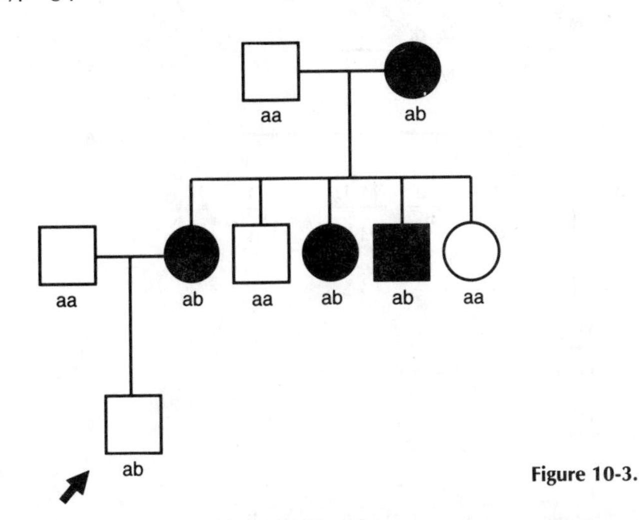

Figure 10-3.

2. It is necessary to determine which allele of the marker locus is on the same chromosome as the abnormal allele of the disease gene. This is called **assigning the phase of the linkage** and can be done in this family by inspection (Figure 10-4). M^a indicates the *a* allele at the marker locus, M^b indicates the *b* allele at the marker locus, *d* indicates the normal allele at the disease locus, and *D* indicates the abnormal allele at the disease locus.

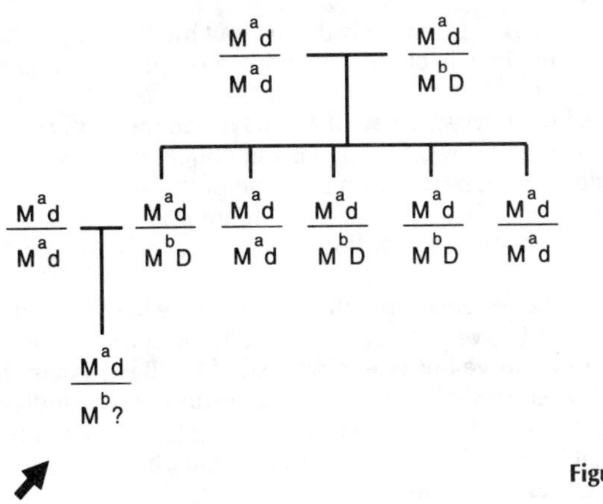

Figure 10-4.

3. If it is also known that the recombination distance between the *M* locus and the *D* locus is 5 cM (i.e., 5%), then the risk of the proband having inherited the disease gene is 95%.
4. Note that when linkage is used, **the 1/2 chance of transmission related to genetic segregation is not included in the calculation**. This is because linkage can only be applied after segregation has occurred and provides information regarding which alleles went to the gamete that produced a given individual.
5. Also note that the **particular alleles involved at the marker locus are irrelevant** as long as the locus is informative. The allele at a marker locus found to be **in coupling** with a linked disease gene varies from family to family.
   a. In this example, the same information could have been obtained if the child's unaffected father and maternal grandfather had both typed M^b/M^b and all of the affected individuals in the family had typed M^a/M^b.

b. In this case, however, the phase of the linkage would be reversed, and M^a would be on the same chromosome as the abnormal allele of the disease gene so that the affected individuals would all be M^aD/M^bd (Figure 10-5).

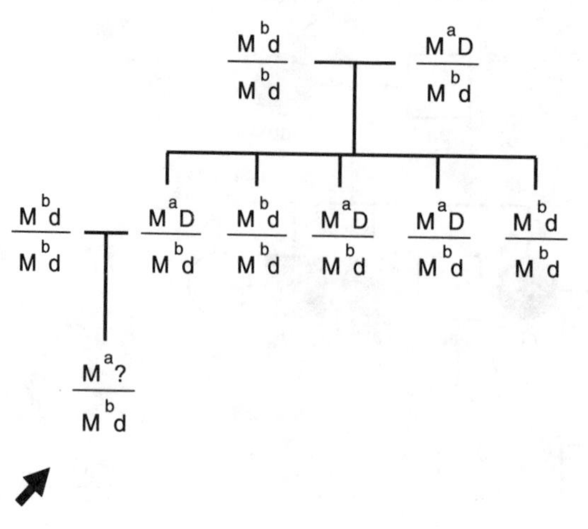

Figure 10-5.

c. The child would still have a 95% change of being affected if found to be M^a/M^b.

C. The **Bayesian method** is the use of information that alters the relative probabilities of prior events to estimate current risks.

1. The Bayesian method is used extensively throughout medicine as well as in everyday life, but its use is generally not explicit. In clinical genetics, an explicit Bayesian approach is employed to modify risk calculations.

a. As **an example of the everyday use of the Bayesian method,** consider a medical student who looks out of her window at 5:30 P.M. and notices that it is dark. She will interpret this observation differently depending on the time of the year.

(1) If it is July, she may be concerned that there is about to be a thunderstorm.

(2) If it is December, she will not find it unusual and is unlikely to be concerned about a thunderstorm.

(3) In both cases, the observation is the same, but the likelihood that darkness at 5:30 P.M. portends a thunderstorm is very different in the summer and winter.

b. As **an example of intuitive but unexplicit use of the Bayesian method in medicine,** consider a laboratory report showing a very high serum potassium level.

(1) If the patient is a healthy man who was being tested as part of a routine physical examination, it is likely that the result is erroneous, and a physician would probably just repeat the test to verify this.

(2) However, if the patient is in renal failure, the elevated potassium may be an ominous sign that hemodialysis is necessary.

(3) Again, the observation is the same, but the interpretation is very different because of additional information.

2. Conditions for use of the Bayesian method in calculating genetic risks

a. The **genotype cannot be determined with certainty** from the pedigree or by testing.

b. Additional information is available that makes one interpretation of the genotype more or less likely.

3. Explicit use of the Bayesian method in clinical genetics is best illustrated by an example question and solution.

a. What is the risk that a sibling of a child with cystic fibrosis (an autosomal recessive condition) will be a **heterozygous carrier** of the abnormal gene?

(1) The answer is straightforward, but for didactic reasons, it is illustrated with a drawing of a **probability space** that indicates all possibilities (Figure 10-6):

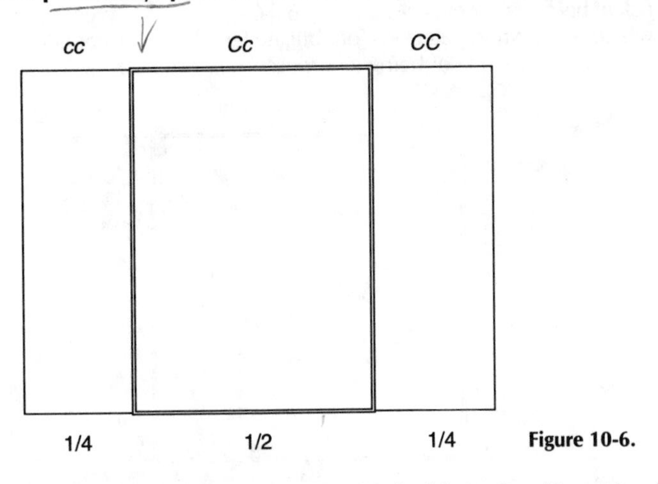

Figure 10-6.

(2) The outcome of interest is shown in the double-outlined box. The chance that the sibling of a child with cystic fibrosis will be a heterozygous carrier is 1/2.

b. If the question is changed slightly and asks what the probability is that a **clinically normal** sibling of a child with cystic fibrosis will be a heterozygous carrier, a Bayesian calculation must be performed.

(1) This is necessary because knowing that the sibling is clinically normal (i.e., does not have cystic fibrosis) changes the probabilities of the other outcomes, as shown in Figure 10-7.

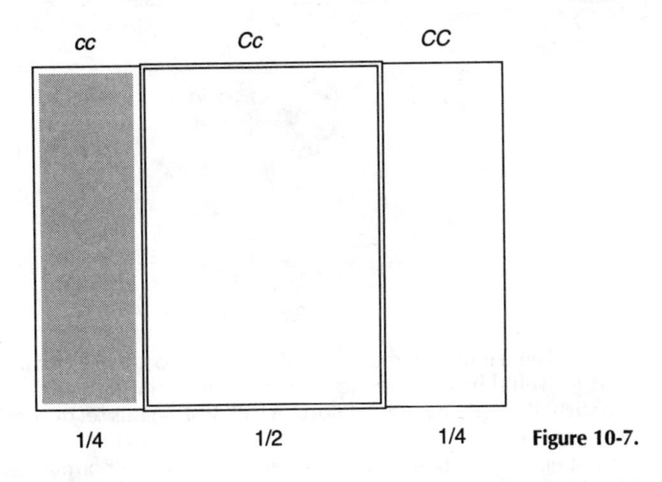

Figure 10-7.

(2) As indicated by the shading in Figure 10-7, it is no longer possible that the sibling has the *cc* genotype, so the only possibilities remaining are that he has a genotype of *Cc* or *CC*.

(3) The **final probability that the sibling is a heterozygous carrier** is thus the chance that he is a heterozygote divided by the sum of all of the probabilities that remain possible (*Cc* or *CC*).

$$P = \frac{1/2}{1/2 + 1/4}$$

$$P = \frac{2/4}{2/4 + 1/4}$$

$$P = \frac{2/4}{3/4}$$

$$P = \mathbf{2/3}$$

4. A somewhat more **realistic example from clinical genetics** is given here. Consider a woman whose only son has DMD, an X-linked recessive disease. If her family history is negative, her a priori risk of being a carrier is 2/3 (see Ch 6 II C 2). If she has creatine phosphokinase (CPK) testing, which is known to detect 3/4 of obligate female carriers of DMD, and the CPK test is normal, what is her chance of being a carrier?

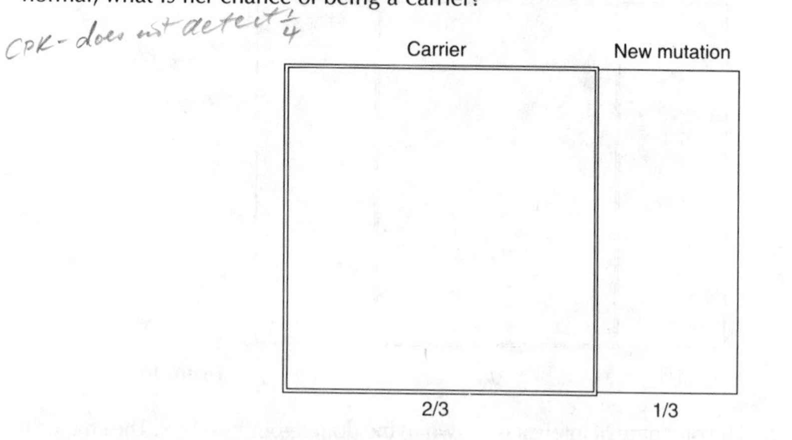

Figure 10-8.

a. Figure 10-8 shows the probability space corresponding to the initial conditions. There are only two possibilities. Either the woman is a carrier ($P = 2/3$) or she is not, in which case her affected son represents a new mutation ($P = 1/3$).

b. Once she has had CPK testing and has had a normal result, the probability space looks like Figure 10-9.

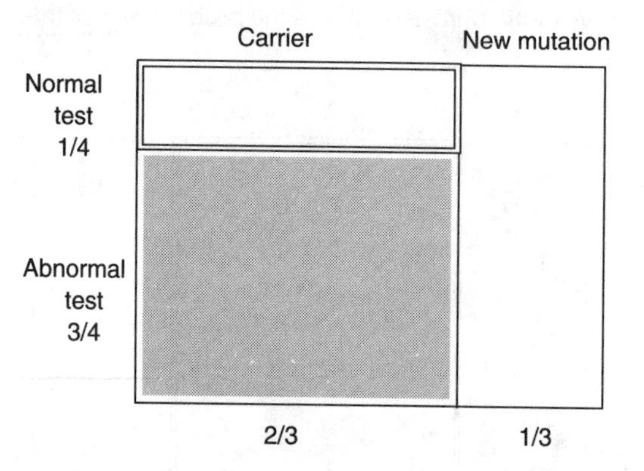

Figure 10-9.

(1) The shaded area indicates that of those women who are in fact carriers (i.e., 2/3 of the total), 3/4 would be expected to have an abnormal CPK test.

(2) This patient does not have an abnormal CPK test, so this set of conditions is no longer possible for her.

(3) She must either be a noncarrier or a carrier who has a normal test.

c. The **probability that this woman is a carrier who has a normal CPK test** is the chance that lies in the double-outlined box divided by the sum of all the probabilities that remain possible:

$$P = \frac{1/4 \times 2/3}{(1/4 \times 2/3) + 1/3}$$

$$P = \frac{2/12}{2/12 + 4/12}$$

$$P = \frac{2/12}{6/12}$$

$$P = 2/6 = \mathbf{1/3}$$

d. The same solution may also be illustrated with a **probability tree** (Figure 10-10). The probability of each alternative event is associated with an arrow.

(1) Probabilities of 1.0 (i.e., certainties) are indicated as arrows without an alternative or an associated number.

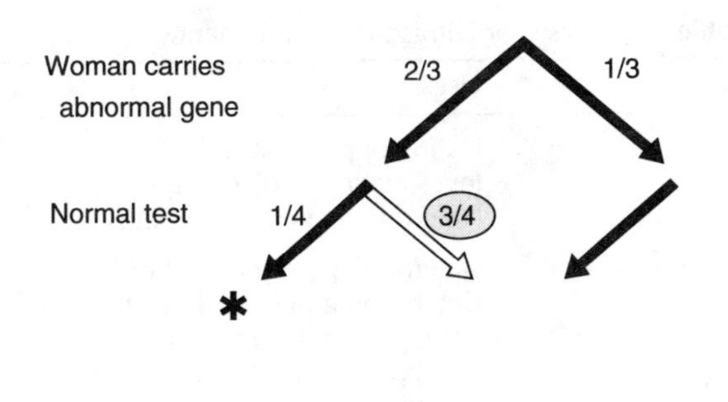

$$P = \frac{2/3 \times 1/4}{(2/3 \times 1/4) + 1/3} = 1/3$$

Figure 10-10.

(2) Possibilities contrary to the facts of the case are covered with stippling, and the corresponding arrows, which are to be excluded, are unfilled.
(3) Each event is listed to the left of the probability tree. Arrows pointing toward the left indicate occurrence of the event. Arrows pointing to the right indicate nonoccurrence.
(4) The outcome of interest is indicated by an asterisk.

III. PRENATAL DIAGNOSIS

A. Introduction. Prenatal diagnosis is a rapidly developing field encompassing both screening [e.g., maternal serum α-fetoprotein (MSAFP)] and definitive testing. In Europe and North America, the use of birth control and the planned reduction in family size have placed a greater emphasis on an optimal outcome of each pregnancy. Also, physicians have recognized the need to assess the genetic and environmental risks facing a pregnancy and the need to be aware of available prenatal diagnostic services. **Before a prenatal diagnostic technique is offered as a service, several criteria need to be met.**

1. **Safety** is related to the experience of the operator of the technique and requires the following assessments:
   a. Maternal and fetal morbidity as compared to a similar background population
   b. Fetal loss rates, both immediate and long-term, following the procedure

2. **Accuracy** requires that the operator of the technique knows false-positives and false-negatives of the test and is aware of the limitations of the technique.

3. **Identification of "at risk" population.** Screening tests may be offered to the total pregnant population, while more invasive or specific analyses are offered only to those with a defined risk factor (e.g., fetal karyotyping to pregnant women over 35 years).

4. **Ongoing quality control** is necessary, including follow-up statistics and standards for procedures and laboratories.

5. **Known diagnostic limitations.** For example, the tissue sample may be obtained easily, but appropriate diagnostic testing on tissue may not be available or may be rarely done.

B. Diagnostic techniques

1. **Ultrasound** was first used in prenatal diagnosis in 1972 for anencephaly. Since then, many fetal structural anomalies have been diagnosed using this technique.
   a. **Level I ultrasound** refers to basic levels of expertise with scans confined to patients where no fetal anomaly is suspected.
   b. **Level II ultrasound** has advanced levels of expertise with technologists and physicians. Specific examinations are done for fetal anomalies.
   c. **Uses of ultrasound** in pregnancy are listed in Table 10-1.
   d. The **detection** by ultrasound **of a placental or fetal anomaly may prompt additional studies,** including karyotyping. A number of subtle fetal features are associated with an increased risk of other abnormalities (Table 10-2).

Table 10-1. Uses of Ultrasound in Pregnancy

Trimester	Uses
First	Dating of pregnancy Investigating of bleeding Determining viability of fetus
Second	Determining presence of twins Determining placental position Screening for fetal anomalies Aiding procedures (e.g., chorionic villus sampling) Screening for suspected fetal anomalies Checking high-risk situations (e.g., a family history of structural abnormalities or maternal disease associated with an increased risk for fetal anomalies)
Third	Measuring interval growth Determining fetal size and positioning Screening for suspected fetal anomaly

2. **Doppler analysis** involves the generation of flow velocity waveforms with a continuous wave unit. The technique is used to evaluate blood flow in the umbilical/placental fetal circulation. **Fetal cardiac pathology** can be defined with Doppler analysis of fetal blood flow patterns.
   a. In **normal pregnancy,** the circulation in the cord and placenta is high flow with low resistance.
   b. In **growth-retarded infants,** placental blood flow may be abnormal with high resistance.
3. **Chorionic villus sampling (CVS)** is a first-trimester method of prenatal diagnosis involving either **transcervical or transabdominal biopsy of the developing placenta** (the chorionic villi), using a flexible catheter or needle (Figure 10-11). First developed in the 1960s, CVS was not commonly used until the availability of ultrasound.
   a. The chorion differentiates into the chorion frondosum (placental site) and the chorion laeve (membranes).
      (1) The frondosum is mitotically very active. The villi are derived from three different cell lineages: polar trophectoderm, extraembryonic mesoderm, and primitive embryonic streak.
      (2) Both the extraembryonic mesoderm and the embryo proper originate from the inner cell mass, but only a small number of cells (3 to 8) become progenitors of the embryo itself.
   b. CVS is routinely done at **9 to 12 weeks of gestation**. Transabdominal placental biopsy is possible throughout the pregnancy and may be an alternative to amniocentesis in selected instances.
   c. Sample tissue can be analyzed directly or cultured.

Table 10-2. Fetal Features Associated with an Increased Risk of Abnormalities

Feature	Abnormality
Large placenta	Triploidy Hydrops fetalis Thalassemia
Early growth retardation	Trisomies Triploidy
Abnormality of body shape	
"Lemon sign" of brain	Neural tube defect
Talipes	Neurologic problems Trisomies
Clefting of lip and palate	Trisomies or triploidy
Nuchal skin thickening	Turner syndrome Down syndrome

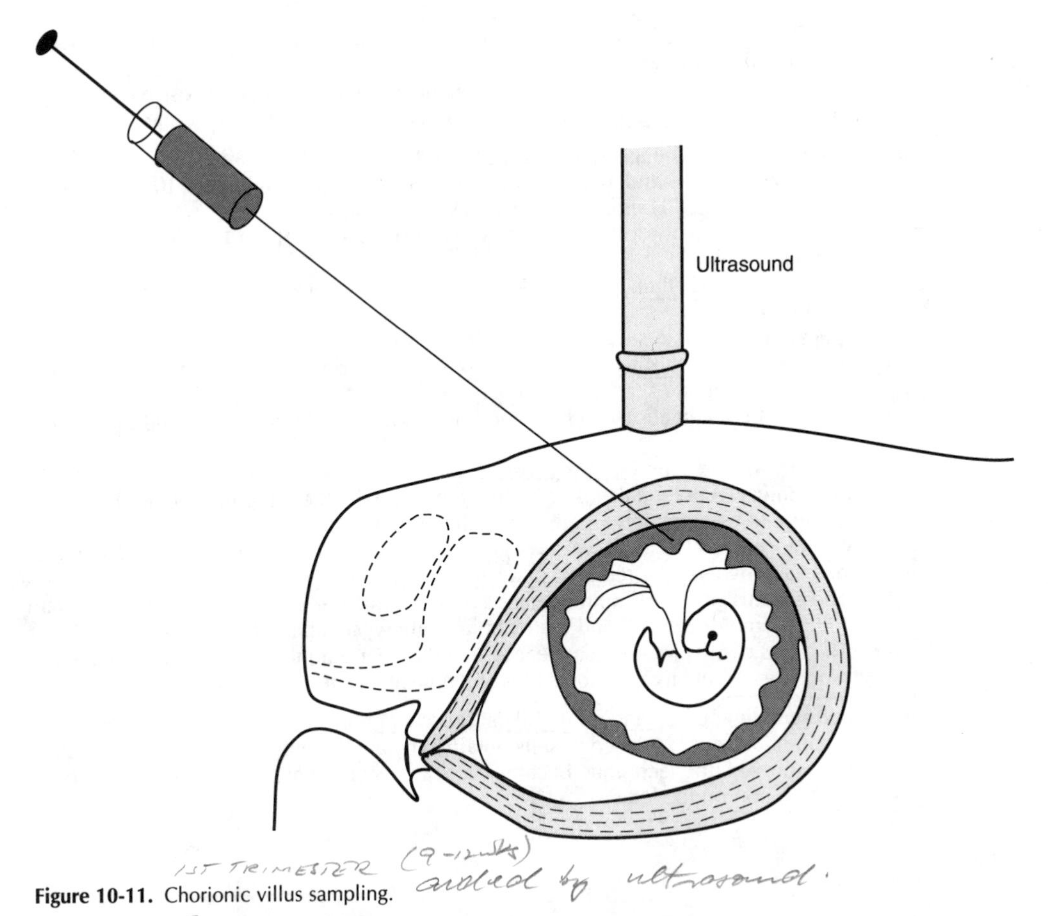

Figure 10-11. Chorionic villus sampling.

(1) **Direct analysis or short-term culture** involves cells from the trophectoderm lineage.

(2) Cells in the **long-term cultures** of chorionic villi represent the extraembryonic and embryonic mesoderm lineages.

d. Results of CVS usually take 2 weeks.

e. Complications of CVS include the following.

(1) **Bleeding.** About 40% of sampling is followed by a minor amount of bleeding, since the sampling is from a vascular site.

(2) **Miscarriage.** A randomized Canadian trial comparing CVS with amniocentesis showed no difference in loss rates. The loss rate due to miscarriage is usually quoted as 1% (the background rate for miscarriage at this stage of pregnancy is 3% to 5%).

(3) **Maternal contamination.** If the villi are not carefully cleaned before being analyzed, maternal cells may be present in sufficient numbers to confuse the results in long-term cultures.

(4) **Confined chorionic mosaicism** is defined as a dichotomy between the chromosomal constitution of the placental cells and the cells of the embryo (fetus). It is reported in 2% of pregnancies studied by CVS.

(a) Mosaicism may originate in early development through nondisjunction, anaphase lag, or structural rearrangement resulting in both a normal diploid and an aneuploid cell line.

(b) If the mutation occurs in trophoblast or extraembryonic mesoderm progenitor cells, the mosaicism may be confined to the placenta and not extend to the fetus.

(c) Most commonly, the mosaicism is confined to the cytotrophoblast (i.e., trophectoderm).

(d) When diagnosed, the usual recommendation is to offer amniocentesis to document normalcy of the fetal karyotype.

(e) The majority of such pregnancies results in normal infants, but some of these pregnancies result in growth-retarded infants or in unexplained perinatal death.

(f) Therefore, pregnancies diagnosed as having confined chorionic mosaicism should be followed carefully.

f. CVS may be done prior to 9 weeks of gestation, but the results indicate a possible increased risk of transverse limb malformations. The basis of this loss may be vascular.

4. Amniocentesis involves the aspiration of amniotic fluid with a fine gauge spinal needle for amniotic fluid cell culture and analysis of the fluid for α-fetoprotein (Figure 10-12). As with CVS, the technique should be done under ultrasound control.

a. Amniocentesis is usually done at **15 to 17 weeks of gestation**. About **15–20 cc of amniotic fluid** are withdrawn.

b. Results from amniotic fluid cell culture take about 2 to 4 weeks, depending on the culture technique used.

c. Complications of amniocentesis include the following.

(1) Failure to aspirate fluid may be due to a uterine contraction, which obliterates the available sampling space, or tenting of the membranes.

(2) Maternal complications include spotting or amniotic fluid leakage and are usually short-lived.

(3) Miscarriage risks are 0.5% in most centers.

(4) Fetal injury. A potential exists to injure the fetus with the sampling needle. The risk is thought to be extremely small, especially with the use of ultrasound.

d. Early amniocentesis (12 to 14 weeks of gestation) trials are now underway. The technique is similar but involves the removal of a smaller amount of fluid.

(1) The **advantage** of early amniocentesis is the timing (i.e., earlier in the pregnancy than routine amniocentesis), and the technique allows an alternative to CVS.

(2) The **risks** of the procedure appear to be no higher than those seen with routine amniocentesis, but large numbers of women have not been studied.

5. Fetal blood sampling [percutaneous umbilical blood sampling (PUBS)] was first developed to diagnose conditions not expressed in cells obtained by amniocentesis or CVS. As ultrasound resolution improved, the technique became more widely used as a faster method to obtain genetic information on the fetus.

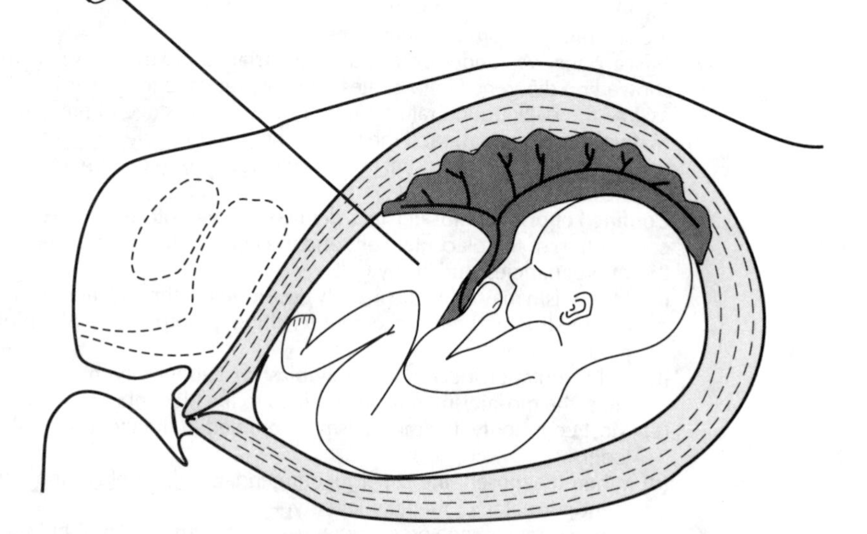

Figure 10-12. Amniocentesis.

a. The **technique** involves percutaneous puncturing of the umbilical cord near the placental insertion with a spinal needle under ultrasound guidance (Figure 10-13). The procedure is done from **17 weeks of gestation to near term**.

b. Usually, no more than **0.5–1.0 cc of fetal blood** is removed, and the sample is immediately analyzed to ensure that the sample is of fetal origin (maternal placental vessels are in close proximity to the sampling site).

c. **Results** are usually available in 48 hours, so the sample is most often used for urgent fetal karyotyping. Fetal blood sampling may also be used for viral or bacterial cultures or analysis of various hematologic indices (e.g., platelet count).

d. **Complications** of fetal blood sampling include the following.

(1) **Fetal loss rates** are difficult to obtain since the fetus being studied may be at increased risk for loss by virtue of the abnormalities present. However, a risk of 2% is seen with experienced centers. In pregnancies with a compromised fetus, the loss risk may be about 5%. Because the complications of fetal blood sampling may lead to rapid delivery of the fetus **after the period of fetal viability, facilities for emergency delivery** should be available.

(2) **Fetal bleeding** from the sampling site may continue for 1 to 2 minutes and is usually not a major problem.

(3) **Fetal bradycardia,** thought to be secondary to vascular spasm, may be documented immediately after the procedure but is usually short-lived.

6. **Maternal serum α-fetoprotein (MSAFP)** involves the use of a unique fetal marker, a protein made by the fetus during early gestation and gradually replaced by albumin.

a. Levels of **α-fetoprotein can be measured in fetal blood, amniotic fluid, and maternal serum**.

(1) The **levels change during pregnancy** and, therefore, must be correlated with gestational age.

(2) Maternal serum levels **increase steadily until about 32 weeks of gestation,** while amniotic fluid levels fall and are generally **unmeasurable after 24 weeks of gestation**.

b. **Neural tube and abdominal wall defects** in the fetus allow leakage of fetal serum into amniotic fluid. In the 1970s, amniotic fluid α-fetoprotein measurement was introduced as a method to diagnose these conditions.

c. **MSAFP screening** was introduced as a presymptomatic population-based process to identify women at increased risk for neural tube defects. Such **screening should meet the following criteria**.

(1) Any woman identified by the screening should be at sufficient risk to justify the use of further invasive confirmatory testing.

(2) Within the identified at-risk group, affected cases should be found. Arbitrarily defined cutoff levels of MSAFP are used to define the at-risk population.

d. The **risks of open neural tube defects must be calculated for specific populations**.

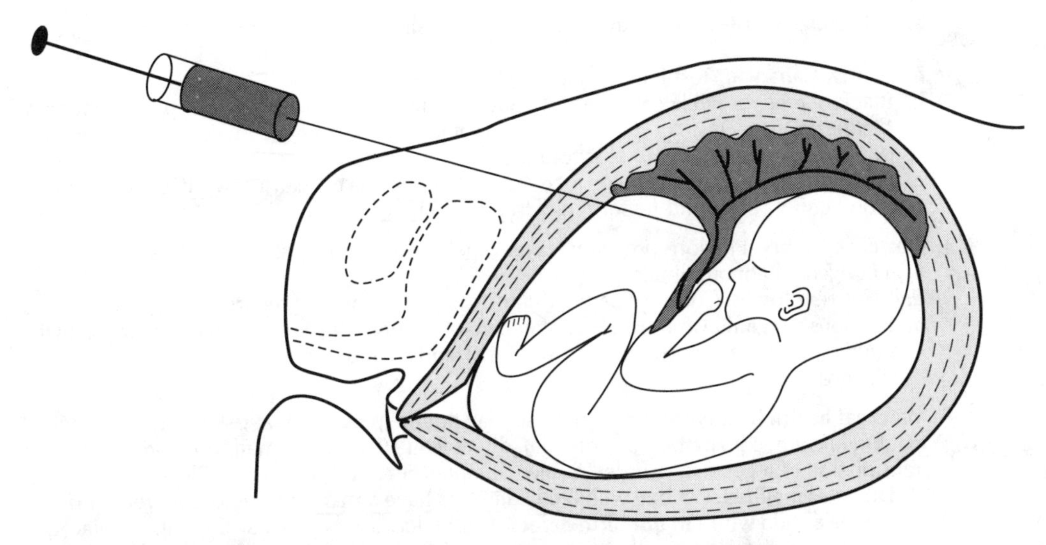

Figure 10-13. Fetal blood sampling [percutaneous umbilical blood sampling (PUBS)].

(1) The rates of open neural tube defects vary both racially (i.e., normal levels of MSAFP in black populations are higher and have a greater dispersal over that seen in other racial groups) and geographically (i.e., the incidence is higher in eastern North America than in western North America).
(2) Other factors altering the normal levels of MSAFP include maternal weight and maternal diabetes.

e. **Screening programs generate patient-specific risk reports,** which give the patient's risk before and after testing and allow both the patient and the physician a more informed decision-making process. These **risk reports require**:
(1) Adequate and independently derived normative data
(2) Prevalence rates of the conditions for which the screening was done
(3) Distributions of MSAFP for those conditions
(4) Appropriate patient data (e.g., gestational age, weight)

f. **Low levels of MSAFP** have been shown to be **associated with an increased risk for Down syndrome**. The mechanism of this association is not clear. Such screening has been calculated to pick up about 20% of the cases of Down syndrome. The addition of other fetal markers, beta-human chorionic gonadotropin (β-hCG) and estriol, may increase the detection rate for Down syndrome to 80%.

g. Because both false-positives and false-negatives exist, MSAFP screening should be offered in a controlled manner. Recommendations include:
(1) Education of health professionals and the patient
(2) Quality control procedures in laboratories
(3) Adequate counseling of the patient before the screening and in the event of an abnormal result

h. After abnormal results, the procedure includes counseling, level II ultrasound, and amniocentesis with measurement of amniotic fluid α-fetoprotein and fetal karyotyping.

C. **The counseling process.** As the field of prenatal diagnosis becomes more complex, the need for a systematic approach to the care of the pregnant patient becomes increasingly necessary. Both genetic and environmental screening have now become a part of routine prenatal care. In this way, the patient requiring specific types of prenatal diagnosis can be identified and appropriately managed. Ideally, such screening should be done before the pregnancy.

1. **Parental age.** Age-related risks for chromosome anomalies should be discussed and prenatal karyotyping by CVS or amniocentesis should be offered to women 35 years of age and older at time of delivery (see Ch 2 II). Paternal age over 55 years is associated with an increased risk of dominant mutations. It may be appropriate to offer detailed ultrasound scanning to such patients.

2. **Race and ethnic background.** A particular ethnic background may indicate an increased risk for certain single gene conditions (see Ch 3 I). Screening can be offered to the following racial groups.
a. **Ashkenazi Jewish.** Test for Tay-Sachs heterozygosity with measurement of hexosaminidase A.
b. **French Canadian.** Test for Tay-Sachs heterozygosity.
c. **Black.** Screen for sickle cell disease heterozygosity with hemoglobin (Hb) electrophoresis.
d. **Mediterranean.** Test for β-thalassemia trait by measuring mean corpuscular volume (MCV) and performing Hb electrophoresis.
e. **Southeast Asian, Indian.** Screen for α- and β-thalassemia by measuring MCV and complete blood count (CBC) and performing Hb electrophoresis.

3. **Obstetric history.** Previous pregnancies may include recurrent losses, stillbirths, or livebirths with congenital abnormalities.
a. **Recurrent losses.** Consider parental translocation in cases of recurrent losses.
b. **Previous congenital abnormality.** After careful review of the autopsy or examination of the child, discuss the appropriate recurrence risk and the most appropriate method of prenatal diagnosis.

4. **Maternal health history.** Maternal disease may put the pregnancy at risk because of general maternal health (e.g., maternal azotemia), direct fetal effects (e.g., malformations), or risk for transmission of a genetic disorder. Some examples of conditions to consider include:
a. **Diabetes mellitus.** Women with type I diabetes have a **two- to threefold increased risk of having a child with a major birth defect**. This includes a specific risk for limb, cardiac, and neural tube defects as well as the increased risk of birth defects overall.

(1) **Poor preconceptional control** may explain some of the increased risk; therefore, optimal control of diabetes beginning prior to conception is important.

(2) **Prenatal diagnosis** includes ultrasound and MSAFP screening.

b. **Epilepsy.** Many commonly used anticonvulsants are associated with an increased risk of craniofacial and cardiac malformations. The mechanism is not clear, nor is the effect of the epilepsy itself. However, seizure control should be optimized with the least number of anticonvulsants, and detailed ultrasound should be offered.

c. **Maternal myotonic dystrophy.** Pregnancies in women with myotonic dystrophy, if the fetus is also affected, are complicated with polyhydramnios and difficulties with labor.

(1) In addition, affected offspring of affected mothers are more likely to have the congenital form of myotonic dystrophy.

(2) Therefore, management involves not only more frequent monitoring of the pregnancy but also may involve prenatal diagnosis.

5. **Exposure history.** The question of whether a maternal exposure increases the risk of congenital malformation must be asked in light of the already existing background risk of malformations of 3% to 5%. Also, the timing of exposure and the dosage must be known. Exposures can be characterized as the following.

a. **Infections.** A number of infections put the pregnancy at increased risk for malformations [e.g., rubella, cytomegalovirus (CMV)] while others have a low or unknown risk.

(1) Prenatal diagnosis is often difficult, as documentation of the presence of the infecting organism may not give information regarding fetal effects.

(2) Detailed ultrasound may be appropriate as a way to assess the fetus.

b. **Medications.** After considering timing, dosage, and maternal effects, information should be sought from either written resources or teratogen information services to assess the fetal risk. Detailed ultrasound may be appropriate to assess particular organ systems known to be involved.

c. **Environmental hazards.** Questions regarding occupational and hobby exposures should be asked of all pregnant women. Exposures to significant radiation, organic solvents, and heavy metals may be of concern and need further assessment.

6. **Assessing the need for prenatal diagnosis.** When considering whether prenatal diagnostic techniques are appropriate for a particular patient, the physician must be aware of recognized indications for prenatal diagnosis (see Table 10-3), and then decide whether the patient

Table 10-3. Indications for Prenatal Diagnosis

Test	Indications
Fetal karyotype	Maternal age ≥35 years at time of delivery Previous stillbirth or livebirth with chromosome abnormality Parental translocation or chromosome abnormality Fetal ultrasound abnormality suggestive of chromosome abnormality Low maternal serum α-fetoprotein (MSAFP)
Amniotic fluid α-fetoprotein	Previous pregnancy with neural tube defect Poorly controlled maternal diabetes mellitus Maternal use of sodium valproate Positive MSAFP screening for neural tube defect Ultrasound finding suggestive of open fetal defect
Fetal blood sampling	Ultrasound abnormality suggestive of chromosome abnormality Hemoglobinopathies or immune deficiencies Diagnosis of fetal infection
Detailed ultrasound	Raised MSAFP at 16–18 weeks of gestation Past history of child with structural malformation Patient affected with, or having family history of, structural malformation Patient affected by medical disorder associated with increased risk for fetal structural malformation Environmental exposure associated with increased risk of structural malformation in fetus

fits these criteria. After review of the previous and present pregnancy history, the maternal health and the family history should be considered.

a. **Some of the diagnostic modalities are used for screening purposes, while others should be considered only with specific indications.** MSAFP screening may be offered to all pregnant patients, but the results give an altered risk figure either for neural tube defects or Down syndrome, not a specific diagnosis.

b. The **decision to have prenatal diagnosis should be made by the patient** after she has heard a description of the procedures and is informed of the risks (Table 10-4) and possible complications. She should be told that prenatal diagnosis is an option and not a requirement to ongoing prenatal care.

7. **Abnormal fetus.** The most difficult situation for both the patient and the physician is the unexpected diagnosis of fetal abnormalities. The patient is referred urgently; counseling and subsequent investigations must, out of necessity, be done in a rapid fashion; and ethical concerns may arise when considering options for the remainder of the pregnancy. The sequence includes the following.

a. **Review of existing information.** This should include a history of the pregnancy to date, details of the abnormal scan or result, and any other relevant information.

b. **Initial counseling of the family.** The couple should be seen together, if possible. A review of the findings, possible diagnoses, and recommended further investigations should be covered. Supportive counseling and follow-up plans can be outlined.

c. **Review of pregnancy options** and discussions of other services likely to be involved in the ongoing care of the mother and infant should take place (e.g., pediatric surgery if a surgically correctable lesion has been identified).

d. **Review of pregnancy outcome.** The infant should be examined, an autopsy should be arranged if appropriate, then all findings should be reviewed with the family. Recurrence risks, if known, and information regarding prenatal diagnosis in subsequent pregnancies can be covered.

IV. GENETIC COUNSELING.

The counselor in medical genetics generally uses nondirective counseling. However, the counseling session is not merely a litany of facts but is colored with the counselor's own experiences and opinions. For the couple or the family seeking the counseling, the information received may have long-term consequences and will often be used as part of the information necessary for decision-making.

A. **Indications for genetic counseling** may change with time as new methods of diagnosis become available or as specific therapies are developed. The most common indications include:

1. **Advanced maternal age** (prenatal diagnosis is offered to pregnant women at age 35 years or older)
2. **Known or suspected hereditary condition in the family**
3. **A fetus or child with birth defects,** including both single and multiple malformations
4. **Mental retardation**
5. **Recurrent spontaneous abortions**
6. **Exposure to known or suspected teratogens** (in many centers, such referrals are handled by telephone)
7. **Consanguinity**

Table 10-4. Risks of Prenatal Diagnostic Techniques

Technique	Time Performed	Risks
Ultrasound	Throughout pregnancy	None known
Chorionic villus sampling	$\leq$9 weeks	Possible transverse limb abnormalities
Chorionic villus sampling	>9 weeks	Loss risk = 1.0%
Amniocentesis	13–14 weeks	Loss risk ~ 0.5%
Amniocentesis	15–17 weeks	Loss risk = 0.5%
Fetal blood sampling	>17 weeks	Loss risk = 2.0%
Doppler	Throughout pregnancy	None known

B. Genetic counseling team

1. The **medical specialist** is often trained in a specific medical specialty (e.g., pediatrics) with additional training as a clinical geneticist. Proposed training may include general training in genetics. The role of the medical or clinical geneticist is to establish the diagnosis and to counsel patients regarding the medical implications of the condition.

2. The **genetic counselor** may be either formally trained or experienced in the area of genetic counseling. Counselors come from a variety of backgrounds, including nursing, social work, and education. Formal training programs involve a master's degree following an undergraduate education with a science/psychology emphasis. The role of the counselor is to be involved in the counseling process, support, and follow-up of the family.

3. **Additional members** include medical specialists consulted for specific investigations and opinions. Ongoing family support may also be given by social workers, religious support persons, and parent and lay groups.

C. Genetic counseling facilities

1. **Teaching or university-based tertiary care hospitals.** These centers are involved in counseling services; the education of other health professionals, fellows, and the lay public; and research into genetic conditions (clinical and basic research).

2. **Component of a specialized clinic.** In such a clinic (e.g., for spina bifida), the provision of genetic counseling is part of the overall service to the family.

3. **Screening programs** (e.g., MSAFP) provide a specific service to a large number of families. Only patients with abnormal results are formally counseled.

4. **Outreach programs** are generally run from a tertiary care center and provide local service on an intermittent basis.

D. Goals of genetic counseling have been defined by the American Society of Human Genetics (1975) and include the following.

1. **Comprehending the medical facts.** After the most precise diagnosis is made (see I), the family is told the diagnosis, prognosis, appropriate investigations, and ongoing management of the condition.

2. **Understanding the mode of inheritance and the recurrence risks.** The inheritance should be clearly explained (the use of diagrams and examples may be helpful) to the couple or to other family members. Recurrence risks may involve standard mendelian genetics, calculated risks, or risks observed in the population (empiric).
   a. The origin of the risk and whether it pertains to a specific diagnosis (e.g., the recurrence risk for cystic fibrosis after one affected child would be 25%) or to a nonspecific diagnosis (e.g., the recurrence risk for nonspecific mental retardation where the cause is unknown would be empiric) should be made clear.
   b. Couples may appreciate risks in different ways.
      (1) **Binary.** Two states are seen: The condition will or will not happen again.
      (2) **Comparison of losses and gains.** The condition is analyzed according to whether there are positive or negative aspects to a recurrence; the balance may determine the decision to have another pregnancy.
      (3) **Numerical risk** is expressed either as a percentage (e.g., 50%) or as a fraction (e.g., 1/2). The percentage may be perceived as being a larger risk than the equivalent fraction, perhaps because the numerator is a number larger than one.

3. **Understanding the options.** When faced with a discrete recurrence risk, the reproductive options would include methods of contraception, adoption, insemination by donor sperm, use of donated ova, prenatal diagnosis with or without termination of the affected fetus, and an unmonitored pregnancy to the couple.

4. **Choosing a course of action.** The couple is encouraged to choose the best course for themselves in light of the recurrence risk, the perceived burden (psychologic, social, economic), the goals of the family, and their religious and ethical standards.

5. **Adjusting to the condition.** The role of the counselor is to assess the family's reaction to the diagnosis and recurrence risk. Follow-up counseling, either by the medical genetics team or a psychologist, may be indicated.

E. Counseling process. To achieve the goals of genetic counseling, the family must be able to receive the information presented and convert the facts into a part of their decision-making. The counseling team (i.e., the medical geneticist and the genetic counselor) interacts with the family most effectively by:

1. Recognizing the **psychologic and emotional burdens associated with the diagnosis.** Families experience degrees of guilt and shame after the diagnosis of a genetic condition in a family member. Either situation can result in a patient who does not keep the counseling appointment, who is angry or belligerent, or who initiates a legal action.
   a. **Guilt** is a type of self-reproach to violations of inner standards. The guilty feelings **may be masked** by repression of "bad behaviors" (e.g., drug or alcohol abuse), by intellectualizing the problem (e.g., the patient who converses in an abstract way about the condition), or by separating the feelings of guilt from other emotions.
      (1) **Realistic guilt feelings** may exist when a woman chooses not to have prenatal diagnosis and delivers an infant with Down syndrome.
      (2) **Unrealistic guilt feelings** are seen when the individual feels responsible for factors over which he or she has no control.
   b. **Shame** is a response to anticipated or actual external disapproval or a response to a failure to reach certain goals and standards. Shame may be masked by denial, relabeling the losses as gains (e.g., "We were chosen to have this child because of our strength."), or by finding fault with others.

2. **Recognizing factors that affect counseling**
   a. North American society is not homogeneous; families come from a **variety of ethnic backgrounds**. The importance of the extended family's involvement in the decision-making process of a young immigrant couple demands that counseling sessions involve all appropriate family members.
   b. **Socioeconomic status** influences a family's access to medical services and their understanding of the information presented.

3. **Providing nondirective counseling.** The family needs support for its decision-making, full disclosure of options, but not paternalistic counseling. A useful method involves using the example of other families while attaching no increased value to a particular option (e.g., "Some families choose prenatal diagnosis while others feel that insemination by donor is the method for them.").

4. **Helping the family relieve feelings of shame and guilt**
   a. The counselor may become a figure of authority, especially early in the process of diagnosis, by stating that a particular condition "is not their fault." Later, the family member may remember this reassurance with comfort.
   b. Directly addressing feelings and normalizing them by comparing the family positively to others in a similar situation may help.
   c. The parents' sense of responsibility for the child's problem may be reframed as "it's normal to feel responsible and to worry and care about the child" while not being the cause of the problem itself.
   d. Helping the parents to define which aspects of their child's condition they can control and which aspects are beyond their control allows the parents a way to feel more positive.

5. **Helping the family identify its personal goals.** Asking the family its expectations of the counseling session helps to set the agenda and allows the counselor to address additional issues.
   a. Goals are often time-related, with an early goal to understand the medical implications and a later goal to decide on another pregnancy.
   b. In this way, the family is helped to formulate the task before them, given the pertinent information and support, and then given the freedom to choose for themselves.

6. **Recognizing counselor biases.** It may be disappointing or frustrating that a family makes choices at variance with those of the counselor. One way to ensure that a couple decides for themselves is to explore their opinions regarding the issues in a nonjudgmental manner. The counselor may list options regarding prenatal diagnosis in a pregnancy in no particular order and emphasize that different families choose different options.

F. Problems in counseling

1. **Genetic heterogeneity** often complicates the establishment of a precise diagnosis. Similar disorders due to mutations at different loci or involving different alleles may have different forms of inheritance. Retinitis pigmentosa may be inherited as an autosomal recessive or dominant as well as an X-linked condition.

2. **Phenocopies** mimic genetic disorders, are caused by environmental factors, and are, therefore, unlikely to recur. Microcephaly may have both genetic and nongenetic causes.

3. The **sporadic case** is common in nuclear families. Here, the pedigree is not helpful, and the counselor must depend on an accurate diagnosis. The advent of molecular methods of diagnosis has helped in some areas (e.g., DMD with an identified deletion) to define a recurrence risk for the family.

4. **Nonpaternity** or questioned paternity may invalidate the risks quoted in a counseling session. Interpretation of molecular analysis is dependent upon known paternity, with the use of both direct and indirect methods (see Ch 5). The inadvertent discovery of nonpaternity during the course of such testing creates ethical concerns regarding whether or not to disclose the fact.

STUDY QUESTIONS

Directions: Each of the numbered items or incomplete statements in this section is followed by answers or by completions of the statement. Select the **one** lettered answer or completion that is **best** in each case.

Questions 1 and 2

A young woman who is 28 weeks pregnant consults a physician because her sister's son was recently found to have hemophilia (an X-linked recessive condition). She recently had an ultrasound examination and was told that her fetus is male. On taking the family history, the physician learns that two of the woman's brothers "bled to death" in early childhood. Further inquiry indicates that they almost certainly had hemophilia. The patient has three normal sons and a normal daughter.

1. What is the chance that the fetus she is carrying has hemophilia?

(A) 1/2
(B) 1/4
(C) 1/9
(D) 1/18
(E) 1/32

The family undergoes testing for linkage to DNA markers. The results of a restriction fragment-length polymorphism (RFLP) that is located 2 cM from the hemophilia gene are shown below.

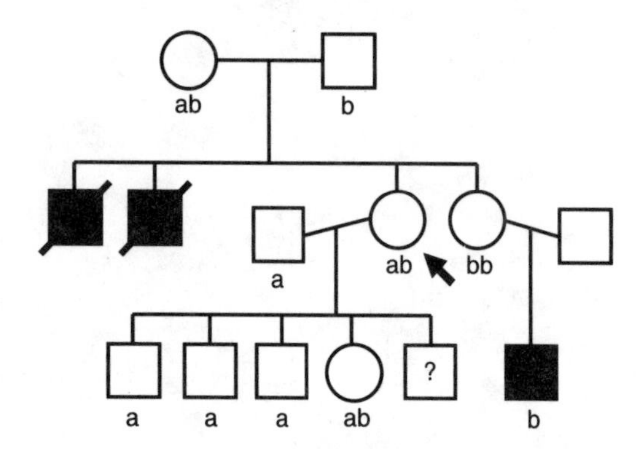

2. On the basis of this additional information, what is the chance that the fetus carried by the proband has hemophilia?

(A) 0
(B) 2%
(C) 25%
(D) 50%
(E) 98%

3. A couple presents for counseling during their pregnancy. The woman, who is 31, is extremely worried because her sister has just delivered a daughter with Down syndrome. The woman is now 6 weeks pregnant and wants amniocentesis. Which would be the most appropriate way to manage and counsel the family?

(A) Offer amniocentesis and schedule the procedure to be done at 15–16 weeks
(B) Offer detailed ultrasound at 8–10 weeks
(C) Ascertain chromosome results on the child affected with Down syndrome
(D) Karyotype the patient—if normal, reassure the patient; if the chromosomes indicate a balanced translocation, offer chorionic villus sampling (CVS) or amniocentesis
(E) Reassure the patient that no prenatal diagnosis is necessary

---

1-D
2-A
3-C

4. A young woman consults a physician because she has a brother and a half-brother who are both profoundly mentally retarded. Further assessment reveals that both have severe hydrocephalus due to stenosis of the aqueduct of Sylvius. In some families, including this one, this kind of hydrocephalus is inherited as an X-linked recessive trait. DNA linkage studies are performed, and a marker 5 cM from the disease gene is found to be informative in the family. The results are shown below.

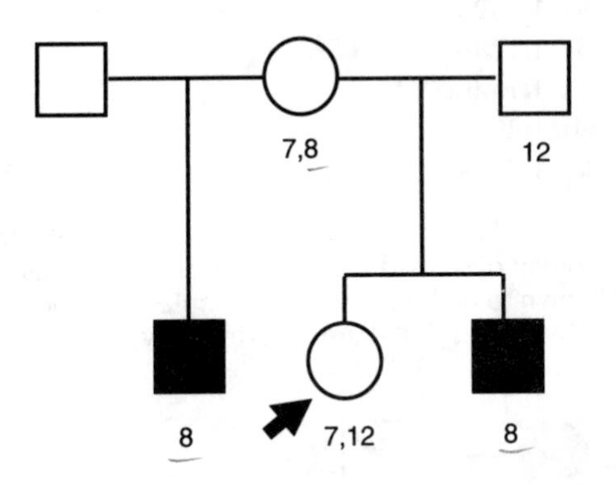

What is the chance that this woman is a heterozygous carrier for the X-linked hydrocephalus gene?

(A) 0
(B) 2.5%
(C) 5%
(D) 50%
(E) 95%

5. A 32-year-old woman has blood drawn for maternal serum α-fetoprotein (MSAFP) screening, and undergoes a routine ultrasound. Her results indicate that she is at significant risk for a neural tube defect. Which of the following actions should be taken to follow up this risk?

(A) Check her dates
(B) Check the patient's weight; was it reported incorrectly as "kg" instead of "lbs"?
(C) Arrange an amniocentesis for measurement of amniotic fluid α-fetoprotein
(D) Schedule a level II ultrasound for assessment of fetal detail
(E) Arrange for measurement of acetylcholinesterase

6. Measurement of the serum creatine phosphokinase (CPK) level can detect about 75% of obligate carriers of Duchenne muscular dystrophy (DMD; an X-linked recessive condition that is lethal in males). A 26-year-old woman has a brother who died of DMD at age 18. There is no other family history of DMD. CPK testing is done on the woman and is normal. What is the risk for DMD in her son?

(A) 1/24 (4.2%)
(B) 1/18 (5.6%)
(C) 1/10 (10%)
(D) 1/8 (12.5%)
(E) 1/4 (25%)

7. All of the following situations are indications for fetal karyotyping EXCEPT

(A) Maternal age of 39
(B) Previous child with trisomy 21
(C) A fetus shown to have a cystic hygroma on ultrasound
(D) A second cousin with spina bifida
(E) A diagnosis of a robertsonian translocation in the father

---

4-C
5-E
6-B
7-D

8. Huntington's chorea is an autosomal dominant disease characterized by progressive dementia and uncontrollable movements. The age of onset is variable. About 10% of individuals carrying the Huntington's chorea gene are clinically affected by age 30, about 50% by age 45, about 80% by age 50, and almost 100% by age 70. What is the chance that the proband (who has just had a completely normal medical examination) carries the gene for Huntington's chorea? (The numbers below the symbols indicate the age in years.)

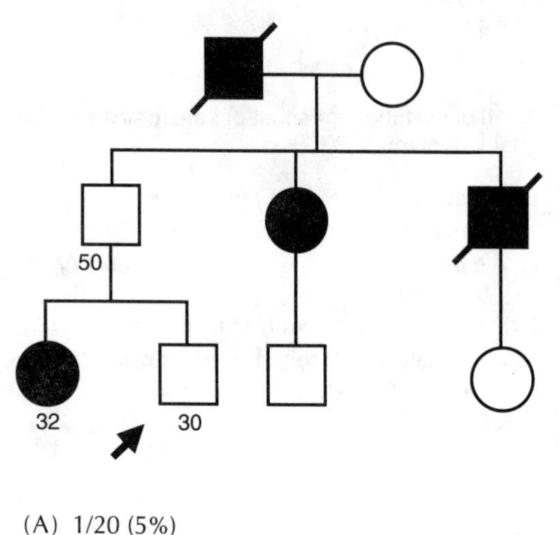

(A) 1/20 (5%)
(B) 9/119 (7.6%)
(C) 1/4 (25%)
(D) 9/19 (47.4%)
(E) 1/2 (50%)

Questions 9 and 10

Retinoblastoma, a malignant tumor of the eye, may be inherited as an autosomal dominant trait. Penetrance in such families is about 4/5 (80%). A woman whose paternal grandmother, father, and brother had retinoblastoma is unaffected. She is 12 weeks pregnant.

9. What is the risk of retinoblastoma in her fetus?

(A) 1/4 (25%)
(B) 1/12 (8.3%)
(C) 1/15 (6.7%)
(D) 1/20 (5%)
(E) 1/25 (4%)

DNA linkage studies are done on the family of the woman described above. The results for a marker known to be linked to the retinoblastoma locus at a distance of 5 cM are shown below.

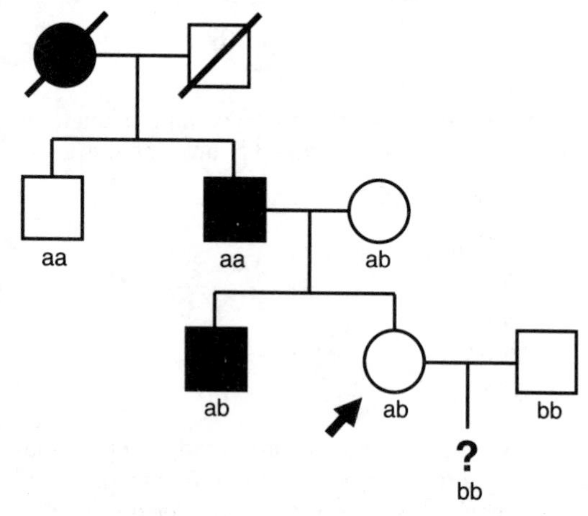

10. What is the risk of retinoblastoma developing in this woman's fetus if the fetus is shown to be *bb* at the marker locus?

(A) 19/1000 (0.19%)
(B) 1/250 (0.4%)
(C) 1/200 (0.5%)
(D) 1/150 (0.67%)
(E) 19/150 (12.7%)

8-D
9-C
10-D

Directions: Each group of items in this section consists of lettered options followed by a set of numbered items. For each item, select the **one** lettered option that is most closely associated with it. Each lettered option may be selected once, more than once, or not at all.

Questions 11–15

Match the following pedigrees with the most likely mode of inheritance.

(A) Autosomal dominant
(B) Autosomal recessive
(C) X-linked recessive
(D) Chromosomal
(E) Any of the above

11.

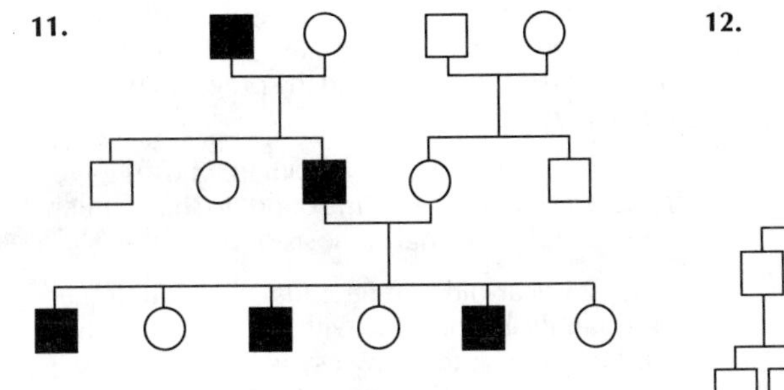

12.

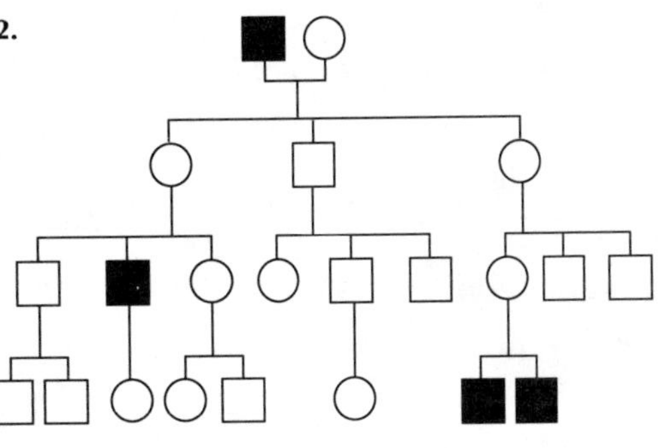

13.

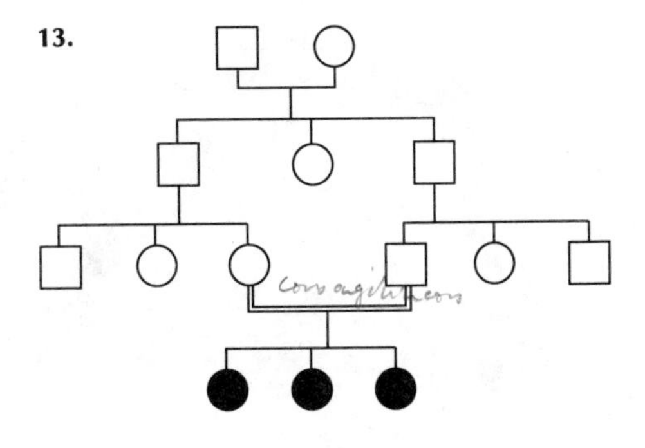

14.

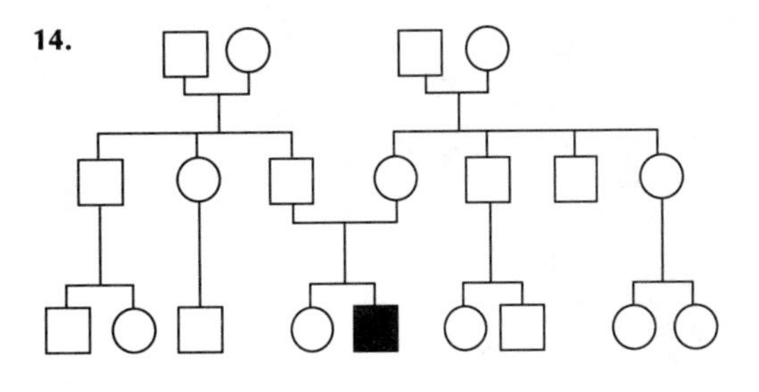

15.

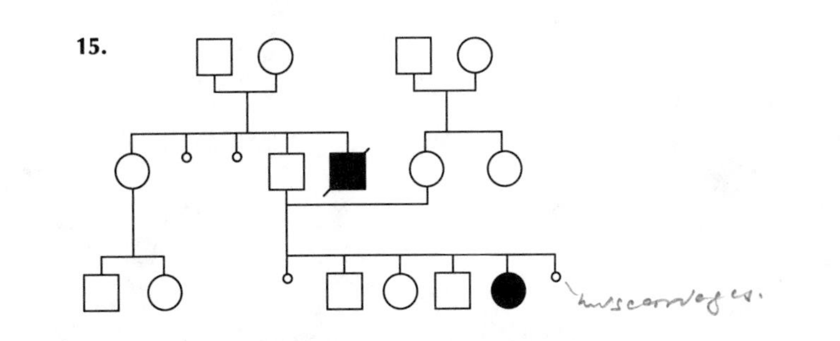

11-A
12-C
13-B
14-E
15-D

Questions 16–20

Match each clinical situation with the most likely etiology.

(A) X-linked inheritance
(B) New dominant mutation
(C) Trisomy 13
(D) Translocation
(E) Autosomal recessive inheritance

16. Two maternal uncles with a clinical picture similar to the affected child

17. Advanced paternal age

18. Advanced maternal age

19. Recurrent early pregnancy losses

20. Consanguinity

Questions 21–25

For each case history that follows, select the most appropriate prenatal diagnostic technique.

(A) Chorionic villus sampling
(B) Maternal serum α-fetoprotein screening
(C) Amniocentesis
(D) Level II ultrasound
(E) Fetal blood sampling

21. A 30-year-old woman with a history of a stillborn infant with normal chromosomes and multiple abnormalities, including congenital heart disease, polydactyly, and cleft palate

22. A family history of Duchenne muscular dystrophy, where affected males have a deletion involving the dystrophin gene and the pregnant woman is a heterozygote

23. An urgent referral of a woman whose19-week fetus was found on ultrasound to have nuchal edema and duodenal stenosis or atresia

24. A 25-year-old primigravida with an unremarkable family history but with extreme anxiety regarding her risk for Down syndrome

25. A 36-year-old healthy woman at 14 weeks of gestation

16-A
17-B
18-C
19-D
20-E
21-D
22-A
23-E
24-B
25-C

ANSWERS AND EXPLANATIONS

1–2. The answers are: 1-D *[II C 2–4]*, **2-A** *[II B]*.
The pedigree is illustrated below.

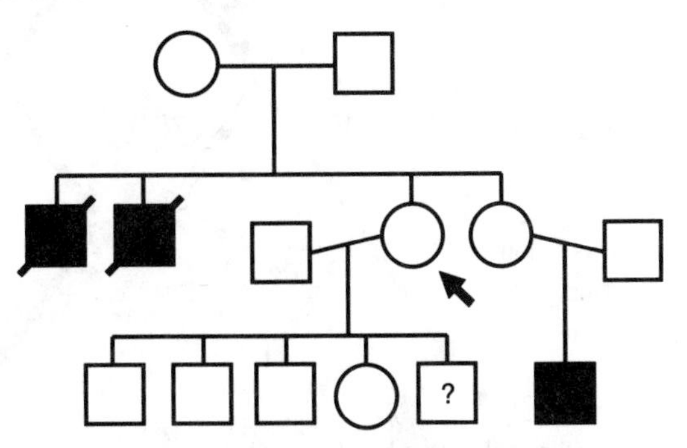

The proband has an a priori risk of 1/2 of having inherited the abnormal hemophilia gene that is segregating in her family. However, the fact that she has had three sons, none of whom has hemophilia, makes it less likely that she is a carrier. The only way that the fetus could have inherited the abnormal gene would be for his mother (the proband) to have inherited the abnormal gene and to have passed it on to the fetus but not to any of her three sons. The other possibilities that must be considered are that the proband did not inherit the abnormal gene from her mother or that the proband did inherit the abnormal gene but did not transmit it to the fetus. Because all three of the proband's sons are unaffected, the chance that she inherited the abnormal gene and transmitted it to one, two, or all three of her sons must be excluded from the calculation.

The chance of a carrier woman transmitting an abnormal X-linked gene to her son is 1/2. If she has three sons, the chance that all three would be affected is (1/2 x 1/2 x 1/2) = 1/8.

The chance that two sons would be affected and one unaffected is 3 x (1/2 x 1/2 x 1/2) = 3/8. There are three ways this could happen: (1) The first son could be unaffected and the other two affected, (2) the second son could be unaffected and the other two affected, or (3) the third son could be unaffected and the other two affected.

The chance that two of the woman's sons would be unaffected and one affected is 3 x (1/2 x 1/2 x 1/2) = 3/8. Again, there are three ways this could happen: (1) only the first son could be affected, (2) only the second son could be affected, or (3) only the third son could be affected.

The chance that all three sons would be unaffected is (1/2 x 1/2 x 1/2) = 1/8. Only this last possibility is compatible with the information given; the others must be excluded from the calculation.

One must perform a Bayesian calculation to determine the overall risk that the proband's fetus is affected with hemophilia. Using a probability space, the problem is illustrated below.

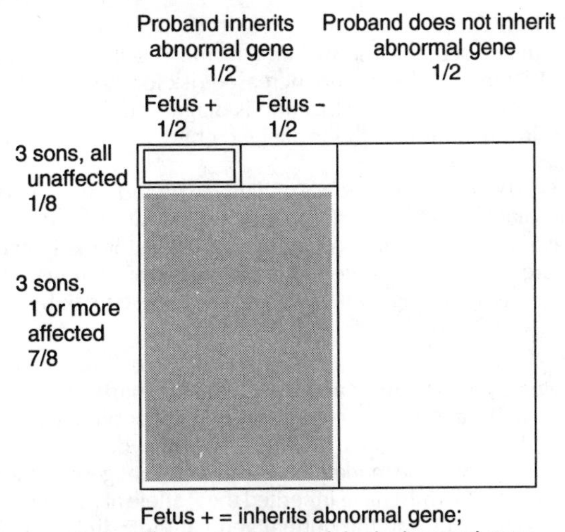

Alternatively, the solution can be illustrated with a probability tree.

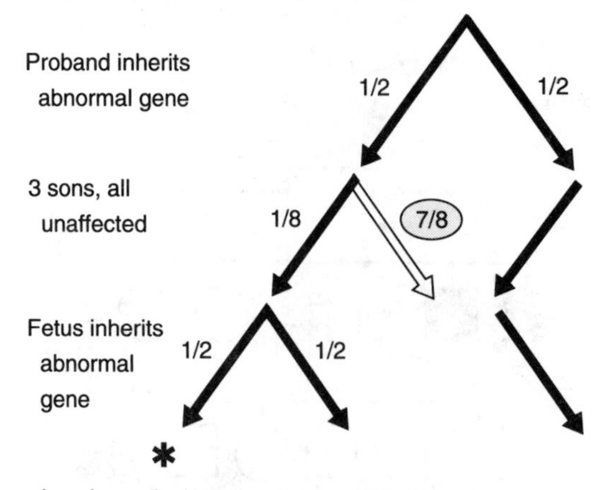

The probability that the proband's male fetus has hemophilia is

$$P = \frac{1/2 \times 1/8 \times 1/2}{(1/2 \times 1/8 \times 1/2) + (1/2 \times 1/8 \times 1/2) + 1/2}$$

$$P = \frac{1/32}{1/32 + 1/32 + 16/32}$$

$$P = \frac{1/32}{18/32}$$

$$P = 1/18$$

The proband's mother is an obligate carrier for hemophilia. Testing for the linked marker reveals that she is heterozygous *ab* at the linked marker locus. The *b* allele is likely to be the one that is in coupling with the hemophilia gene because her grandson who has hemophilia has this allele.

There is essentially no risk that the proband's fetus will have hemophilia. The proband must have inherited a *b* allele at the marker locus and a normal allele at the hemophilia locus from her father. This means that she inherited the *a* allele at the marker locus from her mother. This is likely to be the allele in her mother that is in coupling with the normal allele at the hemophilia locus. Although there was a small chance that the proband inherited the *a* allele at the marker locus and the hemophilia gene from her mother as a result of crossing over, this did not occur in this case because all three of the proband's sons received the *a* allele she received from her mother, and none of them has hemophilia. If the proband did not inherit the hemophilia gene she cannot transmit it (barring a new mutation).

3. The answer is C *[I B 2; III B 1, 6].*
Since the child was newly diagnosed, the chromosome results should be readily available. If the results indicate that the child has trisomy 21, there is no increased risk for the patient (or for other family members) above that associated with age. If a translocation is diagnosed, it will then be appropriate to karyotype the parents of the child. By starting with the affected child, the information is more specific for the entire family.

Detailed ultrasound at 8–10 weeks of gestation is unlikely to provide much information regarding structural detail and would not diagnose Down syndrome at any gestational age.

Since the child may have an inherited translocation as a cause for the Down syndrome, it is not appropriate to simply reassure the patient. Maternal serum α-fetoprotein (MSAFP) may be offered as a screening test, but only after ascertaining that she is not at increased risk because of a translocation.

4. The answer is C *[II B].*
The woman's mother is heterozygous for the *7* and *8* alleles at the marker locus as well as for the X-linked hydrocephalus gene. The *8* allele at the marker locus must be in coupling with the disease gene because both affected boys inherited the marker *8* allele and the X-linked hydrocephalus gene from their mother.

The proband inherited a *12* allele at the marker locus and a normal gene at the X-linked hydrocephalus locus from her unaffected father. She must have inherited the *7* allele at the marker locus from her mother. Since the *7* allele at the marker locus and the normal gene at the X-linked hydrocephalus locus are in coupling and since the two loci are separated by a recombination distance of 5 cM, the proband has only a 5% chance of being a carrier of the abnormal allele for X-linked hydrocephalus. This risk is that of recombination between the marker locus and the X-linked hydrocephalus locus on the X chromosome that she received from her mother.

Note that the usual 50% risk of inheriting the gene for X-linked hydrocephalus does not apply here. The use of linked markers provides information regarding how segregation actually occurred, so the 50% risk related to random segregation is no longer relevant.

5. The answer is E *[III B 1, 6].*
The patient had a routine ultrasound along with her maternal serum α-fetoprotein (MSAFP) assessment. Therefore, there is enough information to check her dates; another dating scan is not necessary. If the weight given is significantly lower, (i.e., 105 lbs instead of 105 kg), the MSAFP value will be falsely elevated.

If the dates are correct and, therefore, the results are valid, it is appropriate to schedule a level II or detailed ultrasound. In some genetic centers this would be done before arranging for amniocentesis, as a reason for the elevation might be seen (e.g., anencephaly); while in other centers, both the level II ultrasound and the amniocentesis are arranged for the same time.

The measurement of acetylcholinesterase is only necessary if the amniotic fluid α-fetoprotein is elevated. This measurement is a more specific indicator of an open neural tube defect.

6. The answer is B *[II C 2–4; Ch 7 II C 2].*
Duchenne muscular dystrophy (DMD) is an X-linked recessive disease that is genetically lethal. In this instance, the family history is negative except for the affected brother. There is a 2/3 chance that the affected boy's mother carries the abnormal gene and a 1/3 chance that he represented a new mutation. If the proband's mother was a carrier, there is a 1/2 chance that she (the proband) inherited the abnormal gene. However, her creatine phosphokinase (CPK) test is normal despite the fact that this test is abnormal in 75% of carriers. This decreases the likelihood that the proband is a carrier.

A Bayesian calculation is necessary. This is illustrated in the following probability tree. The calculation could also be done by drawing a probability space, but probability spaces become more difficult with more complex calculations.

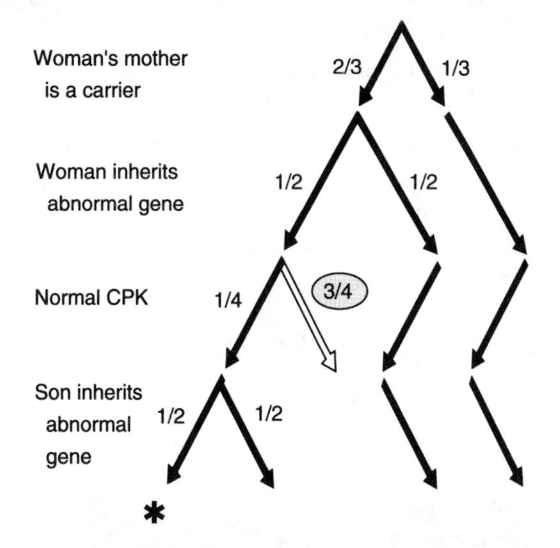

The proband can only have an affected son if her mother was a carrier of the abnormal gene. If the proband's mother transmitted the abnormal gene to the proband, the proband is among the 25% of carriers whose CPK tests are normal, and she transmits the abnormal gene to her son. The other possibilities that must be considered are: (1) the proband's mother was not a carrier (i.e., the affected boy represents a new mutation), (2) the proband's mother was a carrier but did not transmit the abnormal gene to the proband, and (3) the proband's mother was a carrier and transmitted the abnormal gene to the proband who has a normal CPK test, but the proband does not transmit the abnormal gene to her son.

The other possibility, that the proband's mother was a carrier and transmitted the abnormal gene to the proband who has an abnormal carrier test, is not applicable here and must be removed from the calculation.

The chance that the proband's son would have DMD is

$$P = \frac{(2/3 \times 1/2 \times 1/4 \times 1/2)}{(2/3 \times 1/2 \times 1/4 \times 1/2) + (2/3 \times 1/2 \times 1/4 \times 1/2) + (2/3 \times 1/2) + 1/3}$$

$$P = \frac{2/48}{2/48 + 2/48 + 2/6 + 1/3}$$

$$P = \frac{2/48}{2/48 + 2/48 + 16/48 + 16/48}$$

$$P = \frac{2/48}{36/48}$$

$$P = 2/36 = 1/18$$

7. The answer is D *[II C 4].*
Both advanced maternal age (at or over 35 years at the due date) and a previous stillbirth or livebirth with a chromosome abnormality are indications for fetal karyotyping via either chorionic villus sampling (CVS) or amniocentesis.

A cystic hygroma may be associated with a variety of chromosome abnormalities, including Turner syndrome and trisomy 21.

A distant family history of spina bifida does not significantly increase the risk for neural tube defects in this pregnancy, nor does it increase the risk for chromosome abnormalities.

If one parent has a translocation, there will be an increased risk for both early miscarriages and abnormal liveborn infants.

A paternal robertsonian translocation is associated with a 2% risk that an ongoing pregnancy will be chromosomally abnormal.

8. The answer is D *[II C 2–4].*
The proband's father must carry the abnormal gene despite the fact that he does not yet manifest the disease because he has an affected daughter. There is thus a 1/2 chance that he has transmitted the abnormal gene to his son, the proband. The fact that the proband is clinically unaffected at age 30 makes it a little less likely that he carries the abnormal gene. A Bayesian calculation is necessary to determine the risk. This is shown in the probability space below.

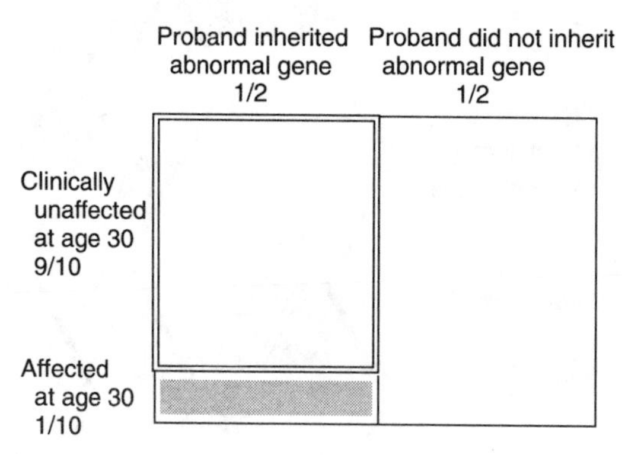

The same solution can be achieved with the probability tree.

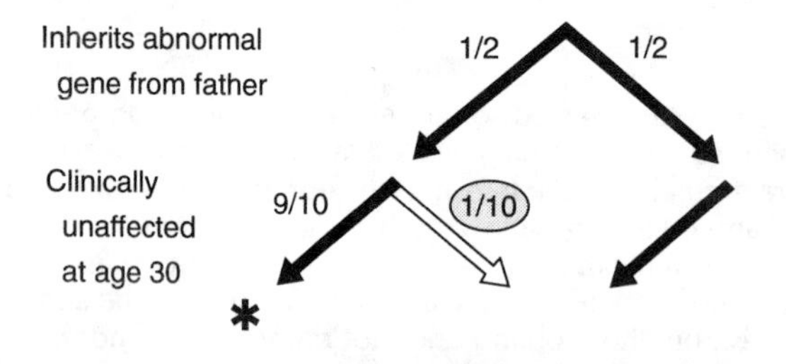

Either the proband inherited the abnormal gene and does not yet express it or he did not inherit the abnormal gene. The remaining possibility, that he inherited the abnormal gene and already expresses it, did not occur and must be removed from the calculation. The chance that the proband carries the gene for Huntington's chorea but does not express it at age 30 is

$$P = \frac{(1/2 \times 9/10)}{(1/2 \times 9/10) + 1/2}$$

$$P = \frac{9/20}{9/20 + 10/20}$$

$$P = \frac{9/20}{19/20}$$

$$P = 9/19$$

9–10. The answers are: 9-C *[II A, C 2–4],* **10-D** *[II A, B, C 2–4].*
The woman's father is an obligate carrier of the abnormal gene for retinoblastoma. She had a 1/2 chance of inheriting the abnormal gene from him, but she shows no evidence of it. This makes it much less likely that she carries the abnormal gene but does not exclude this possibility because 20% of carriers of the retinoblastoma gene do not develop eye tumors. The Bayesian calculation that is necessary is illustrated in the following probability tree.

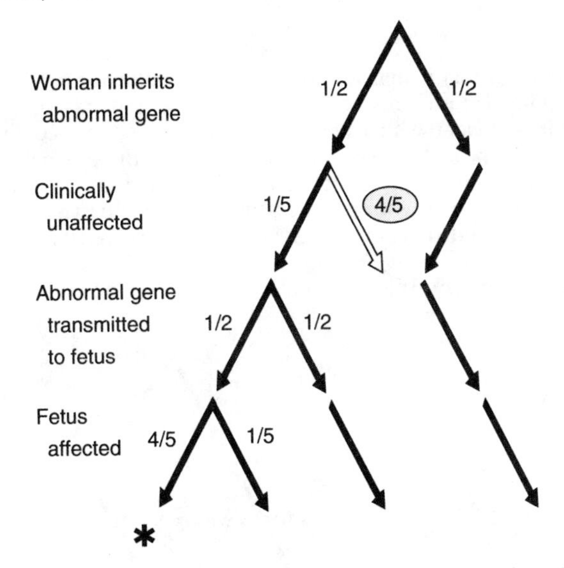

Notice that if the woman is a carrier of the abnormal gene, is unaffected, and transmits the abnormal gene to her fetus, there is only an 80% chance that the fetus will be affected by the disease because of incomplete penetrance. The other possibilities that must be considered are: (1) the woman did not inherit the abnormal gene from her father, (2) she did inherit the abnormal gene despite not showing any sign of it but did not transmit it to her fetus, and (3) she did inherit the abnormal gene despite not showing any sign of it and did transmit it to her fetus who also will not show any sign of it.

The possibility that the woman inherited the abnormal gene from her father and developed retinoblastoma is contrary to the information known in this family and must be excluded from the calculation.

The chance that this woman's fetus will develop retinoblastoma is

$$P = \frac{(1/2 \times 1/5 \times 1/2 \times 4/5)}{(1/2 \times 1/5 \times 1/2 \times 4/5) + (1/2 \times 1/5 \times 1/2 \times 1/5) + (1/2 \times 1/5 \times 1/2) + 1/2}$$

$$P = \frac{4/100}{4/100 + 1/100 + 1/20 + 1/2}$$

$$P = \frac{4/100}{4/100 + 1/100 + 5/100 + 50/100}$$

$$P = \frac{4/100}{60/100}$$

$$P = 4/60 = 1/15$$

In order to interpret the marker data, the physician must determine whether the family is informative for linkage, and, if so, determine the phase. The pedigree can be redrawn as follows to show the genotypes.

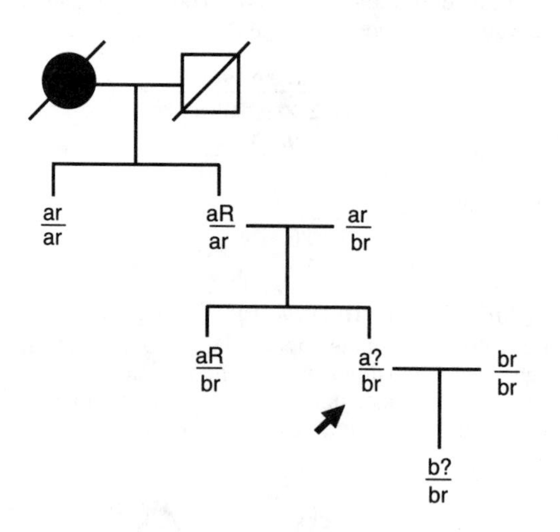

Note that the linked marker is uninformative with respect to whether or not the proband inherited the abnormal gene from her father. He is homozygous for the *a* allele at the marker locus so one cannot determine from the information provided which allele at either the marker locus or the retinoblastoma locus he transmitted to his daughter. The fact that she does not exhibit the disease makes it more likely (but not certain) that she inherited the normal allele at the retinoblastoma locus.

The linked marker is informative with respect to the fetus. The proband transmitted to the fetus the allele at the marker locus that she received from her mother, not the one received from the proband's father with retinoblastoma.

A Bayesian calculation is necessary. This is illustrated in the following probability tree.

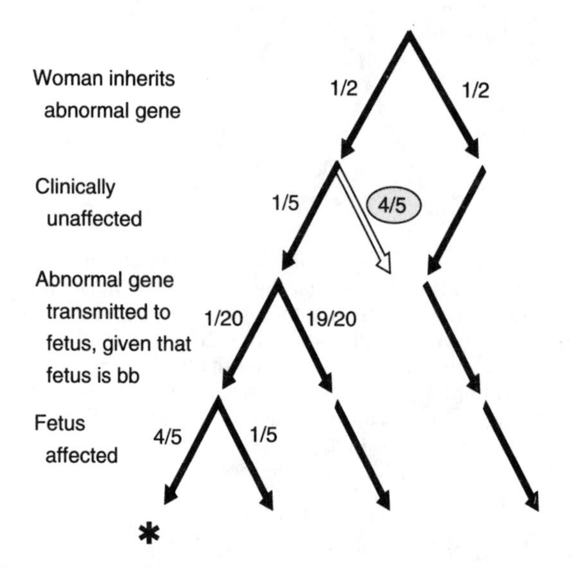

Note that the probability of 1/2 that the proband transmits the abnormal gene to her fetus in the probability tree used in the previous question has been replaced with a probability of 1/20 (5%) in this instance because it is known that the fetus received the proband's *b* marker allele. The only way that the *b* allele at the marker locus could be associated with the abnormal allele at the retinoblastoma locus (assuming the proband inherited it) would be for a cross over event to have occurred. The probability of such an event is 1/20 between two loci that are 5 cM apart.

The fetus would develop hereditary retinoblastoma only if the proband inherits her father's abnormal gene, she does not express it, she transmits it to her fetus with the *b* allele at the marker locus by a cross over event, and expression occurs in the fetus. The other possibilities that need to be considered are: (1) the woman inherits the abnormal gene, she does not express it, she passes it on to her fetus with the *b* allele at the marker locus by a cross over event, but expression does not occur in the fetus; (2) the woman inherits the abnormal gene, she does not express it, but she does not pass it on to her fetus with the *b* allele at the marker locus (i.e., no cross over event occurs between the two loci); and (3) the woman does not inherit the abnormal allele at the retinoblastoma locus from her father.

The other possibility (i.e., that the woman inherits the abnormal allele at the retinoblastoma locus and does express it) is contrary to the information provided in this case and must, therefore, be excluded from the calculation.

The chance that this woman's fetus will develop retinoblastoma, given that it has been found to be *bb* at the marker locus is

$$P = \frac{(1/2 \times 1/5 \times 1/20 \times 4/5)}{(1/2 \times 1/5 \times 1/20 \times 4/5) + (1/2 \times 1/5 \times 1/20 \times 1/5) + (1/2 \times 1/5 \times 19/20) + 1/2}$$

$$P = \frac{4/1000}{4/1000 + 1/1000 + 19/200 + 1/2}$$

$$P = \frac{4/1000}{4/1000 + 1/1000 + 95/1000 + 500/1000}$$

$$P = \frac{4/1000}{600/1000}$$

$$P = 4/600 = 1/150$$

11–15. The answers are: 11-A, 12-C, 13-B, 14-E, 15-D *[II A 4].*
Pedigree 11 shows five affected males, male-to-male transmission, and no affected females, although the pedigree is not large. Y-linked inheritance is a tempting possibility, but all males would be affected, and this is not seen in this pedigree. Therefore, although no females are affected in this particular family, the most likely inheritance is autosomal dominant.

Pedigree 12 shows four affected males, no affected females, and all males are related through unaffected females. This pedigree would be most consistent with X-linked recessive inheritance. If females were also affected and there was no male-to-male transmission, X-linked dominant inheritance would also have to be considered.

Pedigree 13 has a consanguineous couple with three affected daughters. The parents themselves are marked as unaffected. With this history, females affected via autosomal recessive inheritance is most likely. Another less likely possibility is mitochondrial inheritance, since all children are affected.

Pedigree 14 illustrates a common problem for counseling—the singleton case. The important point is that any form of inheritance is possible, or the child may have a nongenetic etiology to the problem.

Pedigree 15 combines two affected children and recurrent miscarriages. The affected children comprise both sexes and are related through the father of one; therefore, neither X-linked nor autosomal recessive inheritance is likely. An inherited chromosomal rearrangement could explain the pedigree.

16–20. The answers are: 16-A, 17-B, 18-C, 19-D, 20-E *[III C; IV A].*
In the family with an affected child and two similarly affected relatives, X-linked recessive inheritance should be considered. If the affected child is a female, this would be much less likely. Since the family history is remote, as much detail as possible should be obtained regarding the uncles.

Advanced paternal age is associated only with new dominant mutations. If the problem is an individual with a condition not otherwise classified, a physician would lean toward a dominant inheritance with advanced paternal age.

Advanced maternal age is associated with an increasing risk of pregnancies with trisomies 13, 18, 21, and perhaps Klinefelter syndrome. If an infant is born with multiple anomalies to a woman over 35, one of the considerations would be a trisomy.

Recurrent pregnancy losses may have many causes, only some of which are genetic (an inherited translocation, aneuploidy in the losses themselves). Generally, single genes have not been identified to cause such losses.

Consanguinity, especially with relationship of first cousins or closer, is associated with an increased risk of autosomal recessive and multifactorial disorders as well as an increased risk of stillbirth. All are secondary to the fetus inheriting homozygous mutant alleles either for single gene or multi-gene conditions.

21–25. The answers are: 21-D, 22-A, 23-E, 24-B, 25-C *[III B 1, 3–6; Table 10-3].*
The woman who has a child with normal chromosomes but multiple structural malformations should be offered level II ultrasound, looking specifically for those malformations seen in the previous infant. Offering fetal karyotyping is not appropriate, since the first infant did not have a chromosome abnormality.

With a family history of a genetic disorder diagnosable by molecular methods, the most efficient method of prenatal diagnosis is chorionic villus sampling. Not only is the sampling done at an earlier time, but also more tissue is available for diagnostic purposes or for faster culture.

A 19-week ultrasound abnormality requires a rapid answer to aid the parents in decision-making. In many areas, the option of pregnancy termination is available only until 20 weeks of gestation. Therefore, the most rapid method would be fetal blood sampling.

The woman with no identified risks but with anxiety could be offered maternal serum α-fetoprotein screening. She should, of course, be aware of the screening nature of this test and what will happen if the result is abnormal.

The 36-year-old woman is at increased risk for chromosome aneuploidies and should be offered either chorionic villus sampling or amniocentesis. However, as she is already at 14 weeks of gestation, the amniocentesis is more appropriate.

11
Clinical Genetics: Part II

J. M. Friedman and Barbara McGillivray

I. GENETIC SCREENING

A. **Genetic screening** is the identification of individuals with a genetic disease, a genetic predisposition to a disease, or a genotype that puts them at increased risk of having a child with a genetic disease.

1. **The purpose of identifying individuals with a genetic disease** is usually **to permit management** of the disease or the complications of the disease in a more effective manner than would otherwise be possible. Examples include:
   a. **Screening for phenylketonuria (PKU) in newborn infants.** Recognition of infants with this condition within the first few weeks of life permits institution of dietary treatment that is effective in preventing severe mental retardation.
   b. **Screening by amniocentesis or chorionic villus sampling for Down syndrome** (trisomy 21) among the fetuses of women who are 35 years of age or older (see IV D 2 c; Ch 10 III B 3, 4). In this case, however, no effective treatment of the condition is possible. Women who are carrying an affected fetus are given the option of terminating the pregnancy or continuing it to term and preparing for the birth of an abnormal child.

2. **The purpose of identifying individuals with a genetic predisposition to a disease** is to enable them to institute measures that will **prevent or delay development** of the disease or enable them to deal with the disease more effectively if it does develop. Examples include:
   a. **Heterozygous carriers of familial hypercholesterolemia (FH)** are at increased risk of developing premature coronary artery disease. While individuals with this genotype cannot alter their genetic endowment, they can alter other factors in their lives that may also predispose them to developing coronary artery disease. Controllable factors include cigarette smoking, diet, and exercise.
   b. Individuals carrying the gene for **Huntington's disease (HD),** an autosomal dominant neurodegenerative disorder with usual onset after 40 years of age, can often be identified prior to the development of symptoms by **linkage studies within a family**. There is no known method of preventing HD from developing or of altering its inexorably progressive course, but many at-risk individuals find presymptomatic genetic screening valuable because it helps them plan their lives more appropriately.

3. **The purpose of identifying individuals at increased risk of having children with a serious genetic disease** is to permit them to take advantage of **reproductive options** that may prevent the birth of affected children. Examples include:
   a. Detection of heterozygous carriers for **Tay-Sachs disease** among Ashkenazi Jewish populations (see Ch 5 II B 2 b (1))
      (1) Identification in this population of couples in which **both partners are heterozygous carriers** of Tay-Sachs disease permits them to **avoid having affected children**.
      (2) Such couples may choose to do so by not having children of their own, by having children via artificial insemination from an unrelated donor, or by prenatal diagnosis with abortion of affected pregnancies.
   b. Use of **DNA testing** to **identify women who carry the gene for Duchenne muscular dystrophy (DMD)** within the family of an affected boy provides another example [see Ch 5 II A 3 a (4)]. Recognition of women who are carriers permits them to avoid having sons with this disease.

B. Criteria for screening. There are thousands of genetic diseases, but screening is used routinely for only a few. Genetic screening is usually undertaken **only when certain conditions relating to the disease, the sceening test, and the system for implementing them are met.**

1. Characteristics of the disease

a. The disease should be **relatively frequent** within the population screened. The more frequent the disease, the more effective will be the screening program.

(1) Some diseases are not common enough to justify screening within the general population but are frequent enough **within an easily identifiable subgroup.** Examples include Tay-Sachs disease among Ashkenazi Jews, sickle cell disease among blacks, and thalassemia among Southeast Asians (see Ch 10 III C 2).

(2) In many uncommon diseases, genetic screening can only be justified **within families in which an affected individual has been born.** DMD is such a condition.

b. The disease must produce a **severe impairment or death.** Screening for a trivial genetic condition such as postminimal polydactyly, which produces a tiny extra digit, cannot be justified because it can be safely and inexpensively treated and produces no long-term morbidity.

c. Some beneficial intervention must be possible if the condition is recognized. This intervention is usually an effective treatment or prevention strategy.

2. Characteristics of the test. An appropriate test must exist that is **capable of identifying people who have the relevant genotype.** Appropriate tests should be **highly sensitive, highly specific, and relatively inexpensive** to perform.

a. Sensitivity is defined as **the proportion of individuals affected with the disease or genotype** in question **who have a positive test.**

(1) False negatives are produced when individuals who have the disease or genotype have a negative test.

(2) A **good test** will have a **very low frequency of false negatives.**

b. Specificity is defined as **the proportion of those who are not affected with the disease or genotype** in question **who have a negative test.**

(1) False positives are produced when individuals who do not have the disease or genotype have a positive test.

(2) A **good test** will produce **few false positives.**

c. Expense

(1) Tests applied to large populations are usually much less expensive to perform if **automation** is possible.

(2) Economy can often be achieved by **combining more than one screening test.** For example, newborn screenings for PKU and congenital hypothyroidism (which is usually not a monogenic disorder) are often done together to decrease costs.

3. Characteristics of the screening system. Effective use of genetic screening requires use of an appropriate test within the context of a **well-organized and comprehensive program for dealing with abnormal results.**

a. There must be **prompt initial testing and follow-up.** A mechanism must exist to assure that abnormal results are acted upon quickly and appropriately.

b. An **ongoing mechanism for determining effectiveness** of the test must be instituted. Although a test may be theoretically valuable, its usefulness must be demonstrated in practice and must be continually reassessed in the face of changing technology and demographics.

c. Education regarding the nature, usefulness, and limitations of the screening must be provided to physicians and other health care professionals. **Public knowledge** regarding the screening should be increased, particularly within the groups for whom the screening is intended.

d. The entire genetic screening program should have a **benefit that exceeds its cost.** There should, at the very least, be a favorable economic benefit-to-cost ratio; social and political costs and benefits must be considered as well.

C. A series of assessments are often involved in genetic screening.

1. Screening tests, to avoid false negatives, frequently produce a substantial number of false-positive results.

a. In such circumstances, all patients who are found to have a positive result on initial screening may require **a more definitive test** to determine whether they are false positives or they do indeed exhibit the condition or genotype of interest.

b. A good example is provided by **maternal serum α-fetoprotein (MSAFP) testing for neural tube defects in pregnancy** (see Ch 10 III B 6).

(1) Most women with abnormal MSAFP screening tests are not carrying a fetus with a neural tube defect (i.e., most positive MSAFP screening tests are false positives).

(2) Additional testing by ultrasound, repeat MSAFP determination, and amniocentesis are required to distinguish true-positive from false-positive results.

2. Concentrating genetic testing in certain **subpopulations known to have a higher frequency** of the condition than the general population can be thought of as **initial screening** that precedes the test in some cases. For example, as mentioned previously, β-thalassemia is much more common among some ethnic groups (e.g., Italians, Greeks, Southeast Asians) than among others. Performing genetic screening only in high-risk ethnic groups increases the efficiency of testing.

3. Screening is often most effective when information obtained from various procedures is combined to determine the result. For example, the risk of a neural tube defect in the fetus of a woman with an elevated MSAFP depends on the absolute value obtained in the test, the woman's ethnic origin, the gestational age, her weight, whether or not she is carrying twins, and whether or not she has diabetes. All of these factors can be included in a single-risk estimate.

D. Ethical and legal concerns regarding genetic screening (see IV)

1. Even though genetic screening programs are developed to benefit patients and families, they sometimes have a **negative impact** on screened populations.

a. For example, people who have been found to have a certain disease or genotype have sometimes been **stigmatized** so that they are unable to obtain employment or insurance.

b. Education, both professional and public, is the most important tool in preventing stigmatization.

2. Genetic screening may be mandated by legislation, which can raise issues of personal conscience, privacy, and consent.

E. Future perspective: the genetic profile

1. Recent advances in molecular genetics have **increased the ability to identify individuals who have genotypes that produce disease** (e.g., in Huntington's chorea), **have genetic predispositions to disease** (e.g., in hyperlipidemias), **or are at high risk for having a child with a given disease** (e.g., detection of heterozygous carriers for cystic fibrosis).

2. Determining the diseases that each individual is likely to develop or pass on to his or her children should be possible if the rate of advances in genetic research continues.

a. Such knowledge would enable patients to:

(1) Alter their lives to reduce nongenetic risk factors of particular relevance and thereby forestall the diseases to which they are genetically predisposed

(2) Take advantage of all available reproductive options if potential offspring carry genetic risks

b. Privacy. Appropriate safeguards will be necessary to make certain that knowledge of a person's genetic endowment is maintained with strictest confidentiality and does not result in discriminatory practices against him or her.

3. Future use of genetic screening. When a sufficiently large number of genotypes can be identified, **every individual will be found to carry some disease-predisposing traits**. Thus, genetic screening will be used to determine the individual genetic predispositions of every person rather than to identify "normal" or "abnormal" people.

II. TREATMENT OF GENETIC DISEASE

A. Genetic disease is treatable. Many people assume that because a disease is determined completely or in large part by genetic factors that it cannot be treated. This is not true.

1. It is true that **genetic diseases generally cannot be cured,** but this is also the case for most other diseases as well.

2. **Diagnostic precision** is often important in planning and evaluating therapeutic interventions in genetic diseases. **Genetic heterogeneity** (see Ch 3 I B 2 f) **is common.**
   a. Different genetic lesions, and thus different pathogenic mechanisms, may be involved in clinically similar conditions.
   b. Effective therapy often requires knowing precisely what gene or protein is altered in an affected patient.
   c. For example, hyperphenylalaninemia, which leads to consequent mental retardation, may be caused by a deficiency of the enzyme phenylalanine hydroxylase (classic PKU) or by certain abnormalities of biopterin metabolism. PKU can be treated successfully by dietary restriction of phenylalanine; abnormal biopterin metabolism cannot.

3. Even within a single genetic entity, there may be substantial **variable expressivity** (see Ch 3 I B 2 c). For example, some patients with neurofibromatosis develop severe complications such as scoliosis, brain tumors, or sarcoma, but most do not. Such variability must be considered in evaluating therapy.

4. Treatment of genetic diseases, particularly if based upon incomplete understanding of pathogenesis, is often associated with **unexpected problems.**
   a. For example, although appropriate therapy can save patients with hereditary retinoblastoma from death from retinal malignancy within the first years of life, these patients may develop osteosarcoma later.
   b. Similarly, early dietary treatment of galactosemia prevents the life-threatening metabolic derangements that occur in affected infants, but "successfully" treated patients often manifest learning disabilities and ovarian failure.

N·B

5. In genetic disease, as in medicine in general, **prevention is better than treatment**. Prevention of genetic disease may sometimes be accomplished by genetic screening (see I A), genetic counseling (see Ch 10 IV), or prenatal diagnosis with elective termination of affected pregnancies.

B. **Amelioration of the clinical phenotype** is the most commonly used method of treatment for genetic disease and is **applicable to all types of genetic disease,** including chromosomal, monogenic, and multifactorial conditions.

1. **Symptomatic treatment** does not require knowledge of either the pathogenesis or the specific nature of the genetic defects, although such knowledge may permit the design of more appropriate and effective treatments.

2. **Examples are numerous** and include the following:
   a. **Surgical correction** of congenital anomalies such as cleft lip, congenital heart disease, or polydactyly
   b. **Special education** in conditions like Down syndrome or fragile X syndrome
   c. **Physical therapy** for congenital hip dislocation or congenital contractural arachnodactyly
   d. **Use of medication,** such as β-blockers, to prevent aortic dissection in Marfan syndrome (see Ch 3 I B 3 b)
   e. **Avoidance of environmental exposures** that trigger a genetic disease [e.g., avoidance of drugs that induce hemolysis in glucose-6-phosphate dehydrogenase (G6PD) deficiency (see Ch 7 III B 1)]

C. **Treatment by amelioration of metabolic abnormalities** is often possible if the pathogenesis of a disease is understood, at least in part, at the biochemical level.

1. Such therapy is restricted almost exclusively to **monogenic disorders** because the necessary pathogenic understanding is not available for most chromosome abnormalities or multifactorial diseases.

2. **Several different approaches** may be employed, depending on the nature of the disease.
   a. **Dietary restriction** can be used to remove the substrate for a deficient enzyme if accumulation of that substrate causes pathology.
      (1) One example is **PKU,** in which limitation of phenylalanine in the diet prevents development of mental retardation.
      (2) In **galactosemia,** removal of milk and dairy products from the diet prevents the metabolic derangements by eliminating galactose, the sugar that cannot be properly metabolized by these patients.

b. Diversion. If the pathology is produced by accumulation of a toxic metabolite, improvement may be achieved by diverting the substance to alternate metabolic pathways. For example, administration of **sodium benzoate to patients with hyperammonemia** due to ornithine transcarbamoylase deficiency **helps eliminate nitrogen through an alternative metabolic pathway**.

c. Removal of the abnormal material may be effective if the pathology is caused by accumulation of a toxic metabolite.

(1) In **hemochromatosis,** removal of excess iron by phlebotomy is useful in preventing progressive organ damage caused by iron deposition.

(2) Similarly, in **Wilson's disease,** removal of excess copper by administration of penicillamine is useful in preventing the neurologic and hepatic damage caused by accumulation of this metal.

d. Activating the defective metabolic pathway can treat some conditions.

(1) **Homocystinuria** is caused by a defect in cystathionine β-synthase, which requires a pyridoxine cofactor. In some patients, the activity of the defective enzyme can be increased substantially by administering very large amounts of pyridoxine (vitamin B_6).

(2) **FH** is caused by abnormalities of cellular receptors for low-density lipoprotein (LDL) cholesterol. Inhibition of HMG-CoA reductase—the cellular enzyme that makes cholesterol—by the drug lovastatin causes cells with LDL receptor deficiency to increase their production of receptors to maintain adequate intracellular cholesterol pools.

D. Replacement therapy

1. Replacement of the deficient substance may be a possible mode of treatment if the pathology is produced by the absence of a product.

a. Several **monogenic disorders** benefit from replacement therapy.

(1) In **classic hemophilia,** replacement of the deficient clotting protein—factor VIII—corrects the bleeding disorder.

(2) Prevention of pulmonary disease in **α_1-antitrypsin deficiency** may be possible by administering a purified protein preparation to affected patients.

b. Some multifactorial conditions also benefit from replacement therapy. For example, type I (insulin-dependent) diabetes mellitus is treated by administration of insulin.

2. Organ transplantation can be considered a method of replacement therapy as well as a method of gene therapy. The donor organ provides the product that is abnormal in the recipient. The donor organ functions because its cells contain the normal gene (see F). A number of genetic diseases have been treated by organ transplantation.

E. Modulation of gene expression has been proposed as a method for treating some monogenic disorders.

1. The approach is to alter the pathogenic process by **activation of a normal gene that has been turned off or by turning off a gene that is active**.

2. For example, patients who have sickle cell anemia and also a condition that leads to abnormal production of fetal hemoglobin after infancy have a much less severe disease than patients with sickle cell anemia alone. This suggests that reactivation of the normal fetal globin gene would be effective therapy in patients with sickle cell disease.

F. Gene therapy involves correction of a genetic defect by altering the genotype.

1. Transplantation of cells, tissues, or organs can be considered to be a form of gene therapy because the cells transferred to the recipient continue to function on the basis of the donor genome.

a. Transplantation **can be used to replace a protein** that is defective or absent in the recipient. This has been done, for example, with the following:

(1) Bone marrow transplantation in severe combined immunodeficiency disease due to adenosine deaminase deficiency

(2) Liver transplantation in homozygous FH.

b. Alternatively, transplantation **can be used to replace an organ that has been damaged** by a genetic disease. This approach has been used to treat renal failure in adult polycystic kidney disease (PKD) and cardiac failure in hereditary cardiomyopathy.

c. In some cases, transplantation serves **to replace both a defective protein and a damaged organ**. This occurs, for example, with the use of liver transplantation for treatment of hepatic failure in α_1-antitrypsin deficiency.

d. There are several **problems with transplantation** that make it a less than ideal method for transferring genetic material.

(1) **Availability of suitable tissue for transplantation is limited,** particularly for organs such as livers and hearts.

(2) Preparation and recovery from transplantation as well as the procedure itself are associated with **significant morbidity and mortality**.

(3) **Tissue rejection or graft-versus-host (GVH) disease** is an ongoing risk; immunosuppressive therapy must be continued throughout life.

(4) Transplantation is **expensive**.

2. **Somatic gene therapy** involves the addition, alteration, or replacement of an abnormal gene in an affected patient.

a. Somatic gene therapy **does not affect the germline,** since only certain somatic tissues or cells of a patient are altered. The change is not passed on to the children of a treated individual.

b. **Several approaches to somatic gene therapy** are being considered.

(1) **Addition of an appropriately functioning normal gene** could be used to treat an individual. This approach might be applicable in diseases such as PKU or galactosemia, in which a person lacks a critical metabolic enzyme.

(2) **Replacement of an abnormal gene by a normal one** may be required in conditions in which a functional alteration produced by the abnormal gene product is pathogenic. This would be the case in diseases such as Marfan syndrome (see Ch 3 I B 3 b) or osteogenesis imperfecta (see Ch 7 II B 3 b), in which an abnormal protein disrupts the structural integrity of tissues even in the presence of a normal gene.

(3) **Correction of an abnormal gene** by targeted mutagenesis or similar manipulations is another potential method for treating genetic diseases in which function of the abnormal gene is pathogenic.

c. **Several technical difficulties** limit the current use of somatic gene therapy, including:

(1) **Transfer of the therapeutic gene to appropriate target tissues.** An effective construct containing the normal gene must be prepared and inserted into the patient's cells.

(a) **Incorporation of the gene.** Insertion of the gene may be accomplished by incorporating it into a retrovirus, plasmid, or other appropriate vector and treating the host cells to induce uptake.

(b) **Site of therapy.** For some cell types (e.g., marrow), the patient's cells are removed, the therapeutic gene is inserted in them, and the cells are returned to the patient. For other types of target cells (e.g., brain cells), uptake of the therapeutic DNA will probably need to be induced in situ.

(c) **Large genes.** Preparation and transfer of a therapeutic gene construct is especially difficult for very large genes, such as those involved in DMD or neurofibromatosis.

(2) **Selection of appropriate target cells**

(a) In some cases, the **cell in which a gene normally functions** can be used as a target cell. This might be the case, for example, in thalassemia, in which therapy would be directed to marrow cells.

(b) In other instances, it might be sufficient to use a more accessible or **convenient target cell**. For example, in a disease such as α_1-antitrypsin deficiency, in which normal protein circulating in the blood is expected to have a beneficial effect, expression of a therapeutic gene might be induced ectopically in marrow cells or implants of skin fibroblasts.

(c) Neurons and other **target cells that do not normally divide** in adults present a particular problem because some transfer vectors are only taken up by dividing cells.

(3) **Induction of appropriate expression of the therapeutic gene in target tissues**

(a) The amount of expression needed to correct the genetic defect will vary in different diseases. In some cases, only a small amount of activity is necessary; in other cases, near normal activity may be required.

(b) Overexpression of some genes may be deleterious.

(4) **Proper regulation of the therapeutic gene.** The necessity for proper regulation is also likely to vary in different diseases. Some genes will need to function within normal regulatory constraints to provide therapeutic benefit; for others, unregulated expression may be adequate.

d. Theoretical analyses of somatic gene therapy have identified many potential risks, but until substantial human experience is obtained, it will not be known which, if any, of these risks are of practical importance. The **theoretical risks** include the following.

(1) **Induction of neoplasia** within the target tissue by action of the vector, activation of a proto-oncogene, or damage to a tumor suppressor gene (see Ch 9 III C 2) is of most concern.

(2) **Damage to other target cell genes** seems unlikely to be a serious problem because the abnormality produced should be restricted to a small clone of cells.

(3) **Alteration of the germline** with risks to children of the treated patient should not occur, as long as therapeutic intervention is limited to somatic cells.

e. Clinical applications of somatic gene therapy **are likely to be limited,** at least in the next few years, even though it is of great interest to scientific researchers. Somatic gene therapy is initially being applied to **monogenic disorders** that exhibit certain particularly favorable characteristics, which include the following.

(1) The mutant gene has been identified and characterized.

(2) A complementary DNA (cDNA) clone of the normal gene is available.

(3) The disease produces very serious morbidity or mortality.

(4) No adequate alternative therapy is available.

(5) Knowledge of disease pathogenesis is sufficient to determine that gene therapy is likely to be a successful intervention.

(6) An appropriate target cell and a suitable method of transfer of the gene are available.

(7) Studies in cultured cells and experimental animals indicate that the specific combination of genetic construct, target cell, and insertion strategy being attempted are likely to be safe and effective.

f. Ethical concerns about gene therapy. Most concerns surround germline gene therapy (see II F 3), which can affect future generations, rather than somatic gene therapy, which does not have an effect on future generations. Some of the concerns regarding the ethics of somatic gene therapy include:

(1) **Safety.** Somatic gene therapies must be subjected to critical assessments of safety and efficacy, as must all new therapies.

(2) **Risks** that may be related to somatic gene therapy (e.g., risk of causing cancer) are not fundamentally different from the risks that may be associated with certain other novel treatments, such as new vaccines or antineoplastic agents. The magnitude of these risks must be evaluated in a similar careful manner.

(3) **Future effects.** Although somatic gene therapies are designed to avoid germline transfer of altered genes, inadvertent involvement of the germline is a possibility. Even the risk of inducing genetic damage that could be transmitted to the offspring is not unique to gene therapy. Similar concerns arise with some forms of cancer radiotherapy and chemotherapy.

3. Germline gene therapy involves the addition, alteration, or replacement of an abnormal gene in such a manner that the gamete-producing cells are also modified.

a. Examples of germline therapy would include the replacement of an abnormal gene in a zygote or the transfer of a therapeutic gene to a patient in a manner that would permit incorporation by the gametes.

b. Safety. Germline gene therapy poses risks and raises important ethical issues beyond those inherent in somatic gene therapy because damage may be caused to future generations.

c. Future. Most geneticists do not envision any role for germline gene therapy in humans.

4. Future of gene therapy. Gene therapy requires substantial knowledge of the nature of the abnormal gene and its function. This will probably limit the applications of gene therapy for the foreseeable future.

a. Gene therapy is likely to be restricted primarily to monogenic diseases in which sufficient pathogenic knowledge is available.

b. Poor understanding of the molecular pathogenesis of multifactorial diseases and chromosome abnormalities creates a substantial impediment to the application of gene therapy (except for organ transplantation).

III. ETHNIC FACTORS IN GENETIC DISEASE

A. Genetic differences. Every human individual is genetically different from every other (with the exception of identical twins).

1. **Polymorphism is the rule,** particularly if one looks at the level of DNA variation (see Ch 4 I).
2. The concept of the **"wild type,"** which is frequently used in genetic studies of experimental organisms, is **not applicable to humans**. The wild type of an organism is one that is homozygous for the normal allele at all loci. Studies in humans indicate that virtually every person carries some genes that could cause or predispose to disease under the appropriate circumstances.

B. Human populations differ genetically from each other.

1. **Physical differences.** Some of these differences are obvious physical traits that humans have traditionally used to define ethnic groups. Examples include body size, skin and hair pigmentation, and facial structure.
2. **Medically important differences.** Many diseases have different frequencies in different populations.
   a. Some examples of **ethnicity of disease** are provided in Table 11-1.
   b. Differences in disease frequencies among populations are most often **apparent for multifactorial and recessive conditions.**
   c. Differences in disease frequencies among populations may also be **due to differences in diet or social practices.**
3. **Natural selection** is responsible for some genetic differences among populations. Examples of natural selection include the following.
   a. **Light skin color** among Northern Europeans probably evolved as an adaptation to a climate in which sunlight was less readily available to form vitamin D in the skin.
   b. The genes for diseases such as sickle cell anemia, G6PD deficiency, and thalassemia have become common in some populations because these genes confer **resistance to malaria** (see Ch 3 II C 2).

Table 11-1. Ethnic Distribution of Disease

Disease	Populations with High Disease Frequency
α-Thalassemia	Southeast Asians
	Africans
Anencephaly	Irish
Cleft lip and cleft palate	Japanese
	Native Americans
	Mexicans
Congenital nephrosis	Finns
Cystic fibrosis	Northern Europeans
Diabetes mellitus	Ashkenazi Jews
	Polynesians
	Native Americans
	Mexicans
Glucose-6-phosphate dehydrogenase deficiency	Italians
	Greeks
	Africans
	Chinese
Hypertension	Africans
Nasopharyngeal cancer	Chinese
β-Thalassemia	Italians
	Southeast Asians
	Greeks
	Africans
Sickle cell anemia	Africans
Tay-Sachs disease	Ashkenazi Jews

4. **Genetic differences among populations arise because mating is not random** within the human species as a whole.
   a. **Genetic isolation.** Populations in which members only mate with other members of the same group are said to be genetically isolated. Genetic isolation may arise because of **geographic factors** or **religious or cultural practices**.
   b. **Small, genetically isolated populations** encounter phenomena that create genetic differences and can be important determinants of gene frequencies.
      (1) **Genetic drift** refers to changes in gene frequencies that occur by chance.
      (2) **Founder effects** (see Ch 3 II C 3). If by chance a particular abnormal gene was present among the small group of founders (i.e., if the initial frequency of the gene in the population was high), the frequency tends to remain high unless the gene is strongly selected against.

5. **Differences in gene frequencies among human populations are usually relative, not absolute.**
   a. Some populations have higher frequencies and some have lower frequencies of various genes, but there are very few genes that are unique to one particular population.
   b. The number of generations humankind has gone through as a species is relatively small. Differences in gene frequencies accumulate with time in genetically isolated populations.
   c. Not many human populations have remained genetically isolated for more than a few generations. Mating between previously isolated populations often occurs as a result of migration, conquest of one group by another, or changes in cultural practices.

IV. ETHICAL ISSUES OF CLINICAL GENETICS

A. Introduction. The goal of genetic counseling is to provide both information and support, so that couples and families can plan decisions about themselves according to their own values. Generally, this is accomplished by **nondirective counseling,** which provides information and empathy but does not direct a course of action for the couple. At times, values may conflict and dilemmas may arise when one can find moral considerations for taking either of two opposing courses of action.

1. **Definition of terms**
   a. **Morality** is behavior according to customs or codes.
   b. **Ethics** is the discipline of reflecting upon what does, and what should, go into formulating morality or what considerations go into sound moral judgement. The considerations should include:
      (1) Analyzing the situation, both in terms of consequences and alternatives
      (2) Weighing the alternatives

2. **Models in common usage.** Ethical theories are bodies of moral principles and rules that justify particular actions. Theories that are still in the process of development should be used as broad models.
   a. **Utilitarianism** is a set of theories based on consequences where the aim is to promote the welfare and protect the interests of most persons.
      (1) Actions may be judged in proportion to their ability to **maximize happiness and minimize pain** (i.e., actions are judged in retrospect). Four criteria are given.
         (a) The obligation is to maximize the good (**utility**).
         (b) Attempt to standardize goodness to reach a **common good,** not an individual personal good (**value**).
         (c) Actions are right or wrong in terms of their consequences (**consequentialism**).
         (d) Consequences affecting all parties should receive equal consideration (**universalism**).
      (2) The **obligation of utilitarianism** is to **benefit the community,** rather than the individual (i.e., individuals are sacrificed to a wider universal good). Therefore, the rights of individuals may be overridden, making utilitarianism inconsistent with the value of **autonomy**. Also, it may not be possible to predict the consequences of decisions.
   b. **Deontology (the theory of duties)** proposes that acts are right or wrong and should not be justified only by their consequences. This model emphasizes truth telling, fidelity to promises, and states that people have a duty to act in certain ways. The duties include the following.
      (1) **Always preserve life** (duty is against suicide or euthanasia).

(2) A **physician's duty is to an individual patient** rather than maximizing the good to others. The doctor–patient relationship has an independent moral significance.

(3) **Acts must be consistently universal.** A procedure performed on one patient with a specific medical problem must be available to every patient who is in that same situation.

c. **Other theories** may be based on **virtues** (i.e., the integrity of the individual and the suppression of self-interest are stressed) or on **rights** (i.e., morality itself is designed to protect the dignity and rights of the individual).

d. **In clinical genetics,** physicians may be concerned with the rights of the fetus or the rights of the malformed newborn. Physicians must recognize non-Western values in patients and their families and also recognize that there may be a difference in female morality (i.e., related more to care and compassion) and male morality (i.e., justice, duty, and rights).

B. Ethical principles

1. **Beneficence** is a duty to confer benefits and to prevent and remove harms. Biomedical research, preventive medicine, and public health interventions could be said to benefit society as a whole. In clinical genetics, somatic gene therapy could benefit by providing nonheritable cures for single gene disorders.

2. **Autonomy** is the principle of being one's own person and choosing one's own course of action, including medical treatment. The individual should not be constrained by others.
   a. The principle **may not apply to those who are incompetent** (i.e., children, those with psychiatric disorders, or those who are unconscious), and **it assumes that the individual has been fully informed**.
   b. An **example of autonomy** is a woman who is counseled about prenatal diagnosis, and who chooses for herself whether to undergo an available procedure.

3. **Justice** is the principle that identical cases are treated in the same manner. **Distributive justice** refers to the fair distribution of both benefits and burdens in society and begins to occur when there are scarcities. For example, if there are limited financial resources to cover all proposed medical programs, a center may decide to offer amniocentesis only to women over a certain age. This decision treats those with the greater risk equally but does not make the procedure available to those with the lesser risk.

4. **Nonmaleficence** is a duty derived from that of beneficence and proposes that one ought not to inflict evil or harm. This duty assumes that a physician will be competent. Withholding or withdrawing treatment to an infant or adult may be seen as harmful, except in the case where the individual was judged to be in the dying process.

5. **Veracity**—the duty to tell the truth and refrain from lying to others—implies that the patient has a right to information about himself.
   a. **Nondisclosure** may be considered when information is thought to be harmful to the patient, but health care professionals should disclose what a reasonable patient would want to know and should fulfill the individual patient's need for information.
   b. For example, when researching a new lipid-lowering drug, patients should be made aware of all the risks and benefits of such drugs before being asked to participate.

6. **Fidelity** is the duty to keep contracts and promises. The physician has a duty not to neglect patients once accepting them into care. If a couple presents for prenatal diagnosis and the fetus is found to have a chromosome abnormality, the counselor has a duty to impart the information and continue support, regardless of how difficult it may be to divulge the diagnosis.

C. Making decisions in the presence of ethical dilemmas. As part of providing genetic counseling, the physician may be faced with having to choose between two or more incompatible courses of action and must make a moral judgement to proceed or to be effective with counseling. It may be helpful to have a **framework for such problems**.

1. **Identify the facts,** which include accurate medical information, details about the family, support for the family, and the physician's own biases and values regarding the situation.

2. **Identify the ethical principles involved.** If one course of action is clearly dominant, then the moral problem is solved. If not, the principles should be ordered in terms of their importance. For example, is the patient's autonomy the most important or is the physician's duty of beneficence to the patient overriding? If the choice is not clear, other steps may be necessary.

3. **Compare with similar cases.** If a similar case can be found where one principle was found to be most important, the physician may use that example as a precedent to make a similar decision. To use such comparisons, a paradigm must exist without morally relevant differences from case to case.

4. **Use an ethics committee.** Most hospitals have an ethics committee consisting of representatives from medicine, nursing, social work, philosophy, law, religion, and the lay public. The purpose of these committees is to examine such problems around an ethical framework and help the clinician. Most committees of this type **do not make decisions for the physician; instead they offer advice and support**.
   a. **Case 11-A. Pregnancy affected by abnormal results.** The ethics committee of a large teaching hospital is asked to meet urgently to provide direction on a clinical problem. A 24-year-old woman is pregnant (26 weeks) for the second time. She presented to her family physician only 1 week earlier for her first prenatal visit. The woman was concerned because she lost a brother with spina bifida and severe hydrocephalus. Because of this concern, fetal ultrasound was arranged, and this unfortunately indicated that the fetus had spina bifida.

      The young couple returned 1 week later obviously distraught and demands to terminate the pregnancy. The family physician explained that this would not be possible, at which time the woman tearfully stated she could neither sleep nor eat and would end her life if she could not terminate the pregnancy.

      (1) What principles are important in this moral dilemma?

      (2) Through what process should the committee go to evaluate this situation?
   b. **Explanation of Case 11-A.** The main principles in conflict here are **autonomy** (the woman's choice to discontinue the pregnancy) and **beneficence** (the best course for the fetus is continuation of the pregnancy until judged the most appropriate time to balance treatment of the spina bifida and the prematurity).

      The committee must start with adequate information. Important issues include: (1) prognosis for the fetus—need the help of the local spina bifida treatment group; (2) the couple's understanding of the current treatment and prognosis for neural tube defect; (3) psychiatric assessment and possible treatment of the woman; and (4) exploration of the couple's family situation and supports.

      The committee needs to explore the moral conflicts and the likely outcomes of alternate courses of action. For instance, if labor is induced, the infant will be extremely premature in addition to having spina bifida and may ultimately have induced complications. If the woman can be convinced to continue the pregnancy, she may decompensate further emotionally and feel that she has no choice in the situation.

      What may be helpful would be for the committee to recommend counseling for the couple, meeting with the spina bifida team as well as the physicians managing prematurity, and then the committee meeting again to discuss the couple's request. Often, such requests are made while the patients are still in shock after receiving abnormal results and without having full information or support.

5. **Realize decisions may be wrong and learn from each mistake.** Identifying and making moral decisions should be a continuous process, both as the physician learns to analyze situations and as applied ethics copes with the rapid advances in medicine.

D. Areas of conflict in clinical genetics

1. **Confidentiality and the rights of the family.** Prior to counseling a family, confidential information is gathered regarding medical conditions, relationships, and use of medications. Such information may not be divulged without the permission of its owner (or guardian), even though knowledge of a diagnosis (e.g., HD) may be deemed important for those at risk to know. **The autonomy of the affected individual is in conflict with beneficence to the patient.**
   a. **Case 11-B. Blood samples and family studies.** A family (a widowed mother, her two children, and their spouses) requests counseling and predictive testing for an adult genetic disorder of which the children's father died 5years ago. His father and two sisters were also affected. The sisters are still alive and are living in long-term care facilities. Linkage analysis for the condition is available, and blood samples are requested from all appropriate individuals, including the two affected sisters.

      The extended family has not been close, and, in fact, the husband of one affected sister refuses to allow a blood sample to be taken from his wife. The lab has determined that linkage analysis will not be informative without her blood.

(1) **What are the responsibilities in this situation?**
(2) **Are there ever compelling reasons to insist that a blood sample from an affected individual be made available for family studies?**

b. **Explanation of Case 11-B.** This situation may certainly occur when trying to arrange family studies. The reasons may include feelings of guilt on the part of affected individuals, earlier distancing of the normal individuals from the affected individuals, and a feeling that confidentiality will not be preserved.

Responsibilities of the genetic counseling service are to the couple seeking counseling and, at the same time, to suggest ways to end the stalemate regarding the testing. Outlining predictive testing, suggesting what individuals might do with the information, determining whether specific investigations or monitoring is available to the individual at high risk, or determining whether the information would make a difference in life planning are all items that are essential to the family seeking such testing. Reassuring the family that results are given individually and that there will be no discussion of another's results may alleviate the concerns regarding loss of control over personal information.

It may be argued that an essential blood sample for linkage (or for specific molecular analysis) belongs to the family, but patient autonomy should always be considered first. Attempts at restoring communication within the family may help. There are no quick solutions to this dilemma and no one right course.

2. **Prenatal diagnosis**
   a. If a woman chooses not to have prenatal diagnosis but is at high risk (25%) to have another child with a severe disorder, **maternal autonomy may be in conflict with beneficence for society at large or nonmaleficence to the fetus.**
   b. **Maternal and fetal autonomy may be in conflict** when termination is considered upon diagnosis of a chromosome abnormality.
   c. If a woman has a prenatal diagnosis because she is over 35 years of age and elects to abort a fetus with normal female chromosomes, she may have a legal right to do so, but this morally pits **maternal autonomy against fetal autonomy**.

3. **Molecular or other predictive testing**
   a. Either **direct or indirect molecular methods** (see Ch 5 II, III) may allow predictive testing for single gene conditions (e.g., HD).
   b. **All concerns regarding confidentiality of results** for predictive testing must be considered. These include:
      (1) Not conveying results to other family members without permission
      (2) Not conveying results to outside agencies
      (3) Autonomy of the individual to either refuse to give blood to enable family studies or to release results
      (4) Obtaining truly informed consent for studies
      (5) The use of minors for predictive testing

4. **Gene therapy.** As methods develop to introduce genes into human somatic cells, the public becomes more concerned about long-term implications. While the aim of such therapy is to alleviate the effects of a particular genetic condition, the lay public often wonders if such methods will be used to "further the human race." Both public education and ethical principles are essential.

STUDY QUESTIONS

Directions: Each of the numbered items or incomplete statements in this section is followed by answers or by completions of the statement. Select the **one** lettered answer or completion that is **best** in each case.

1. A good genetic screening test should have which one of the following characteristics?

(A) High sensitivity
(B) A high rate of false-negative results
(C) A high rate of false-positive results
(D) Low specificity
(E) None of the above

2. Of the following statements, which one describes the essential difference between somatic gene therapy and germline gene therapy?

(A) Somatic gene therapy involves the replacement of an abnormal gene by a normal gene; germline gene therapy involves the addition of a normal gene
(B) Somatic gene therapy does not affect the genetic makeup of the children of a treated patient; germline gene therapy may affect future generations
(C) Somatic gene therapy is performed postnatally; germline gene therapy is performed prenatally
(D) Somatic gene therapy requires lifelong immunosuppression; germline gene therapy does not
(E) Somatic gene therapy is only applicable to monogenic diseases; germline gene therapy can be applied to chromosome abnormalities and multifactorial disorders as well

3. Of the following statements regarding treatment of genetic disease by organ or tissue transplantation, which one is correct?

(A) Transplantation is considered to be an experimental therapy when applied to genetic disease
(B) Although transplantation is useful in replacing an organ damaged by genetic disease, the disease usually recurs in the donor organ
(C) Transplantation in a genetic disease may involve removal of a healthy organ
(D) Bone marrow transplantation from a sibling unaffected by a genetic disease to a sibling affected by the disease can be considered to be germline gene therapy
(E) Tissue rejection is rarely a problem in transplantation for genetic disease because the defective genotype prevents graft-versus-host disease

4. Which one of the following statements about genetic differences between human populations is correct?

(A) Genetic differences rarely affect genes for multifactorial diseases
(B) Genetic differences arise because of genetic isolation among groups
(C) Genetic differences are selected against by evolution
(D) Genetic differences are of no medical significance
(E) Genetic differences often are reflected by genes that occur only within a particular group

---

1-A 2-B 3-C 4-B

5. Recently, a clinic has been made aware of a patient choosing to terminate a female fetus on the basis of sex. The genetic counseling staff is upset about this event and decides to give information pertaining only to the normalcy of the karyotype. They will no longer divulge information regarding fetal sex, because they feel this may prevent other women from choosing to terminate a female fetus. Of the following sets of ethical principles, which set contains two conflicting principles?

(A) Fidelity and nonmaleficence
(B) Justice and nonmaleficence
(C) Autonomy and beneficence
(D) Autonomy and veracity
(E) Beneficence and veracity

6. A good genetic screening program should have all of the following characteristics EXCEPT

(A) the overall cost of having the program must be substantially less than the cost of not having the program
(B) the program must include provision for professional education regarding the screening
(C) the diseases for which patients are screened must be lethal or have serious adverse consequences
(D) the diseases for which patients are screened must be curable if appropriately treated
(E) the program must incorporate an effective mechanism to follow up abnormal results

Directions: Each group of items in this section consists of lettered options followed by a set of numbered items. For each item, select the **one** lettered option that is most closely associated with it. Each lettered option may be selected once, more than once, or not at all.

Questions 7–9

Match the following examples or descriptions with the most appropriate kind of genetic disease treatment.

(A) Dietary restriction
(B) Diversion to an alternate metabolic pathway
(C) Removal of an abnormal metabolite
(D) Replacement of an inactive or defective product
(E) None of the above

7. Administration of growth hormone to a child with hereditary dwarfism who lacks pituitary function

8. Administration of phenobarbital to prevent seizures in a child with hyperammonemia, resulting from ornithine transcarbamoylase deficiency

9. Administration of large doses of vitamins in a child with mental retardation due to a chromosome abnormality

Questions 10–14

Match the following definitions with the correct ethical principle.

(A) Autonomy
(B) Beneficence
(C) Veracity
(D) Fidelity
(E) Justice

10. The duty to keep contracts and promises

11. Disclosing what a reasonable patient would want to know

12. Respecting a patient's decision not to seek medical care

13. Not allowing harm to come to a patient

14. Treating like cases alike

5-C 6-D 7-D 8-E 9-E 10-D 11-C 12-A 13-B 14-E

ANSWERS AND EXPLANATIONS

1. The answer is A *[I B 2].*
A good screening test should have high sensitivity so that all or almost all of the target group will be identified. The specificity should also be high so that most of those who are not in the target group will have a negative test. The rate of false positives should be low, and the rate of false negatives should be very low.

2. The answer is B *[II F 2-4].*
The essential difference between germline and somatic gene therapy is that in germline gene therapy the manipulation affects germ cells and can be transmitted to the offspring, while in somatic gene therapy, the germline is unaffected. Gene addition or replacement could, in principle, be done either in somatic tissues or in the germline. Both germline and somatic gene therapy could be performed either postnatally or prenatally. An advantage of any type of gene therapy over tissue transplantation is that recipient cells that have been cured by gene therapy should be completely compatible with the host. Thus, immunosuppression should be unnecessary. It is likely that somatic gene therapy will be applied to monogenic diseases almost exclusively in the foreseeable future because the understanding of pathogenesis in other kinds of genetic diseases is currently insufficient to permit rational gene therapies to be designed. It is unlikely that there will be any need to use germline gene therapy in humans.

3. The answer is C *[II F 1].*
In genetic disease, an organ without apparent damage may have to be replaced with an organ that provides a function lacking in the recipient. For example, removal of a histologically normal liver from a patient with cardiac failure due to homozygous hypercholesterolemia may be required to permit transplantation of a donor liver with normal cholesterol metabolism. Transplantation is a standard treatment in genetic conditions such as adult polycystic kidney disease (PKD) or infantile hepatic failure due to α_1-antitrypsin deficiency. Although genetic disease may recur in a transplanted organ if transplantation does not correct the essential genetic defect, the disease does not recur in situations in which the essential defect is corrected. For example, if the marrow of a patient with β-thalassemia is completely replaced with normal marrow, the disease would not recur. Marrow transplantation between siblings is considered somatic, not germline gene therapy. Tissue rejection after transplantation is no less a problem in genetic diseases than in others, unless the disease being treated is an immunodeficiency disease.

4. The answer is B *[III B 4, 5].*
Differences in gene frequencies among populations arise largely because of nonrandom mating between them. Multifactorial and autosomal recessive diseases most commonly exhibit different frequencies in different populations. Certain genes may become common in particular groups because of natural selection. Genetic differences among populations are important medically because they predispose some patients to different diseases than other patients. Genetic differences among populations are usually relative rather than absolute because it is very uncommon for human populations to remain genetically isolated for more than a few generations.

5. The answer is C *[IV B].*
The main conflict here is whether the women having prenatal diagnosis can make autonomous decisions regarding the results and continuation of the pregnancy or whether the staff can decide that the best interests of the patient (which could be the mother and is certainly the fetus) are served by withholding information. Fidelity is of lesser importance in this case, but it is involved because a person could claim the contract between the patient and the counselor is such that all information must be divulged. Nonmaleficence is involved in terms of preventing harm to the fetus but does not address the good intended to the woman or to female fetuses in general. Justice is not directly involved since none of the women will be given information regarding fetal sex. Veracity could be involved since the women will not be told the sex, but conversely, one could claim that no lies are told. Veracity does have the same weight as autonomy and beneficence in this situation.

6. The answer is D *[I B 1, 3].*
Good genetic screening programs do not need to be limited to diseases that can be cured. Screening may also be valuable to help delay the onset or improve the management of genetic disease. Alternatively, screening may be justified to help couples avoid having children with a serious and untreatable genetic disease. Good genetic screening programs must have a net benefit that exceeds their cost and should deal only with a disease that causes serious morbidity or mortality. Timely and effective follow-up of abnormal results and both professional and public education are essential to a good program.

7–9. The answers are: 7-D, 8-E, 9-E *[II D 1]*.
Administration of growth hormone to a pituitary dwarf is an example of amelioration of a metabolic abnormality by replacement of a missing product.

Anticonvulsant therapy in a child with an inborn metabolic error is symptomatic treatment but does not alter the metabolic imbalance.

Giving large doses of vitamins to a child with a chromosome abnormality is unlikely to produce any of the effects listed. Pathogenesis of the phenotypic alterations is not understood well enough in such conditions to permit rational biochemical therapy.

10–14. The answers are: 10-D, 11-C, 12-A, 13-B, 14-E *[IV B]*.
Fidelity refers to the often unspoken contract between the physician and the patient to be ethical and to provide information. Veracity involves truth telling, which includes the provision of all the information that a reasonable patient is expected to ask. The principle of autonomy is one of the strongest, and it concerns the individual's right to make decisions concerning his or her own life, including health care and reproductive decisions. Beneficence not only involves the provision of benefit to the patient but also the prevention and removal of harm. Justice involves matters of fair provision of services and is concerned with both macro- and microallocation of such resources. Generally, the physician is not expected to provide all minute details, especially regarding rare or unlikely events, unless such events are potentially serious.

Comprehensive Exam

Introduction

One of the least attractive aspects of pursuing an education is the necessity of being examined on what has been learned. Instructors do not like to prepare tests, and students do not like to take them.

However, students are required to take many examinations during their learning careers, and little if any time is spent acquainting them with the positive aspects of tests and with systematic and successful methods for approaching them. Students perceive tests as punitive and sometimes feel that they are merely opportunities for the instructor to discover what the student has forgotten or has never learned. Students need to view tests as opportunities to display their knowledge and to use them as tools for developing prescriptions for further study and learning.

A brief history and discussion of the National Board of Medical Examiners (NBME) examinations [now the United States Medical Licensing Examination (USMLE)] are presented here, along with ideas concerning psychological preparation for the examinations. Also presented are general considerations and test-taking tips, as well as ways to use practice exams as educational tools. (The literature provided by the various examination boards contains detailed information concerning the construction and scoring of specific exams.)

Before the various NBME exams were developed, each state attempted to license physicians through its own procedures. Differences in the quality and testing procedures of the various state examinations resulted in the refusal of some states to recognize the licensure of physicians licensed in other states. This made it difficult for physicians to move freely from one state to another and produced an uneven quality of medical care in the United States.

To remedy this situation, the various state medical boards decided they would be better served if an outside agency prepared standard exams to be given in all states, allowing each state to meet its own needs and have a common standard by which to judge the educational preparation of individuals applying for licensure.

One misconception concerning these outside agencies is that they are licensing authorities. This is not the case; they are examination boards only. The individual states retain the power to grant and revoke licenses. The examination boards are charged with designing and scoring valid and reliable tests. They are primarily concerned with providing the states with feedback on how examinees have performed and with making suggestions about the interpretation and usefulness of scores. The states use this information as partial fulfillment of qualifications upon which they grant licenses.

Students should remember that these exams are administered nationwide and, although the general medical information is the same, educational methodologies and faculty areas of expertise differ from institution to institution. It is unrealistic to expect that students will know all

---

The author of this introduction, Michael J. O'Donnell, holds the positions of Assistant Professor of Psychiatry and Director of Biomedical Communications at the University of New Mexico School of Medicine, Albuquerque, New Mexico.

the material presented in the exams; they may face questions on the exams in areas that were only superficially covered in their classes. The testing authorities recognize this situation, and their scoring procedures take it into account.

The Exams

The first exam was given in 1916. It was a combination of written, oral, and laboratory tests, and it was administered over a 5-day period. Admission to the exam required proof of completion of medical education and 1 year of internship

In 1922, the examination was changed to a new format and was divided into three parts. Part I, a 3-day essay exam, was given in the basic sciences after 2 years of medical school. Part II, a 2-day exam, was administered shortly before or after graduation, and Part III was taken at the end of the first postgraduate year. To pass both Part I and Part II, a score equalling 75% of the total points available in each was required.

In 1954, after a 3-year extensive study, the NBME adopted the multiple-choice format. To pass, a statistically computed score of 75 was required, which allowed comparison of test results from year to year. In 1971, this method was changed to one that held the mean constant at a computed score of 500, with a predetermined deviation from the mean to ascertain a passing or failing score. The 1971 changes permitted more sophisticated analysis of test results and allowed schools to compare among individual students within their respective institutions as well as among students nationwide. Feedback to students regarding performance included the reporting of pass or failure along with scores in each of the areas tested.

During the 1980s, the ever-changing field of medicine made it necessary for the NBME to examine once again its evaluation strategies. It was found necessary to develop questions in multidisciplinary areas such as gerontology, health promotion, immunology, and cell and molecular biology. In addition, it was decided that questions should test higher cognitive levels and reasoning skills.

To meet the new goals, many changes have been made in both the form and content of the examination. Changes include reduction in the number of questions to approximately 800 in Step 1 and Step 2 of the USMLE to allow students more time on each question, with total testing time reduced on Step 1 from 13 to 12 hours and on Step 2 from 12.5 to 12 hours. The basic science disciplines are no longer allotted the same number of questions, which permits flexible weighing of the exam areas. Reporting of scores to schools includes total scores for individuals and group mean scores for separate discipline areas. Only pass/fail designations and total scores are reported to examinees. There is no longer a provision for the reporting of individual subscores to either the examinees or medical schools. Finally, the question format used in the new exams is predominately multiple-choice, best-answer.

The New Format

New questions, designed specifically for Step 1 are constructed in an effort to test the student's grasp of the sciences basic to medicine in an integrated fashion—the questions are designed to be interdisciplinary. Many of these items are presented as vignettes, or case studies, followed by a series of multiple-choice, best-answer questions.

The scoring of this exam is altered. Whereas in the past the exams were scored on a normal curve, the new exam has a predetermined standard, which must be met in order to pass. The exam no longer concentrates on the trivial; therefore, it has been concluded that there is a common base of information that all medical students should know in order to pass. It is anticipated that a major shift in the pass/fail rate for the nation is unlikely. In the past, the average student could only expect to feel comfortable with half the test and eventually would complete approximately 67% of the questions correctly, to achieve a mean score of 500. Although with the standard setting method it is likely that the mean score will change and become higher, it is unlikely that the pass/fail rates will differ significantly from those in the past. During the first testing in 1991, there was not differential weighing of the questions. However, in the future, the NBME will be researching methods of weighing questions based on both the time it takes to answer questions vis-à-vis their difficulty and the perceived importance of the information. In addition, the NBME is attempting to design a method of delivering feedback to the student that will have considerable importance in discovering weaknesses and pinpointing areas for further study in the event that a retake is necessary.

Materials Needed for Test Preparation

In preparation for a test, many students collect far too much study material only to find that they simply do not have the time to go through all of it. They are defeated before they begin because either they leave areas unstudied, or they race through the material so quickly that they cannot benefit from the activity.

It is generally more efficient for the student to use materials already at hand; that is, class notes, one good outline to cover or strengthen areas not locally stressed and to quickly review the whole topic, and one good text as a reference for looking up complex material needing further explanation.

Also, many students attempt to memorize far too much information, rather than learning and understanding less material and then relying on that learned information to determine the answers to questions at the time of the examination. Relying too heavily on memorized material causes anxiety, and the more anxious students become during a test, the less learned knowledge they are likely to use.

Positive Attitude

A positive attitude and a realistic approach are essential to successful test taking. If concentration is placed on the negative aspects of tests or on the potential for failure, anxiety increases and performance decreases. A negative attitude generally develops if the student concentrates on "I must pass" rather than on "I can pass." "What if I fail?" becomes the major factor motivating the student to **run from failure rather than toward success**. This results from placing too much emphasis on scores rather than understanding that scores have only slight relevance to future professional performance.

The score received is only one aspect of test performance. Test performance also indicates the student's ability to use information during evaluation procedures and reveals how this ability might be used in the future. For example, when a patient enters the physician's office with a problem, the physician begins by asking questions, searching for clues, and seeking diagnostic information. Hypotheses are then developed, which will include several potential causes for the problem. Weighing the probabilities, the physician will begin to discard those hypotheses with the least likelihood of being correct. Good differential diagnosis involves the ability to deal with uncertainty, to reduce potential causes to the smallest number, and to use all learned information in arriving at a conclusion.

The same thought process can and should be used in testing situations. It might be termed **paper-and-pencil differential diagnosis**. In each question with five alternatives, of which one is correct, there are four alternatives that are incorrect. If deductive reasoning is used, as in solving a clinical problem, the choices can be viewed as having possibilities of being correct. The elimination of wrong choices increases the odds that a student will be able to recognize the correct choice. Even if the correct choice does not become evident, the probability of guessing correctly increases. Just as differential diagnosis in a clinical setting can result in a correct diagnosis, eliminating incorrect choices on a test can result in choosing the correct answer.

Answering questions based on what is incorrect is difficult for many students since they have had nearly 20 years experience taking tests with the implied assertion that knowledge can be displayed only by knowing what is correct. It must be remembered, however, that students can display knowledge by knowing something is wrong, just as they can display it by knowing something is right. **Students should begin to think in the present as they expect themselves to think in the future.**

Paper-and-Pencil Differential Diagnosis

The technique used to arrive at the answer to the following question is an example of the paper-and-pencil differential diagnosis approach.

A recently diagnosed case of hypothyroidism in a 45-year-old man may result in which of the following conditions?

(A) Thyrotoxicosis
(B) Cretinism
(C) Myxedema
(D) Graves' disease
(E) Hashimoto's thyroiditis

It is presumed that all of the choices presented in the question are plausible and partially correct. If the student begins by breaking the question into parts and trying to discover what the question is attempting to measure, it will be possible to answer the question correctly by using more than memorized charts concerning thyroid problems.

- The question may be testing if the student knows the difference between "hypo" and "hyper" conditions.
- The answer choices may include thyroid problems that are not "hypothyroid" problems.
- It is possible that one or more of the choices are "hypo" but are not "thyroid" problems, that they are some other endocrine problems.
- "Recently diagnosed in a 45-year-old man" indicates that the correct answer is not a congenital childhood problem.
- "May result in" as opposed to "resulting from" suggests that the choices might include a problem that **causes** hypothyroidism rather than **results from** hypothyroidism, as stated.

By applying this kind of reasoning, the student can see that choice **A,** thyroid toxicosis, which is a disorder resulting from an overactive thyroid gland ("hyper") must be eliminated. Another piece of knowledge, that is, Graves' disease is thyroid toxicosis, eliminates choice **D**. Choice **B,** cretinism, is indeed hypothyroidism, but it is a childhood disorder. Therefore, **B** is eliminated. Choice **E** is an inflammation of the thyroid gland—here the clue is the suffix "itis." The reasoning is that thyroiditis, being an inflammation, may **cause** a thyroid problem, perhaps even a hypothyroid problem, but there is no reason for the reverse to be true. Myxedema, choice **C,** is the only choice left and the obvious correct answer.

Preparing for Board Examinations

1. **Study for yourself.** Although some of the material may seem irrelevant, the more you learn now, the less you will have to learn later. Also, do not let the fear of the test rob you of an important part of your education. If you study to learn, the task is less distasteful than studying solely to pass a test.

2. **Review all areas.** You should not be selective by studying perceived weak areas and ignoring perceived strong areas. This is probably the last time you will have the time and the motivation to review **all** of the basic sciences.

3. **Attempt to understand, not just memorize, the material.** Ask yourself: To whom does the material apply? Where does it apply? When does it apply? Understanding the connections among these points allows for longer retention and aids in those situations when guessing strategies may be needed.

4. **Try to anticipate questions that might appear on the test.** Ask yourself how you might construct a question on a specific topic.

5. **Give yourself a couple days of rest before the test.** Studying up to the last moment will increase your anxiety and cause potential confusion.

Taking Board Examinations

1. In the case of the USMLE, be sure to **pace yourself** to use the time optimally. Each booklet is designed to take 2 hours. You should use all your allotted time; if you finish too early, you probably did so by moving too quickly through the test.

2. **Read each question and all the alternatives carefully** before you begin to make decisions. Remember the questions contain clues, as do the answer choices. As a physician, you would not make a clinical decision without a complete examination of all the data; the same holds true for answering test questions.

3. **Read the directions for each question set carefully.** You would be amazed at how many students make mistakes in tests simply because they have not paid close attention to the directions.

4. It is not advisable to leave blanks with the intention of coming back to answer the questions later. Because of the way Board examinations are constructed, you probably will not pick up any new information that will help you when you come back, and the chances of getting numerically off on your answer sheet are greater than your chances of benefiting by skipping around. If you feel that you must come back to a question, mark the best choice and place a note in the margin. Generally speaking, it is best not to change answers once you have made a decision, unless you have learned new information. Your intuitive reaction and first response are correct more often than changes made out of frustration or anxiety. **Never turn in an answer sheet with blanks.** Scores are based on the number that you get correct; you are not penalized for incorrect choices.

5. **Do not try to answer the questions on a stimulus–response basis.** It generally will not work. Use all of your learned knowledge.

6. **Do not let anxiety destroy your confidence.** If you have prepared conscientiously, you know enough to pass. Use all that you have learned.

7. **Do not try to determine how well you are doing as you proceed.** You will not be able to make an objective assessment, and your anxiety will increase.

8. **Do not expect a feeling of mastery** or anything close to what you are accustomed to. Remember, this is a nationally administered exam, not a mastery test.

9. **Do not become frustrated or angry** about what appear to be bad or difficult questions. You simply do not know the answers; you cannot know everything.

Specific Test-Taking Strategies

Read the entire question carefully, regardless of format. Test questions have multiple parts. Concentrate on picking out the pertinent key words that might help you begin to problem-solve. Words such as "always," "never," "mostly," "primarily," and so forth play significant roles. In all types of questions, distractors with terms such as "always" or "never" most often are incorrect. Adjectives and adverbs can completely change the meaning of questions—pay close attention to them. Also, medical prefixes and suffixes (e.g., "hypo-," "hyper-," "-ectomy," "-itis") are sometimes at the root of the question. The knowledge and application of everyday English grammar often is the key to dissecting questions.

Multiple-Choice Questions

Read the question and the choices carefully to become familiar with the data as given. Remember, in multiple-choice questions there is one correct answer and there are four distractors, or incorrect answers. (Distractors are plausible and possibly correct or they would not be called distractors.) They are generally correct for part of the question but not for the entire question. Dissecting the question into parts aids in discerning these distractors.

If the correct answer is not immediately evident, begin eliminating the distractors. (Many students feel that they must always start at option A and make a decision before they move to B, thus forcing decisions they are not ready to make.) Your first decisions should be made on those choices you feel the most confident about.

Compare the choices to each part of the question. **To be wrong,** a choice needs to be **incorrect for only part** of the question. **To be correct,** it must be **totally** correct. If you believe a choice is partially incorrect, tentatively eliminate that choice. Make notes next to the choices regarding tentative decisions. One method is to place a minus sign next to the choices you are certain are incorrect and a plus sign next to those that potentially are correct. Finally, place a zero next to any choice you do not understand or need to come back to for further inspection. Do not feel that you must make final decisions until you have examined all choices carefully.

When you have eliminated as many choices as you can, decide which of those that are left has the highest probability of being correct. Remember to use paper-and-pencil differential diagnosis. Above all, be honest with yourself. If you do not know the answer, eliminate as many choices as possible and choose reasonably.

Vignette-Based Questions

Vignette-based questions are nothing more than normal multiple-choice questions that use the same case, or grouped information, for setting the problem. The NBME has been researching question types that would test the student's grasp of the integrated medical basic sciences in a more cognitively complex fashion than can be accomplished with traditional testing formats. These questions allow the testing of information that is more medically relevant than memorized terminology.

It is important to realize that several questions, although grouped together and referring to one situation or vignette, are independent questions; that is, they are able to stand alone. Your inability to answer one question in a group should have no bearing on your ability to answer other questions in that group.

These are multiple-choice questions, and just as with single best-answer questions, you should use the paper-and-pencil differential diagnosis, as was described earlier.

Single Best-Answer—Matching Sets

Single best-answer—matching sets consist of a list of words or statements followed by several numbered items or statements. Be sure to pay attention to whether the choices can be used more than once, only once, or not at all. Consider each choice individually and carefully. Begin with those with which you are the most familiar. It is important always to break the statements and words into parts, as with all other question formats. **If a choice is only partially correct, then it is incorrect.**

Guessing

Nothing takes the place of a firm knowledge base, but with little information to work with, even after playing paper-and-pencil differential diagnosis, you may find it necessary to guess the correct answer. A few simple rules can help increase your guessing accuracy. Always guess consistently if you have no idea what is correct; that is, after eliminating all that you can, make the choice that agrees with your intuition or choose the option closest to the top of the list that has not been eliminated as a potential answer.

When guessing at questions that have choices in numerical form, you will often find the choices listed in an ascending or descending order. It is generally not wise to guess the first or last alternative, since these are usually extreme values and are most likely incorrect.

Using the Challenge Exam to Learn

All too often, students do not take full advantage of practice exams. There is a tendency to complete the exam, score it, look up the correct answers to those questions missed, and then forget the entire thing.

In fact, great educational benefits can be derived if students would spend more time using practice tests as learning tools. As mentioned earlier, incorrect choices in test questions are plausible and partially correct or they would not fulfill their purpose as distractors. This means that it is just as beneficial to look up the incorrect choices as the correct choices to discover specifically why they are incorrect. In this way, it is possible to learn better test-taking skills as the subtlety of question construction is uncovered.

Additionally, it is advisable to go back and attempt to restructure each question to see if all the choices can be made correct by modifying the question. By doing this, four times as much will be learned. By all means, look up the right answer and explanation. Then, focus on each of the other choices and ask yourself under what conditions they might be correct? For example, the entire thrust of the sample question concerning hypothyroidism could be altered by changing the first few words to read:

"Hyperthyroidism recently discovered in . . ."
"Hypothyroidism prenatally occurring in . . ."
"Hypothyroidism resulting from . . ."

This question can be used to learn and understand thyroid problems in general, not only to memorize answers to specific questions.

In the practice exams that follow, every effort has been made to simulate the types of questions and the degree of question difficulty in the USMLE Step 1. While taking these exams, the student should attempt to create the testing conditions that might be experienced during actual testing situations. Approximately 1 minute should be allowed for each question, and the entire test should be finished before it is scored.

Summary

Ideally, examinations are designed to determine how much information students have learned and how that information is used in the successful completion of the examination. Students will be successful if these suggestions are followed:

- Develop a positive attitude and maintain that attitude.
- Be realistic in determining the amount of material you attempt to master and in the score you hope to obtain.
- Read the directions for each type of question and the questions themselves closely and follow the directions carefully.
- Guess intelligently and consistently when guessing strategies must be used.

- Bring the paper-and-pencil differential diagnosis approach to each question in the examination.
- Use the test as an opportunity to display your knowledge and as a tool for developing prescriptions for further study and learning.

The USMLE is not easy. It may be almost impossible for those who have unrealistic expectations or for those who allow misinformation concerning the exam to produce anxiety out of proportion to the task at hand. It is manageable if it is approached with a positive attitude and with consistent use of all the information that has been learned.

Michael J. O'Donnell

STUDY QUESTIONS

Directions: Each of the numbered items or incomplete statements in this section is followed by answers or by completions of the statement. Select the **one** lettered answer or completion that is **best** in each case.

1. On which one of the following tissue specimens can cytogenetic analysis be performed?

(A) Products of conception that have been preserved frozen from a miscarriage
(B) Chorionic villus cultured for biochemical analysis
(C) A fixed tissue block from a malignant tumor
(D) Dried blood from a routine neonatal screening card
(E) Clotted blood from which the serum has been removed

2. Of the following comparative statements about linkage and linkage disequilibrium, which one is correct?

(A) Linkage is measured in family studies; linkage disequilibrium is measured in population studies
(B) Linkage is restricted to loci on a single chromosome; linkage disequilibrium may affect loci on different chromosomes
(C) Linked loci must be polymorphic; linkage disequilibrium may be demonstrated in both polymorphic and nonpolymorphic loci
(D) Linkage refers to alleles; linkage disequilibrium refers to loci
(E) Linkage is usually measured by restriction fragment length polymorphism (RFLP) studies; linkage disequilibrium is measured by cytogenetic studies

3. Which one of the following statements regarding the effects of medical interventions on disease gene frequencies is correct?

(A) Successful treatment of an autosomal dominant disease is likely to decrease the disease frequency substantially in future generations
(B) Successful treatment of an X-linked recessive disease is likely to decrease the disease frequency substantially in future generations
(C) Successful treatment of an autosomal recessive disease is likely to increase the disease frequency substantially in future generations
(D) Genetic counseling is likely to decrease the frequency of disease genes substantially in future generations
(E) None of the above

4. Which one of the following statements about normal phenotypic variation is true?

(A) Most normal characteristics in boys are inherited from their fathers and most normal characteristics in girls are inherited from their mothers
(B) The genes involved are the same in all populations
(C) Environmental factors play a role
(D) Most normal characteristics such as eye or hair color are transmitted as autosomal recessive traits
(E) The occurrence in a child of a more extreme phenotype than in either parent results from a new mutation

5. Which one of the following statements about teratogens and mutagens is correct?

(A) An agent cannot act as both a teratogen and a mutagen
(B) Exposures of either the mother or the father may produce teratogenic effects
(C) Mutagens produce their effects by alteration of the proteins of the ova or sperm
(D) Mutagens affect single cells; teratogens affect groups of cells
(E) Birth defects due to mutagens are more common than birth defects due to teratogens

---

1-B
2-A
3-E
4-C
5-D

6. Major events of early mammalian embryogenesis include which one of the following occurrences?

(A) Establishment of clonal cell lineages with predetermined fates
(B) Segmentation
(C) Amplification of preformed maternal messenger RNA (mRNA)
(D) Activation of proto-oncogenes

7. Measurement of the serum creatine phosphokinase (CPK) level can detect about 75% of obligate carriers of Duchenne muscular dystrophy (DMD), an X-linked recessive condition that is lethal in males. A 26-year-old woman had a brother who died of DMD at age 18. There is no other family history of DMD. CPK testing is done on the woman and results are normal. What is the risk for DMD in her son?

(A) 1/4
(B) 1/8
(C) 1/10
(D) 1/18

8. The father of a 24-year-old male patient has a diagnosis of Huntington's disease and is cared for at home. Several paternal relatives are similarly affected, and most are now in long-term care facilities. The patient and his wife request predictive testing. All appropriate blood samples are obtained, and the results obtained in duplicate indicate that there is nonpaternity; that is, the patient's biological father is not the legal and affected father. The mother is interviewed separately and gently but does not confirm your genetic interpretation. Which one of the following solutions would best aid this couple in planning their future?

(A) Respect the autonomy of the mother and tell the patient the molecular testing was not informative
(B) Impress upon the mother the importance of divulging the results to her son
(C) Respect the autonomy of the mother, say nothing about the nonpaternity, but tell the patient the results indicate he has not inherited the Huntington's allele
(D) Call the couple in and review all the results, including the nonpaternity; ask the young man not to tell his mother about the nonpaternity

9. Of the following statements regarding genetic polymorphisms, which one is *always* correct? They

(A) alter protein structure or function
(B) allow tissues from two different people to be distinguished
(C) reflect alterations of DNA sequence
(D) are inapparent without special laboratory testing
(E) are useful in linkage studies

10. Which one of the following statements regarding multifactorial congenital anomalies is true?

(A) Defects typically occur with other embryologically unrelated malformations in an affected patient
(B) Defects usually have a characteristic appearance that distinguishes them from similar defects due to other causes
(C) Concordance for defects is similar in monozygotic and dizygotic twins
(D) Recurrence risks for defects are similar in the siblings and children of an affected patient
(E) Defects typically occur with equal frequencies in both sexes

11. Which one of the following statements about teratogenesis is correct?

(A) The likelihood that an agent will produce a teratogenic effect can usually be predicted on the basis of its pharmacologic action in an infant
(B) The teratogenic effect of exposure to a combination of agents can usually be predicted by "adding" the effects of the individual agents in the combination
(C) The period during which the embryo is most susceptible to teratogenic effects is usually the first 2 weeks after conception
(D) Congenital anomalies are unlikely to be produced by exposure to teratogenic agents later than 10 weeks after conception
(E) The circumstances, route, and dosage of exposure are as important in determining teratogenic risk as the chemical nature of the agent

---

6-B
7-D
8-C
9-C
10-D
11-E

12. A woman, who was affected by Rh hemolytic disease as a newborn, married a man who did not have Rh hemolytic disease as a newborn but who had an elder brother and sister who died of this condition. What is the chance that their children will have Rh hemolytic disease?

(A) 0
(B) 12.5%
(C) 25%
(D) 50%
(E) 100%

13. Which one of the following factors is not an indication for fetal karyotyping?

(A) A maternal age of 39 years
(B) A previous child with trisomy 21
(C) A fetus shown to have a cystic hygroma on ultrasound
(D) A second cousin with spina bifida
(E) A diagnosis of a robertsonian translocation in the father

14. Which one of the following statements regarding treatment of genetic diseases is true?

(A) Genetic diseases cannot be treated if they are congenital
(B) Genetic diseases that are etiologically heterogeneous cannot be treated
(C) Genetic diseases can only be treated if they are understood at least in part at the biochemical level
(D) Diagnostic precision is often critically important in designing effective treatment for genetic diseases
(E) Treatment of genetic diseases requires proper regulation of the therapeutic gene

15. A female patient was treated throughout pregnancy with Kulout, a newly introduced tranquilizer. Her baby has cleft lip and cleft palate. Which one of the following observations provides the strongest evidence that this association is causal?

(A) The frequency of cleft palate is increased among the offspring of rats treated during pregnancy with 100 times the human dose of Kulout
(B) Kulout is metabolized through the same metabolic pathway as thalidomide
(C) Six other cases of congenital anomalies among children of women who took Kulout during pregnancy have been reported: One child has congenital heart disease, one has club feet, one has hypospadias, one has Down syndrome, one has spina bifida, and one has a cavernous hemangioma
(D) The infant has several other congenital anomalies in addition to cleft lip and cleft palate, and the same pattern of anomalies has been noted among three other children whose mothers also took Kulout early in pregnancy
(E) A clinical geneticist has examined the infant and says that she does not have a recognized genetic syndrome

16. Of the following approaches to treatment of genetic disease, which one is most commonly used?

(A) Somatic gene therapy
(B) Organ transplantation
(C) Modulation of gene expression
(D) Amelioration of metabolic abnormalities
(E) Amelioration of the clinical phenotype

17. Prenatal diagnosis for neural tube defects is indicated in a pregnant woman who

(A) has long-standing insulin-dependent diabetes mellitus
(B) had rubella during the first trimester of pregnancy → PDA + deafness
(C) used marijuana daily throughout gestation
(D) had an abdominal x-ray early in gestation
(E) took androgenic hormones for bodybuilding before finding out that she was pregnant

---

12-A
13-D
14-D
15-D
16-E
17-A

18. Which of the following statements regarding the teratogenic effects of alcohol is true?

(A) Maternal binge drinking a few times early in pregnancy does not appear to pose a risk to the fetus
(B) Maternal drinking of less than 3 ounces of absolute alcohol (i.e., the equivalent of about 6 beers) per day during pregnancy is unlikely to produce an adverse effect on the fetus
(C) Behavioral abnormalities often occur in children with fetal alcohol syndrome
(D) Limb reduction defects (phocomelia) are common features of fetal alcohol syndrome
(E) Mental retardation may occur in fetal alcohol syndrome but is uncommon

19. Differences between meiosis and mitosis include all of the following EXCEPT

(A) meiosis results in a reduction of chromosome number in the cell from 46 to 23, while mitosis does not result in a change in chromosome number
(B) each mitotic cell division and each second meiotic division is preceded by a complete round of DNA synthesis
(C) all somatic cells undergo mitosis, but meiosis is restricted to germ cells
(D) recombination is much more common in meiosis than in mitosis
(E) chromosome pairing (synapsis) occurs in meiosis but not in mitosis

20. A physician is about to see a young couple of which one person has been identified as having a reciprocal translocation. The important factors to consider for counseling include all of the following EXCEPT

(A) the number of spontaneous abortions
(B) how the couple was ascertained
(C) the sex of the affected partner
(D) whether the anomaly is de novo or familial
(E) the particular chromosome segments involved

21. All of the following statements about genetic markers are true EXCEPT they

(A) are inherited as typical multifactorial traits
(B) have alternate forms that are readily recognizable
(C) may be detected as alterations in DNA sequence or protein structure or function
(D) can be used clinically in linkage studies
(E) are used to identify individuals in forensic medicine

22. Of the following statements, all are assumptions on which the Hardy-Weinberg law is based EXCEPT

(A) mating within the population is completely random
(B) the genes involved are autosomal dominant
(C) there is no mutation occurring at the locus
(D) there is no selection for any of the genotypes at the locus
(E) there is no migration into or out of the population

23. A couple has a child with Tay-Sachs disease, which is a recessive condition associated with lack of the enzyme hexosaminidase A. All of the following scenarios could be predicted EXCEPT

(A) the heterozygotes in the family should be mildly affected
(B) the heterozygotes can be detected by having intermediate levels of the enzyme
(C) prenatal diagnosis is available by measuring the level of the enzyme in amniocytes
(D) the siblings of the affected child will be at low risk to have affected children
(E) there may be a milder disease involving the same enzyme

24. All of the following infectious agents are potentially teratogenic in humans EXCEPT

(A) rubella virus
(B) *Gonococcus*
(C) chickenpox (varicella) virus
(D) cytomegalovirus (CMV)
(E) *Toxoplasma gondii*

18-C	21-A	24-B
19-B	22-B	
20-A	23-A	

25. A physician is called to the nursery to examine a newborn with ambiguous genitalia. The infant has a small phallic structure with hypospadias, bilateral cryptorchidism, but no other obvious problems. All of the following immediate studies will help to establish the diagnosis quickly EXCEPT

(A) karyotyping
(B) 17-ketosteroid levels
(C) lower abdominal ultrasound
(D) androgen-binding studies on genital skin cells
(E) electrolyte determinations

26. A 32-year-old woman has blood drawn for maternal serum α-fetoprotein (MSAFP) testing and undergoes a routine ultrasound. The results indicate that her fetus has a significantly increased risk for a neural tube defect. All of the following procedures should be done to follow up this risk EXCEPT

(A) determine the gestational age precisely
(B) check that the patient's weight was correctly reported
(C) arrange an amniocentesis for measurement of amniotic fluid α-fetoprotein
(D) schedule a level II ultrasound for assessment of fetal detail
(E) arrange for measurement of acetylcholinesterase in the patient's blood

27. Genetic screening is justified in each of the following cases EXCEPT to

(A) detect female carriers of a lethal X-linked recessive disease within the families of affected boys
(B) identify couples at high risk of having children with a life-threatening malformation
(C) identify individuals who have a low genetic risk of heart attack so they can be provided life insurance at lower rates
(D) identify relatives likely to be affected within families of patients with a late-onset autosomal dominant neurodegenerative disease before development of symptoms
(E) detect carriers of a dominant gene that predisposes to a serious complication of anesthesia within an ethnic community with a high incidence of the disorder

28. True statements regarding direct and indirect approaches to DNA diagnosis include all of the following EXCEPT

(A) for direct testing, at least part of the normal gene must be cloned
(B) direct approaches are more accurate
(C) DNA from other family members is needed for indirect approaches (i.e., linkage) to diagnosis
(D) genetic heterogeneity may complicate indirect approaches to DNA testing
(E) DNA polymorphisms due to variations in noncoding DNA are predominantly used in direct approaches to DNA testing

29. A 49-year-old woman is being assessed preoperatively. She informs the anesthesiologist that her mother died during an anesthetic procedure and that her mother's brother also had significant medical problems, including high fever, during an operation. All of the following statements regarding this woman's condition are correct EXCEPT

(A) she should be considered at increased risk for the autosomal dominant condition malignant hyperthermia
(B) she should be encouraged to wear a medic alert bracelet stating she is at risk for malignant hyperthermia
(C) the medical records of her relatives should be obtained to learn more about the reason for problems during anesthesia
(D) she should undergo surgery only in case of an emergency
(E) problems in surgery can be reduced by medications or by use of different anesthetics

30. Each of the following statements regarding studies that are used to identify teratogenic effects in humans is true EXCEPT

(A) most associations observed in individual case reports are causal
(B) a causal relationship is suggested by a recurrent pattern of congenital anomalies among infants born to women with similar exposures during pregnancy
(C) most human teratogens have been recognized initially in a clinical series
(D) both cohort studies and case–control studies are often subject to serious biases of ascertainment
(E) animal studies alone cannot prove that an exposure is teratogenic in humans

---

25-D
26-E
27-C
28-E
29-D
30-A

31. Twin studies are important tools in the analysis of behavioral genetics, but they have several inherent problems. All of the following are important factors in such studies EXCEPT

(A) zygosity
(B) self-selection
(C) sex of study participants
(D) source of study participants
(E) sample size

32. Examples of symptomatic treatment of genetic disease include all of the following EXCEPT

(A) restriction of phenylalanine in the diet of patients with phenylketonuria (PKU)
(B) use of hearing aids in hereditary deafness
(C) administration of antibiotics for pulmonary infections in cystic fibrosis (CF)
(D) speech therapy for a child with cleft palate
(E) plastic surgery to correct facial malformation in a dominantly inherited multiple congenital anomaly syndrome

33. All of the following statements regarding genetic mutations are true EXCEPT

(A) mutations underlying most genetic diseases are known
(B) major rearrangements are, in general, uncommon causes of mutation
(C) deletions are common causes of mutation in Duchenne muscular dystrophy (DMD)
(D) the majority of Jewish persons with Tay-Sachs disease have the same mutation
(E) the majority of persons with cystic fibrosis (CF) have the same mutation

34. All of the following statements regarding an association observed in a birth defects epidemiology study are true EXCEPT the

(A) association may occur because both the child's anomaly and the mother's exposure are related to another factor
(B) association is more likely to be causal if pregnant women exposed to a much higher dose of the agent have a higher risk of having affected children than pregnant women exposed to a lower dose
(C) association is more likely to be causal if similar congenital anomalies can be induced in the offspring of rats subjected to similar exposures during pregnancy
(D) agent is likely to be an important cause of birth defects if the relative risk is less than 0.05
(E) association may be due to chance even if it is statistically significant

35. Characteristics of acquired chromosome abnormalities that develop in association with neoplasms include all of the following EXCEPT

(A) oncogene activation
(B) loss of a tumor suppressor gene
(C) demonstration of clonal evolution
(D) transmission to the offspring of an affected individual

36. Sickle cell disease, thalassemia, and glucose-6-phosphate dehydrogenase deficiency (G6PD) are genetic diseases of the red blood cells that occur commonly among Africans. Each of the following statements about this association is true EXCEPT

(A) the genes are thought to provide protection against malaria in some circumstances
(B) the genes were probably selected for in some African populations
(C) the genes are rarely seen in people whose ancestry is not African
(D) the genes may cause disease in people living in areas in which malaria occurs
(E) the genes may cause disease in people living in areas in which malaria does not occur

31-C
32-A
33-A
34-D
35-D
36-C

37. Certain abnormalities of the sex chromosomes have been associated with behavioral or psychiatric problems. All of the following statements regarding such associations are true EXCEPT

(A) XYY males tend to be immature and impulsive
(B) Turner syndrome females tend to be immature, with an increased incidence of mood disorders
(C) Klinefelter syndrome (XXY) is associated with poor judgment and learning problems in children
(D) Klinefelter syndrome (XXY) is associated with an increased incidence of schizophrenia
(E) XXX females show an increased incidence of schizophrenia

38. A 35-year-old woman decides to have amniocentesis after a counseling session at a clinic. She is extremely concerned about Down syndrome because a close friend of hers recently delivered an affected child. The results of her amniocentesis come back, and, although the fetus does not have Down syndrome, the results are abnormal (47, XYY). The physician is aware of the literature that reports the majority of such males are normal and elects to keep the results of the amniocentesis to himself. All of the following ethical principles are involved in this situation EXCEPT

(A) veracity
(B) autonomy
(C) beneficence
(D) fidelity
(E) justice

Questions 39–41

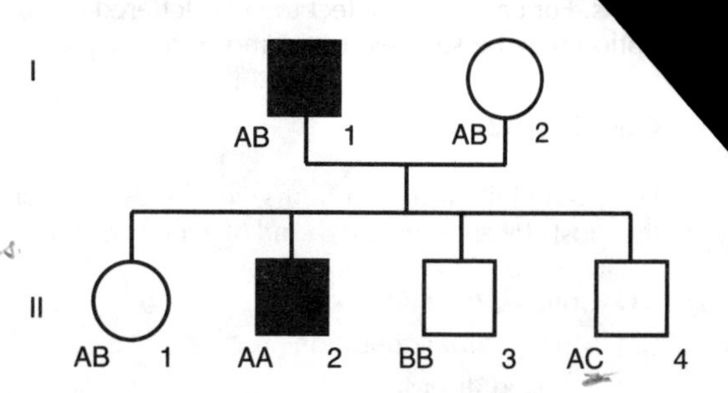

The figure above shows the DNA typing for a family that has Huntington's disease (HD). I-1 has HD, as does his son II-2. Family members II-1, II-3, and II-4 want to know what their risk is of having inherited the abnormal gene. The family members have had DNA analysis with a marker approximately 2 million base pairs from the mutation causing HD.

39. The risk for HD in II-1 is

(A) significantly increased
(B) significantly decreased
(C) mildly increased
(D) not changed because the markers are uninformative
(E) nonexistent

40. The risk for HD in II-3 is

(A) significantly increased
(B) significantly decreased
(C) mildly increased
(D) not changed
(E) nonexistent

41. The risk for HD in II-4 is

(A) significantly increased
(B) significantly decreased
(C) mildly increased
(D) not changed
(E) nonexistent

37-B
38-E
39-D
40-B
41-E

items in this section consists of lettered options followed by a set of numbered ct the **one** lettered option that is most closely associated with it. Each lettered nce, more than once, or not at all.

ror each clinical description given below, select the most closely associated kind of genetic disease.

(A) Chromosome abnormality
(B) Multifactorial condition
(C) X-linked disease
(D) Autosomal recessive disease
(E) Autosomal dominant disease

42. Due to a combination of genetic and nongenetic influences

43. A mendelian disorder carried on the X chromosome

44. Occurs only when both genes in a pair are mutant and located on a chromosome other than X or Y

45. The cause of most common diseases of adulthood, such as coronary artery disease and non–insulin-dependent diabetes

Questions 46–50

Match the following clinical descriptions with the most likely chromosome abnormality.

(A) Triploidy
(B) Trisomy 13
(C) Turner syndrome (45,X)
(D) Trisomy 18
(E) Trisomy 21

46. Flat occiput, Brushfield's spots, atrioventricular canal cardiac defect, single palmar creases

47. Small-for-age infant with small facies, congenital heart disease, peculiar hand positioning, rocker-bottom heels

48. Stillborn infant with holoprosencephaly, polydactyly, midline cleft lip and palate

49. Second trimester abortus with cystic hygroma and massive hydrops

50. Spontaneous abortus with large cystic placenta and a growth-disorganized embryo

Questions 51–55

Genetic polymorphisms can be identified in a variety of ways. Match the most appropriate description to the genetic polymorphism it characterizes.

(A) Restriction fragment length polymorphisms (RFLPs)
(B) Variable number of tandem repeats (VNTRs)
(C) Enzyme activity variants
(D) Antigenic protein variants
(E) Chromosome heteromorphisms

51. It is so polymorphic that two unrelated people are very unlikely to share the same alleles

52. Important in blood transfusion

53. May produce illness if a patient is exposed to certain drugs

54. Used in DNA fingerprinting in forensic medicine

55. Often reflect an alteration of a single nucleotide that produces gain or loss of a restriction endonuclease recognition site

Questions 56–59

Cystic fibrosis (CF), an autosomal recessive disease, occurs with a frequency of 1/2500 among people of northern European origin. What is the approximate risk of CF in a child of each of the following matings within this population?

(A) 1/25
(B) 1/50
(C) 1/100
(D) 1/400
(E) None of the above

56. A man and woman, neither of whom has a family history of CF

57. A woman with CF whose husband has no family history of CF

58. A woman whose first cousin has CF and a man who has no family history of CF

59. A man who has a child with CF by a previous marriage and a woman with no family history of CF

42-B	45-B	48-B	51-B	54-B	57-B
43-C	46-E	49-C	52-D	55-A	58-D
44-D	47-D	50-A	53-C	56-E	59-C

Questions 60–64

A 15-year-old boy develops insulin-dependent diabetes mellitus (IDDM). His parents wish to know the risk of their other children developing diabetes. The pedigree below was developed for the family. Match each of the affected son's siblings in the pedigree to the best estimate of risk.

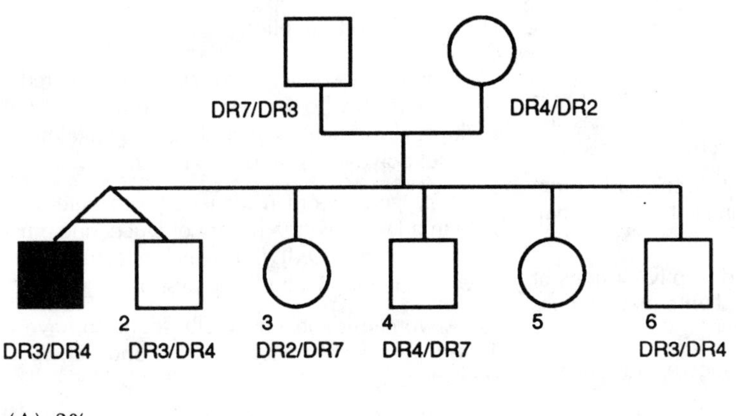

(A) 2%
(B) 5%
(C) 20%
(D) 50%
(E) 100%

60. The proband's monozygous twin (II-2) D

61. The proband's sister with human leukocyte antigen (HLA) DR2/DR7 (II-3) A

62. The proband's brother with HLA DR4/DR7 (II-4) B

63. The proband's sister for whom no HLA typing is available (II-5) B

64. The proband's brother with HLA DR3/DR4 (II-6) C

60-D
61-A
62-B
63-B
64-C

Questions 65–69

Match each of the following statements describing a potential teratogenic effect in humans with the most appropriate agent.

(A) Heroin
(B) Valproic acid
(C) Cocaine
(D) Isotretinoin
(E) Diethylstilbestrol

65. Maternal use of this agent during pregnancy often produces brain and cardiac abnormalities and craniofacial malformations in fetuses

66. Maternal use of this agent during early pregnancy frequently produces genital tract abnormalities in female fetuses

67. Placental abruption and disruptive lesions of the brain have been associated with maternal use of this agent during pregnancy

68. This agent has not been shown to be teratogenic in humans

69. Prenatal diagnosis for neural tube defects should be offered to women who have used this agent during early pregnancy

Questions 70–74

A deranged former employee snuck into the hospital nursery and clipped the identification bracelets from five newborn boys. In order to straighten the situation out, blood typing has been done on the infants and their parents. Match the following parents to the correct infant.

(A) Infant A: O, MN, Rh^+
(B) Infant B: A, MN, Rh^-
(C) Infant C: A, M, Rh^+
(D) Infant D: AB, N, Rh^+
(E) Infant E: B, MN, Rh^+

	Mother Blood Group			Father Blood Group		
Family	ABO	MN	Rh	ABO	MN	Rh
70.	B	N	Rh^-	A	MN	Rh^-
71.	B	N	Rh^+	AB	N	Rh^-
72.	AB	MN	Rh^+	O	MN	Rh^+
73.	O	M	Rh^+	A	M	Rh^+
74.	A	M	Rh^+	A	N	Rh^+

Questions 75–79

For each of the following patients, match the best form of prenatal diagnosis.

(A) Chorionic villus sampling (CVS)
(B) Maternal serum α-fetoprotein (MSAFP)
(C) Amniocentesis
(D) Level II ultrasound
(E) Fetal blood sampling

75. A pregnant woman who previously delivered a stillborn infant who had normal chromosomes but multiple abnormalities, including congenital heart disease, polydactyly, and cleft palate

76. A pregnant woman identified as a heterozygote in a family with a history of Duchenne muscular dystrophy (DMD), in which affected males have a deletion involving the dystrophin gene

77. A woman identified by ultrasound to have a 19-week fetus with nuchal edema and duodenal stenosis or atresia

78. A 25-year-old primigravida with an unremarkable family history, who is extremely concerned regarding her risk for a fetus with Down syndrome

79. A 36-year-old healthy woman who is 14 week's pregnant

Questions 80–84

Match each of the following methods for treatment of genetic disease with the appropriate situation.

(A) When the pathology is produced by accumulation of a toxic metabolite
(B) When the pathology is produced by absence of an essential product
(C) Both
(D) Neither

80. Dietary restriction

81. Activation of the defective metabolic pathway

82. Diversion of a substance to an alternate biochemical route

83. Organ transplantation

84. Somatic gene therapy

65-D	68-A	71-D	74-A	77-E	80-A	83-C
66-E	69-B	72-E	75-D	78-B	81-C	84-C
67-C	70-B	73-C	76-A	79-C	82-A	

Questions 85–89

Match each description of cytogenetic techniques with the most appropriate term.

(A) Q banding
(B) G banding
(C) Replication banding
(D) In situ hybridization
(E) NOR staining

85. A procedure for demonstrating the location of transcriptionally active ribosomal RNA (rRNA) synthesis

86. The most commonly used method for chromosome analysis; the standard in most laboratories

87. A method of cytogenetic preparation in which bands that stain darkly with the standard method fluoresce brightly

88. A method that can be used to demonstrate the position of single-copy genes on chromosomes

89. A method that stains only the short arms of chromosomes 13, 14, 15, 21, and 22

Questions 90–94

Match each genetic disease with the type of mutation that is most commonly associated with it.

(A) Deletion
(B) Insertion
(C) Duplication
(D) Point mutation
(E) Frameshift mutation

90. Cystic fibrosis (CF)

91. α-Thalassemia

92. β-Thalassemia

93. Duchenne muscular dystrophy (DMD)

94. Growth hormone deficiency

Questions 95–99

The locus for Huntington's disease (HD), an autosomal dominant condition, is near the tip of the short arm of chromosome 4 (4p). A marker locus that is 4 centimorgans (cM) from the HD locus has two alleles, called *1* and *2*. In a particular family, a woman with HD has an affected brother, sister, and father. The woman's genotype at the marker locus is *1,2,* her father's genotype is *1,1,* her brother's genotype is *1,1,* and her sister's genotype is *1,2*. The woman's mother and husband and her brother's wife, all of whom are unaffected, have a *1,2* genotype at the marker locus. Match each of the following offspring to the closest risk of developing HD.

(A) 2%
(B) 4%
(C) 48%
(D) 50%
(E) 96%

95. A son of the woman who has a *1,1* genotype at the marker locus

96. A daughter of the woman who has a *2,2* genotype at the marker locus

97. A fetus of the woman who has not yet been genotyped at the marker locus

98. A son of the woman's brother who has a *1,1* genotype at the marker locus

99. A daughter of the woman's brother who has a *1,2* genotype at the marker locus

85-E	88-D	91-A	94-A	97-D
86-B	89-E	92-D	95-E	98-D
87-A	90-A	93-A	96-B	99-D

Questions 100–103

For each of the following diseases, match the approximate frequency of heterozygous carriers.

(A) 1/25
(B) 1/50
(C) 1/1250
(D) 1/5000
(E) None of the above

100. An X-linked recessive disease with a frequency of 1/10,000

101. An autosomal recessive disease with a frequency of 1/10,000

102. An X-linked recessive disease with a frequency of 1/2500 males

103. An autosomal recessive disease with a frequency of 1/2500

Questions 104–108

Phenylketonuria (PKU) is associated with elevated levels of phenylalanine, low tyrosine levels, and severe mental retardation, if untreated. Match each description of children with PKU with the most appropriate reason for their having the disease.

(A) Because the disease was transferred via mitochondrial inheritance
(B) Because PKU is a heterogeneous disorder that may be caused either by enzyme deficiency or abnormal cofactor metabolism
(C) Because PKU was transferred transplacentally to the fetus
(D) Because a treatment goal is prevention of substrate accumulation
(E) Because the initial enzyme deficiency impacts on melanin production

104. A boy who does not respond to dietary restriction of phenylalanine

105. A severely affected son of an untreated woman with PKU

106. A blond child who experienced an increase in pigmentation with treatment

107. An infant is tested in whom the diagnosis was missed because of testing performed immediately postpartum

108. A child whose treatment involved restriction, not total removal, of phenylalanine

Questions 109–111

Several research methods are used to study the genetic contribution to psychiatric illness. Match each description below to the study method it best fits.

(A) Twin study
(B) Pedigree analysis
(C) Adoption study
(D) Family interview
(E) Linkage analysis

109. Allows distinction between genetic and environmental components in psychiatric illness

110. Tends to underestimate psychiatric morbidity

111. Allows identification of mild psychiatric illness

Questions 112–116

Match each genetic term to the correct clinical situation.

(A) Advanced paternal age
(B) Consanguinity
(C) Recurrent early miscarriage
(D) Advanced maternal age
(E) Two maternal uncles with a clinical picture similar to that of the affected child

112. X-linked inheritance

113. New dominant mutation

114. Trisomy 13

115. Translocation

116. Autosomal recessive inheritance

100-D 103-A 106-E 109-C 112-E 115-C
101-B 104-B 107-C 110-B 113-A 116-B
102-C 105-C 108-D 111-D 114-D

Questions 117–121

A 5-year-old boy who presents with muscle weakness is diagnosed as having Duchenne muscular dystrophy (DMD). The physician approaches the family because she has heard about a trial on a new drug that promises to slow the rate of loss of muscle bulk. The physician insists that the trial coordinator review the consent process with her before discussing the trial with the family. Match each of the following aspects of informed consent with the ethical principle it illustrates.

(A) Veracity
(B) Autonomy
(C) Beneficence
(D) Justice
(E) Fidelity

117. The right to refuse entry into the trial

118. Full discussion of the risks and benefits of the drug

119. The signatures of both the patient or guardian and the investigator on the consent form indicating acceptance

120. The initial approach to the family by their physician

121. The right to full care should the patient refuse to enter the trial

Questions 122–126

Match each of the following descriptions of chromosome structure with the most appropriate term.

(A) Nucleosome
(B) Euchromatin
(C) Heterochromatin
(D) Centromere
(E) Telomere

122. A DNA sequence required to maintain the stability of chromosome ends

123. The last segment of the chromosome to replicate and separate during mitosis

124. A complex of eight histones wrapped with two turns of the chromosomal DNA

125. A segment of highly condensed genetic material that is largely devoid of active genes

126. A complex associated with specific proteins that attach to microtubules involved in chromosome movement during cell division

---

117-B	120-C	123-D	126-D
118-A	121-D	124-A	
119-E	122-E	125-C	

Questions 127–132

The most common type of color blindness, green-shift, is seen in 5% of Caucasian males and is inherited as an X-linked recessive trait. The mutant allele (*g*) causes a shift in the absorption of a green-sensitive pigment. Females can have three different genotypes, *GG, Gg,* and *gg,* while males are either *GY* or *gY*. Match the most likely outcomes to the following possible crosses.

(A) *GG* × *GY*
(B) *GG* × *gY*
(C) *Gg* × *GY*
(D) *Gg* × *gY*
(E) *gg* × *gY*
(F) *gg* × *GY*

127. Risk of having affected daughters is 50%

128. May have both normal and affected sons and both homozygous normal and heterozygous daughters

129. All offspring are clinically affected

130. All daughters are homozygous normal

131. All offspring are clinically normal, but daughters are all heterozygous

132. All daughters are heterozygous; all sons are affected

Questions 133–136

For each clinical situation described below, select the corresponding congenital anomaly.

(A) Malformation
(B) Disruption
(C) Deformation
(D) Dysplasia

133. A child has short legs because of abnormal formation of cartilage (as reflected by histologically abnormal growth plates) in the long bones

134. A child is born with only her right kidney because of complete loss of blood supply to the left kidney at 14 weeks of gestation

135. A twin has an indentation in his skull because his head and his twin's head were both tightly jammed into the mother's pelvis during the last month of pregnancy

136. A child is born with severe chorioretinitis and blindness of her right eye because of an in utero rubella infection

---

127-D	130-A	133-D	136-B
128-C	131-B	134-B	
129-E	132-F	135-C	

Questions 137–141

For each clinical presentation listed below, match the correct condition.

(A) Pure gonadal dysgenesis
(B) 21-Hydroxylase deficiency
(C) 5α-Reductase deficiency
(D) Mixed gonadal dysgenesis
(E) Androgen insensitivity

137. A phenotypic male infant is brought into the emergency room with vomiting and dehydration; he has no dysmorphic features, but the intern notices that the infant does have bilateral cryptorchidism

138. A 6-month-old female infant presents with an inguinal hernia, which at surgery is found to contain a gonad that is a normal testis

139. While an infant is being assessed for a suspected coarctation, she is noted to have neck webbing, a shield-shaped chest, and an enlarged clitoris; her chromosomes show the presence of two cell lines

140. A 17-year-old female presents with primary amenorrhea and is noted to have normal stature, mild neck webbing, and, on ultrasound, a normal-sized uterus but a mass in the right ovarian area; she has no breast development as yet; chromosomes are 46,XY

141. An infant is noted to have ambiguous genitalia and severe hypospadias, with the urethral opening on the perineum; gonads are present in a bifid scrotum; there is no evidence of a uterus on ultrasound; serum testosterone levels are normal

Questions 142–146

A woman has cystic fibrosis (CF) and is the only person in her family affected with the disease. The family comes in for counseling and is told that the carrier frequency in the background population is 1/30. Match the woman's unaffected relatives to the following carrier risks. Assume that all partners of family members are unrelated and have no family history of CF.

(A) 1/30
(B) 1/4
(C) 1/2
(D) 2/3
(E) 1

142. Mother

143. Daughter

144. Granddaughter

145. Brother

146. Husband's brother

137-B 138-E 139-D 140-A 141-C 142-E 143-E 144-C 145-D 146-A

Questions 147–150

For each clinical description given below, select the corresponding pattern of anomalies.

(A) Sequence
(B) Developmental field defect
(C) Syndrome
(D) Association

147. A child has congenital heart disease, abnormal external ears, and hydrocephalus as a result of maternal treatment with isotretinoin during pregnancy

148. A child has spina bifida and club feet, which have resulted from little normal fetal movement related to the lack of innervation produced by the spina bifida

149. A stillborn female infant has lumbar vertebral defects, renal agenesis, sirenomelia (fusion of the legs into a single midline appendage), and absence of the vagina, uterus, and fallopian tubes, all of which result from abnormal development of the caudal mesoderm prior to the fourth week of gestation

150. A child has thrombocytopenia, bilateral absence of the radius, ulnar hypoplasia, and congenital heart disease, all of which form a pattern of anomalies known to be transmitted as an autosomal recessive trait

Questions 151–154

Match each description below with the term most closely related.

(A) Positive eugenics
(B) Negative eugenics
(C) New mutations
(D) Genetic lethals
(E) Lethal equivalents

151. Rare in the general population but common among patients with autosomal dominant diseases that reduce fertility

152. Each person carries an average of three to five of these

153. Accounts for the increased risk of some dominant diseases among the children of older fathers

154. Seeks to decrease the propagation of "less desirable" types of people

---

147-C
148-A
149-B
150-C
151-C
152-E
153-C
154-B

ANSWERS AND EXPLANATIONS

1. The answer is B *[Ch 1 IV C 1; Ch 10 III B 3 b-c].*
Cytogenetic studies require dividing cells and thus can only be performed on living tissue. Frozen, fixed, or dried material of any sort is unsatisfactory. Chromosome analysis is done on heparinized blood specimens; clotted blood cannot be used. Cytogenetic studies can be done on a variety of growing tissues, including specimens obtained via amniocentesis, chorionic villus sampling, or skin biopsy or from fresh autopsy materials.

2. The answer is A *[Ch 4 II C 1].*
Linkage is the occurrence of two or more genetic loci in such close proximity on a chromosome that they are likely to segregate together at meiosis. Since linkage is a result of meiosis, it can only be measured in studies in which the products of meiosis are determined (i.e., in family studies). Linkage disequilibrium is the tendency for certain alleles at linked loci to occur together more often than expected by chance. Linkage disequilibrium is demonstrated with respect to allele frequencies in a population and can only be measured in population studies.

3. The answer is E *[Ch 6 III A 1–3].*
Neither treatment of genetic diseases nor genetic counseling is likely to have a substantial effect on the frequencies of disease genes in future generations. Prevention and treatment of genetic disease benefit individual patients and families. The effect on the frequencies of the disease genes in future pregnancies is generally negligible.

4. The answer is C *[Ch 7 IV B].*
Normal variation usually results from the combined effects of many genetic and nongenetic factors; therefore, it is multifactorial. On the average, each parent contributes half of the genetic factors involved in any particular trait to a child, regardless of sex. The genes involved in normal variation may differ in different populations. While the genetic endowment of an individual determines the range of potential development of a normal trait such as intelligence, full achievement of the genetic potential depends on an optimal environment. Extreme values of a phenotype are uncommon on the basis of normal variation but may occur. Thus, a child may be taller or lighter skinned or more intelligent than her parents if she happens by chance to have inherited more of the extreme predisposing factors from both her mother and father.

5. The answer is D *[Ch 8 II D 3; III A, B].*
Mutagens cause mutation, or alteration of the genetic material (DNA or chromosomes), in a single cell. In contrast, teratogens affect the development of a tissue, organ, or structure. Many mutagens, such as x-irradiation and some cancer chemotherapeutic agents, are also potentially teratogenic. Since teratogens affect the developing embryo or fetus, they typically act via exposure of the mother and not the father. About 10% of congenital anomalies are thought to be due to teratogenic effects. Mutagens are rarely recognized as a cause of congenital anomalies.

6. The answer is B *[Ch 9 III A 2; IV A 2].*
The major events of early mammalian embryogenesis are cell lineage specification (which does not occur clonally and does not predetermine precise developmental fates), segmentation, and regional specialization along the body axis. Maternal gene products in the zygote are not critically important in mammalian embryogenesis. Activation of proto-oncogenes occurs during neoplasia, not normal embryogenesis.

7. The answer is D *[Ch 10 II C 2–4].*
Duchenne muscular dystrophy (DMD) is an X-linked recessive disease that is lethal. In this instance, the family history is negative except for the affected brother. There is a 2/3 chance that the woman in the question (the proband) carries the abnormal gene and a 1/3 chance that her son has a new mutation. If the proband's mother was a carrier, there is a 1/2 chance that the proband inherited the abnormal gene. However, the proband's creatine phosphokinase (CPK) test showed normal results despite the fact that the results are abnormal in 75% of carriers. This decreases the likelihood that the proband is a carrier.

A Bayesian calculation is necessary. This is illustrated in the following probability tree. The calculation could also be done by drawing a probability space, but use of probability spaces becomes more difficult with more complex calculations.

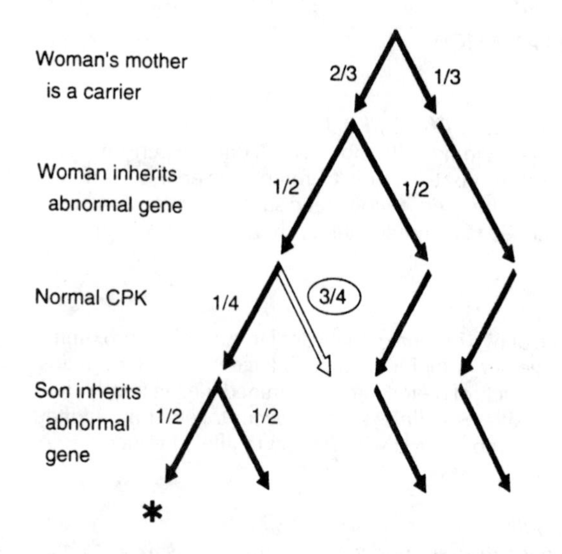

The proband can only have an affected son if: (1) her mother was a carrier of the abnormal gene; (2) her mother transmitted the abnormal gene to the proband; (3) the proband is among the 25% of carriers whose CPK test results are normal; and (4) she transmits the abnormal gene to her son. The other possibilities that must be considered are: (1) the proband's mother was not a carrier (i.e., the affected boy has a new mutation); (2) the proband's mother was a carrier but did not transmit the abnormal gene to the proband; and (3) the proband's mother was a carrier and transmitted the abnormal gene to the proband, but the proband did not transmit the abnormal gene to her son.

The chance that the proband's son has DMD is

$$P = \frac{(2/3 \times 1/2 \times 1/4 \times 1/2)}{(2/3 \times 1/2 \times 1/4 \times 1/2) + (2/3 \times 1/2 \times 1/4 \times 1/2) + (2/3 \times 1/2) + 1/3}$$

$$P = \frac{2/48}{2/48 + 2/48 + 2/6 + 1/3}$$

$$P = \frac{2/48}{2/48 + 2/48 + 16/48 + 16/48}$$

$$P = \frac{2/48}{36/48}$$

$$P = 2/36 = 1/18$$

8. The answer is C *[Ch 11 IV B, D 3].*
There is no easy or perfect answer to this problem. The physician has been made privy to information through the linkage analysis that has the potential to do great damage to this family. On the other hand, the physician would like to reassure the patient that he is not at risk for Huntington's disease. If the physician chooses option A, the patient still believes he is at risk when information is to the contrary. Option B impinges on the mother's autonomy, as she has chosen not to disclose the paternity, even after a private session. Going ahead in spite of this could jeopardize her relationship with the family. Option D may be the most destructive as it betrays the mother's autonomy while placing a burden on the son, and likely altering his relationship with his mother. Option C is not perfect but does allow the mother her privacy and gives correct information to the patient. It contravenes the principle of veracity, but there may be some justification for this.

9. The answer is C *[Ch 4 I B 1–2].*
All genetic polymorphisms ultimately reflect alterations of DNA sequence, but these alterations may be demonstrated in a variety of ways. Some polymorphisms affect protein structure or function, but many do not. Demonstration of most polymorphisms requires special studies of the DNA, proteins, or antigens, but some polymorphisms are apparent as variations in gross physical features. Highly polymorphic variants [such as some variable number of tandem repeats (VNTRs)] can be used to distinguish tissues from different individuals, but most genetic polymorphisms are not sufficiently variable for this purpose. Moreover, monozygotic twins cannot be distinguished by polymorphisms because they are genetically identical. Although polymorphisms are used extensively in linkage studies, limited variability or difficulty of detection of some polymorphisms limits their usefulness.

10. The answer is D *[Ch 7 IV C 1 c, d, f].*
Many common congenital anomalies (e.g., cleft palate, anencephaly, ventricular septal defect) may be inherited as multifactorial traits. However, these same defects also occur as features of malformation syndromes and other generalized patterns of congenital anomalies. Multifactorial congenital anomalies usually do not differ in their appearance from similar defects seen as components of various dysmorphic syndromes, but the multifactorial anomalies are usually isolated (i.e., they occur in a child who is otherwise normal). Like other multifactorial diseases, multifactorial congenital anomalies characteristically exhibit a much higher concordance rate in monozygotic twins (who share all of their genes) than in dizygotic twins (who share only half of all genes). The recurrence risk for multifactorial congenital anomalies is similar in an affected patient's children and siblings, who share about 50% of their genes with the patient.

11. The answer is E *[Ch 8 II C 1–3, 5].*
The risk of teratogenic effects in a pregnancy depends not only on the chemical or pharmacologic characteristics of the agent but also upon the dose, route, and other circumstances of exposure. Knowledge of the pharmacologic action of an agent in a child is usually insufficient to predict its teratogenic potential, because different biochemical and physiologic mechanisms are often involved in the embryogenesis of a tissue and in its function once it is formed. Simultaneous exposures to several teratogenic agents may produce effects that are similar to, greater than, or less than those of the same agents individually. The embryo is most sensitive to teratogenic effects between 2 weeks and 10 weeks after conception. Malformations are unlikely to be produced after this time, but disruptive, cytotoxic, or growth-inhibiting effects may occur.

12. The answer is A *[Ch 9 II D 3 a–c].*
The woman must have been Rh^+ and her mother Rh^- because the woman's red cells were attacked by her mother's anti-Rh antibody. The husband's mother must have been Rh^- and he must be Rh^- as well, because he did not get Rh hemolytic disease despite the fact that his mother was highly sensitized. Since the woman is Rh^+ and her husband is Rh^-, they are not at risk to have a child with hemolytic disease of the newborn due to the major Rh antigen.

13. The answer is D *[Ch 10 IV A].*
A distant family history of spina bifida does not significantly increase the risk for neural tube defects in a fetus, nor does it increase the risk for chromosome abnormalities.

Both advanced maternal age (i.e., the mother is at or over 35 years of age at the due date) and a previous stillbirth or livebirth with a chromosome abnormality are indications for fetal karyotyping via either chorionic villus sampling (CVS) or amniocentesis.

A cystic hygroma may be associated with a variety of chromosome abnormalities, including Turner syndrome and trisomy 21.

If one parent has a translocation, there will be an increased risk for both early miscarriages and abnormal liveborn infants. A paternal robertsonian translocation is associated with a risk of 2% that the fetus of an ongoing pregnancy will be chromosomally abnormal.

14. The answer is D *[Ch 11 II A 1–2].*
Many genetic diseases with similar phenotypic manifestations are etiologically and pathogenically heterogeneous. Effective therapy often requires knowing precisely what gene or protein is altered in an affected patient. Even if the precise diagnosis or biochemical alteration is unknown and regardless of the age of onset, symptomatic treatment of genetic diseases is often possible. Therapeutic genes are employed only in somatic gene therapy, an infrequently used experimental approach.

15. The answer is D *[Ch 8 II C 1, 2, 6; IV B 1, 2].*
Most human teratogens produce a characteristic pattern of congenital anomalies rather than a single isolated defect. The observation of this pattern repeatedly within a series of infants born to women exposed to a given agent suggests that the agent is teratogenic, especially if the exposure is uncommon. Experimental animal teratology studies, particularly when performed at vastly greater doses than those used in pregnant women, are difficult to interpret in terms of human risk. Pharmacologic similarities among agents do not necessarily predict similarities in teratogenic effects, because the metabolism of embryos differs from the metabolism of adults. Isolated case reports of infants with a variety of congenital anomalies who were delivered from women who took a medication during pregnancy are more likely to represent a chance association than a causal one, especially if use of the medication among pregnant women is common. Most congenital anomalies are of multifactorial or unknown cause. Ruling out a recognizable mendelian or chromosomal cause eliminates only a small proportion of the congenital anomalies that are not due to teratogenic factors.

16. The answer is E *[Ch 11 II B–F].*
Symptomatic treatment of genetic disease by medical, surgical, or other means is by far the most commonly used approach. Somatic gene therapy has very limited application and is considered to be experimental. Organ transplantation is more widely used but only in certain special circumstances. Modulation of gene expression is a theoretically attractive approach, but knowledge of gene regulation is currently insufficient for substantial clinical application. Amelioration of metabolic abnormalities is used for inborn metabolic errors, a relatively uncommon group of disorders, in which the pathogenesis is more or less well understood.

17. The answer is A *[Ch 8 II D 1 a, 2 c, 3 a, 5 b, 6 a (4)].*
Neural tube defects occur in 1% to 2% of infants of women with insulin-dependent diabetes mellitus. Prenatal diagnosis for neural tube defects is, therefore, indicated in these women's pregnancies. Although first-trimester rubella infection may be teratogenic, it does not cause neural tube defects. Maternal use of marijuana during pregnancy has not been shown to be teratogenic. Although ionizing radiation may produce teratogenic effects, the dose involved in an abdominal x-ray is unlikely to pose a significant risk to the fetus. Maternal use of androgenic hormones during pregnancy may cause virilization of a female fetus but is not associated with an increased risk for neural tube defects.

18. The answer is C *[Ch 8 II D 5 a (1)].*
Common features of fetal alcohol syndrome include behavioral abnormalities, mental retardation, growth deficiency, and characteristic facial appearance. Limb reduction defects are not common features. Although women who drink only in occasional binges or who drink less than 3 ounces of absolute alcohol daily during pregnancy do not have children with classic fetal alcohol syndrome, the risk of less severe damage to their fetuses may still be substantial. The only amount of maternal drinking during pregnancy that is known to be safe is none.

19. The answer is B *[Ch 1 III A, B 2].*
Every cell division in mitosis is preceded by a complete round of DNA synthesis, but only the first division of meiosis is preceded by a complete round of DNA synthesis. Meiosis, which only occurs in germ cells, results in haploid cells (a reduction of chromosome number in the cell from 46 to 23). There is no change in chromosome number in mitosis, which all somatic cells undergo. Recombination occurs more frequently in meiosis than in mitosis. Chromosome pairing (i.e., synapsis) occurs in meiosis, but it does not occur in mitosis.

20. The answer is A *[Ch 2 III F; Table 2-3; Case 2-A].*
If the couple was ascertained because of recurrent spontaneous abortions, the initial investigation may have been prompted by the number of spontaneous abortions. However, once a translocation has been discovered, the precise number of miscarriages is not important to risk counseling.

If ascertainment was because of a stillborn or liveborn infant with abnormalities, it is clear that subsequent affected pregnancies might also not abort early but progress to term instead. If, on the other hand, the pregnancies have all ended in early miscarriages, it is more likely that unbalanced chromosome combinations will not be viable.

For some reciprocal translocations, the sex of the parent should be considered before giving recurrence risk figures. For example, a reciprocal translocation with a 3:1 segregation may have a higher risk in the female carrier. This information is not currently available for all rearrangements, but can be reviewed before counseling.

If a translocation is found in a dysmorphic infant, counseling the parents will depend on whether the rearrangement is de novo (i.e., found only in the child) or familial. If a parent is a balanced carrier, counseling to the couple will take into account the previously mentioned factors and risks will not change if other family members are also affected. However, although the couple's risk does not change, the identification of other possibly affected relatives is extremely important, as counseling can then be provided to the extended family.

The degree of potential imbalance as a consequence of a translocation, as well as the intrinsic nature of the rearrangement and its behavior at meiosis, affects the pregnancy outcomes. Risk figures should be sought from the literature or from the family itself, if enough members are affected with the anomaly.

21. The answer is A *[Ch 4 I A 2, B 1 a, b (2), 2].*
Genetic markers are polymorphisms that are easily detected clinically or by laboratory testing. Genetic markers are inherited in a simple mendelian pattern, not as multifactorial traits. Alternative alleles must be easily differentiated in order for marker loci to be useful. Genetic markers may be detected by DNA-based methods [e.g., restriction fragment length polymorphisms (RFLPs)], as protein structural variants

[e.g., human leukocyte antigen (HLA) antigens], or as functional alterations of proteins (e.g., the ability to taste phenylthiocarbamide). Genetic markers are useful in linkage studies and in forensic medicine.

22. The answer is B *[Ch 6 I B 2].*
The Hardy-Weinberg law can be applied regardless of whether the genes involved are dominant, recessive, autosomal, or X-linked. The Hardy-Weinberg law is based on the following assumptions: (1) Mating within the population is completely random, (2) there is no mutation occurring at the locus, (3) there is no selection for any of the genotypes at the locus, (4) there is no migration into or out of the population. These assumptions are probably never entirely correct for any one locus, but if they are nearly correct, the Hardy-Weinberg law will be applicable.

23. The answer is A *[Ch 5 II B 2 b (1); Ch 7 II B 1 c].*
Most biochemical disorders are inherited as autosomal recessive traits. Heterozygotes are clinically normal, although most have intermediate levels of the enzyme. Normal siblings have a 2/3 chance of being heterozygotes, but their partners are unlikely to be carriers of the same abnormal gene, so the risk that their children will be affected is low. Like other enzyme deficiencies, allelic mutations may cause disease of lesser severity (some mutations may lead to nonproduction of the enzyme, while others may lead to a less efficient enzyme).

24. The answer is B *[Ch 8 II D 2].*
Transplacental infections with the viruses that cause rubella and varicella, cytomegalovirus (CMV), or *Toxoplasma gondii* may cause permanent damage to the developing human embryo. Maternal gonorrheal infection during pregnancy is not known to be teratogenic, although perinatal exposure can cause neonatal ophthalmia.

25. The answer is D *[Ch 9 IV C 4, D 3].*
Androgen-binding studies on cultured genital skin cells are appropriate if the suspected diagnosis is androgen insensitivity. However, such studies often take several months, and the diagnosis may be inferred in other ways before then. Results of karyotyping can be back in 48 hours. If the newborn is karyotyped XX, the fetus is a masculinized female or could be the rare XX male with genital ambiguity. If the infant is chromosomally male, he is much more likely to have a form of incomplete androgen insensitivity.

Levels of 17-ketosteroids are elevated in many forms of congenital adrenal hyperplasia and can be determined quickly, but the clinician should be aware that the levels may not be greatly elevated until several days after birth.

An ultrasound of the lower abdomen allows assessment of the urinary tract, which may be abnormal in any infant with ambiguous genitalia. Also, the presence or absence of a uterus is crucial to evaluate. If a uterus is present and the chromosomes are 46,XX, the most likely diagnosis is congenital adrenal hyperplasia.

If the differential diagnosis includes congenital adrenal hyperplasia, the infant may have life-threatening electrolyte abnormalities with low sodium and elevated potassium levels. Electrolytes are also a concern in an infant who has renal failure, especially if the urinary tract is malformed.

26. The answer is E *[Ch 10 III B 6].*
The measurement of acetylcholinesterase in the amniotic fluid, not the blood, is useful as a more specific indicator of a neural tube defect.

The patient had a routine ultrasound along with her maternal serum α-fetoprotein (MSAFP) assessment. Therefore, enough information should already be available to verify the gestational age, but the physician should make certain that it was recorded accurately when the MSAFP was analyzed. Also, if the weight used in the calculation is too low (e.g., if instead of 105 kilograms, 105 pounds is used as the weight), the MSAFP value will be falsely elevated.

If the gestational age and other information used to interpret the MSAFP are correct, and therefore the interpretation is valid, it is appropriate to schedule a detailed (level II) ultrasound. This is sometimes done before arranging the amniocentesis, since it may be possible to detect a reason for the elevated MSAFP (e.g., anencephaly) via the detailed ultrasound. In some cases, both the level II ultrasound and the amniocentesis are arranged for the same time.

27. The answer is C *[Ch 11 I A 1–3].*
Genetic screening is justified to improve the management of a genetic condition or its complications in patients who have it, to prevent or delay the manifestations of a disease in genetically predisposed individuals (as is the case in D and E), or to provide reproductive options to people at high risk of having children with a serious genetic disease (as in A and B). Screening to obtain a better rating on life insurance

does not fall into any of these categories and implies that those found not to be in the genetically low-risk group will have to pay more for their insurance. This is discriminatory and therefore inappropriate.

28. The answer is E *[Ch 5 II A 2; III A 1, 3].*
Direct testing requires that part of the normal gene has been cloned. Direct approaches to DNA diagnosis utilize a probe that is part of, or complementary to, the gene in question. As such, the results are more accurate. In contrast, for linkage analysis (indirect testing), which utilizes DNA polymorphisms close to the gene of interest, DNA from numerous family members is needed. Genetic heterogeneity can complicate indirect approaches to DNA diagnosis.

29. The answer is D *[Ch 7 III B 2].*
Necessary surgery for this woman can be undertaken, but appropriate medications, such as dantrolene sodium, should be available in surgery to prevent or reduce the severity of a hyperthermic response. Clearly, this woman should be considered at an increased risk for malignant hyperthermia, and wearing a medic alert bracelet may prevent a catastrophic outcome in an emergency.

30. The answer is A *[Ch 8 IV B 1–5].*
Case reports are useful in raising suspicion about the teratogenic potential of an agent, but associations observed in individual case reports are usually spurious. Most human teratogens have been recognized in a clinical series when a recurrent and unusual pattern of congenital anomalies was observed among infants born to women who had similar exposures during pregnancy. Epidemiologic studies of both the cohort and case-control type may be subject to a variety of serious biases. This must be considered when interpreting such studies. Experimental animals differ from humans in many important ways. Consequently, a teratogenic effect observed in experimental animals is not necessarily predictive of a similar effect occurring in women, even if similar conditions of exposure occur.

31. The answer is C *[Ch 9 V A 2 b].*
Twin studies are valuable research tools for studying complicated genetic disorders. Twin studies allow conclusions to be drawn regarding the strength of genetic contributions to disease. However, the study method may have several inherent problems. Accurate zygosity data are crucial to distinguish monozygotic from dizygotic pairs and to allow comparisons between the two groups. Self-selection (i.e., the likelihood for twins who are concordant for a disorder to come to attention) and small sample size tend to bias the results of twin studies. If the twins who participate in a study are drawn from different sources (e.g., a hospital versus a community setting), they may not be comparable and results may be biased. However, sex of the study participants has not been an important factor in twin studies.

32. The answer is A *[Ch 11 II B].*
Dietary restriction in phenylketonuria (PKU) is designed to improve the metabolic abnormality that occurs, not just to assuage a symptom. The other therapies are examples of symptomatic treatments involving an electronic device, a medication, speech therapy, and surgical intervention, respectively.

33. The answer is A *[Ch 5 II A 1 a, 3 a, B 2 b (1)].*
Mutations underlying most genetic diseases are unknown. Of 4000 genetic diseases, less than 10% have been defined at the DNA level. Major rearrangements are uncommon causes of mutation. However, exceptions to this rule are Duchenne and Becker muscular dystrophies, in which deletions are common causes of mutation. Most persons with Tay-Sachs disease are of Ashkenazi Jewish descent and have the same point mutation. Most persons with cystic fibrosis (CF) have a common mutation.

34. The answer is D *[Ch 8 IV C 1–2].*
Relative risk (RR) is the ratio of the occurrence rate of a condition among individuals who were exposed to an agent and the occurrence rate among individuals who were not exposed. An RR of 0.05 indicates that infants born to women exposed to an agent during pregnancy have a rate of birth defects that is one-twentieth that of infants of unexposed women (i.e., the exposure exerts a protective effect).

An association may be seen in a birth defects epidemiology study if congenital anomalies and maternal exposures are not directly related to each other, but both are associated with a common third factor.

Observing a dosage effect supports the interpretation that an association between a maternal exposure and congenital anomalies in infants is causal.

Statistical significance (p) indicates that an observed association is unlikely to be due to chance. In studies in which many comparisons are made, some statistically significant associations are expected to occur by chance. For example, if $p < 0.05$ is chosen as the significance level, 1/20, or 5/100, comparisons would be expected to be statistically significant purely on the basis of chance.

35. The answer is D *[Ch 9 III B 3 b, C 1 d, 2 b (3)].*
Chromosome rearrangement is a common means of activating oncogenes in neoplastic cells. Deletions and chromosome loss are common methods of loss of tumor suppressor genes. Several different cytogenetic lines may be apparent within a neoplastic population, but all of the lines can usually be shown to have arisen from a common progenitor (i.e., they are clonally related). The chromosome abnormalities that are acquired during the neoplastic process arise somatically and do not involve the germ cells. Such chromosome abnormalities cannot, therefore, be transmitted to the children of an affected person.

36. The answer is C *[Ch 3 II C; Ch 11 III B].*
Thalassemia and glucose-6-phosphate dehydrogenase deficiency (G6PD) are frequently seen in African as well as non-African populations that have historically been exposed to malaria, such as Greek, Italian, Chinese, and Southeast Asian populations. The genes responsible for these diseases are thought to have been selected for because they provide protection against malaria. Protection is seen in both heterozygous carriers and affected individuals, but sickle cell disease and thalassemia occur only in homozygotes. G6PD deficiency, an X-linked recessive trait, usually produces disease only in affected males. Sickle cell disease, thalassemia, and G6PD deficiency can produce disease whether or not malaria happens to occur where the affected person lives.

37. The answer is B *[Ch 9 V D 3].*
Although Turner syndrome females often are immature, they are not mentally handicapped and do not exhibit an increased incidence of psychiatric disease. Psychiatric illness (e.g., schizophrenia) is increased in males who are affected by Klinefelter syndrome and females who are affected with trisomy X. As children, XXY males demonstrate poor judgment and learning problems. XYY males tend to be immature and impulsive.

38. The answer is E *[Ch 11 IV B].*
By acting in a paternalistic manner, even though his intentions are good, the physician has removed this woman's autonomy to make a choice regarding the results of her amniocentesis. Beneficence (the physician's decision to withhold the diagnosis from the woman to avoid worrying her) is in conflict with autonomy. The physician has not been truthful with his patient (veracity) and has not fulfilled his contract with her to inform her of results of the testing so that she can choose whether or not to continue her pregnancy (fidelity). Justice is not an issue, since the woman was offered a prenatal diagnosis, as would be every other 35-year-old pregnant woman.

39–41. The answers are: 39-D, 40-B, 41-E *[Ch 5 III B 1; Case 5-C].*
Evidently, the gene causing Huntington's disease (HD) in this family is inherited together with the A marker. However, II-1 has AB markers, and it is impossible to know whether she has inherited the A or the B marker from the affected father; therefore, her risk of having inherited the HD gene has not changed.

II-3 has inherited marker B from his affected father and marker B from his mother. In view of the fact that the gene for HD is inherited together with marker A, this man's risk of being affected with HD has significantly decreased.

II-4 has inherited an A marker from his mother. However, II-4 has a C marker. The supposed father (I-1) does not have a C marker. Therefore, it is highly unlikely that I-1 is the father of II-4. This being the case, II-4 has no risk of having inherited anything from I-1. The frequency of undisclosed illegitimacy is high in the population (3% to 5%) and can dramatically alter risk estimates.

42–45. The answers are: 42-B, 43-C, 44-D, 45-B *[Ch 1 I B].*
Multifactorial conditions are caused by a combination of many factors, some of which are genetic and some of which are not. In contrast, mendelian diseases are caused by single abnormal genes or gene pairs (i.e., they are unifactorial).

Mendelian diseases caused by genes located on chromosomes other than the X or Y are said to be autosomal. Single gene disorders are termed X-linked if the responsible gene is on the X chromosome.

Conditions that are expressed when just a single gene of a pair is abnormal are called dominant. Diseases that require both genes of a pair to be abnormal for expression are said to be recessive.

Most common diseases of middle life are caused by a combination of many interacting genetic and nongenetic predisposing factors. Mendelian disorders and chromosome abnormalities occur in adults but are relatively uncommon.

46–50. The answers are: 46-E *[Ch 2 II B 2 a]*, **47-D** *[Ch 2 II B 2 c]*, **48-B** *[Ch 2 II B 2 b]*, **49-C** *[Ch 2 II C 1 b, 2]*, **50-A** *[Ch 2 II A 2 a].*
Down syndrome (trisomy 21) may be suspected initially in an infant on the basis of hypotonia and typical

facial features. The head is brachycephalic with a small occiput, the ears are small and squared, there are often epicanthic folds, as well as Brushfield's spots involving the irides. About 40% of the children have congenital heart disease, with the most characteristic lesion being an atrioventricular canal cushion defect. The hands are short and broad, and half of the infants have associated single palmar creases.

The trisomy 18 infant is usually significantly growth-retarded, with tiny facial features and prominent occiput. Because of abnormal tendon placements, there may be overlapping of the second and fifth fingers. Almost all affected infants have congenital heart disease, and many have other major malformations as well (e.g., diaphragmatic hernia).

The trisomy 13 infant, unlike the trisomy 18 infant, may be well-grown. The infant has a variety of midline abnormalities, including clefting of the lip and palate and midline scalp lesions, as well as congenital heart disease and omphalocele. Many affected infants also have postaxial polydactyly. The brain is frequently malformed, with the most serious lesion being holoprosencephaly.

The Turner syndrome conceptus may abort spontaneously early in pregnancy, or a fetus with Turner syndrome may have a cystic hygroma (secondary to delayed development of lymph channels from the posterior neck) or overt hydrops fetalis. Internally, the fetus may have a coarctation of the aorta and horseshoe kidneys.

The triploid embryo is usually the result of fertilization of a haploid oocyte by two sperm. The placenta is large and shows cystic changes like those seen with molar pregnancies. The embryo itself may be growth-disorganized and tiny.

51–55. The answers are: 51-B, 52-D, 53-C, 54-B, 55-A *[Ch 4 I B 1].*
RFLPs (restriction fragment length polymorphisms) are inherited DNA sequence variations that can result in the gain or loss of a restriction endonuclease recognition site, often by alteration of a single nucleotide.

VNTRs (variable number of tandem repeats) are a special kind of RFLP that involve differences in the number of short repeated DNA segments between restriction sites. Some VNTR loci are so polymorphic that two unrelated people are very unlikely to share the same alleles. This extreme polymorphism makes such VNTRs useful as "DNA fingerprints" in forensic medicine.

Polymorphic enzyme activity variants may be involved in pharmacogenetic interactions. For example, some variants of the enzyme glucose-6-phosphate dehydrogenase (G6PD) predispose people to developing anemia when they take certain medications.

Polymorphic antigenic variants expressed on the surface of blood cells are responsible for transfusion reactions. Common examples of such polymorphisms include the ABO and Rh blood groups.

Chromosome heteromorphisms are heritable differences in chromosome appearance that are useful in determining the origin of chromosome rearrangements.

56–59. The answers are: 56-E, 57-B, 58-D, 59-C *[Ch 6 I B 6].*
Since cystic fibrosis (CF) is an autosomal recessive disease, it occurs in homozygotes for the mutant gene. The frequency of such homozygotes in this population is 1/2500. In terms of the Hardy-Weinberg law,

$$q^2 = 1/2500 \text{ and } q = \sqrt{\frac{1}{2500}}\text{, or } 1/50$$

The frequency of heterozygous carriers in the general population is

$$2pq\text{, or about } 2/50 = 1/25$$

A man and a woman, neither of whom has a family history of CF, would each have a 1/25 chance of being carriers for this condition. In order for the couple to have an affected child, both would have to be carriers and both would have to transmit the abnormal gene to their child. The chance of this is:

$$\frac{1}{25} \times \frac{1}{25} \times \frac{1}{4} = \frac{1}{2500}$$

A woman with CF is homozygous for the abnormal gene. All of her children must inherit this gene from her. If her husband has no family history of CF, his chance of being a carrier is 1/25 and his chance of transmitting the gene, if he is a carrier, is 1/2. Thus, the chance that this couple would have an affected child is:

$$1 \times \frac{1}{25} \times \frac{1}{2} = \frac{1}{50}$$

A woman whose first cousin is affected with CF has a 1/4 chance of carrying the CF gene herself. Her aunt or uncle (her affected cousin's parent) must carry the abnormal gene, so her parent has a 1/2 chance of being a carrier. The woman's husband has a 1/25 chance of being a carrier for CF. If both the woman and her husband are carriers, their child has a 1/4 chance of being homozygous for the CF gene. Thus, the risk that the child will have CF is:

$$\frac{1}{4} \times \frac{1}{25} \times \frac{1}{4} = \frac{1}{400}$$

A man who has a child with CF by a previous marriage must be a carrier of the abnormal gene. His new partner has a 1/25 chance of being a CF carrier. In order for them to have an affected child, both must be carriers and both must transmit the abnormal gene. The chance that their child would have CF is:

$$1 \times \frac{1}{25} \times \frac{1}{4} = \frac{1}{100}$$

60–64. The answers are: 60-D, 61-A, 62-B, 63-B, 64-C *[Ch 7 V A 1].*
About 95% of patients with insulin-dependent diabetes mellitus (IDDM) have human leukocyte antigen (HLA) DR4. HLA DR3/DR4 heterozygotes are particularly susceptible to IDDM, and the DR3/DR4 antigens account for more than half the genetic contribution underlying IDDM.

If a proband with IDDM and his sibling share no haplotype, the risk to the sibling of developing IDDM is approximately 2%. The risk is 5% to a sibling who does share one HLA haplotype with a proband affected with IDDM. About half of the normal population has HLA DR3 or HLA DR4; the empiric risk for developing IDDM is approximately 5% to a sibling for whom no HLA typing is available.

If the proband and sibling both share DR3 and DR4, the risk to the sibling for developing IDDM is approximately 20%. Clearly, the HLA haplotype alone does not underlie the genetics of IDDM, and there must be other genes that also contribute to the development of the disease.

65–69. The answers are: 65-D, 66-E, 67-C, 68-A, 69-B *[Ch 8 II D 5 a (2), (3) (b), (5), (8), b].*
Treatment of a woman with isotretinoin during early pregnancy can cause serious craniofacial malformations and brain and cardiovascular abnormalities in her offspring.

The daughters of mothers who were treated with diethylstilbestrol (DES) during pregnancy often exhibit cytologic abnormalities of the vagina and cervix and have a greatly increased risk of developing cancer of these organs.

Maternal use of cocaine during pregnancy is associated with the occurrence of placental abruption and disruptive lesions of the fetal brain.

Although children born to women who abuse heroin during pregnancy have symptoms of withdrawal in the newborn period, they do not appear to have an increased risk of malformations.

An increased frequency of neural tube defects has been observed among the children of women treated during pregnancy with valproic acid, a commonly used anticonvulsant. Women who are taking this medication during pregnancy should be offered prenatal diagnosis for neural tube defects.

70–74. The answers are: 70-B, 71-D, 72-E, 73-C, 74-A *[Ch 9 I B 1–2, D 1–3].*
The couple in family 70 can only be the parents of infant B. Infant A, C, D, or E cannot be their child because those children are Rh^+ and both the mother and father are Rh^-.

The couple in family 71 can only be the parents of infant D. The child of family 71 must have just the N antigen in the MN group. Infant D is the one who has only the N antigen.

The couple in family 72 could be the parents of infant B, C, or E. However, infant B must belong to family 70, and infant C must belong to family 73, since infants A, B, D, and E all have the N blood group antigen, which neither the mother nor the father in family 73 has. Therefore, infant E must belong to family 72. The couple in family 72 cannot be the parents of infant A, who has blood type O, because the child must inherit the gene for either the A or the B antigen from his mother. The couple in family 72 cannot be the parents of infant D, because he must have inherited the gene for either the A or the B blood group antigen from his father, and the father has the O antigen.

The couple in family 74 could be the parents of infant A or B, but infant B must belong to family 70, so infant A must be family 74's child. The couple in family 74 cannot be the parents of infant C because he did not inherit the gene for the N blood group antigen, which the father must have transmitted to his son. The couple in family 74 cannot be the parents of infant D or E because both have a gene for the B blood group, which neither the mother nor father in family 74 has.

75–79. The answers are: 75-D, 76-A, 77-E, 78-B, 79-C *[Ch 10 III B 1, 3–6].*
The woman who previously delivered a stillborn infant with normal chromosomes but multiple structural

malformations should be offered level II ultrasound, looking specifically for the malformations that were seen in the previous infant. Offering fetal karyotyping is not appropriate, since the first infant did not have a chromosome abnormality.

With a family history of a genetic disorder diagnosable by molecular methods [e.g., Duchenne muscular dystrophy (DMD)], the most efficient method of prenatal diagnosis is chorionic villus sampling (CVS). Not only is the sampling done at an earlier time, but more tissue is available for diagnostic purposes or for faster culture.

The 19-week fetus shown by ultrasound to have an abnormality requires a rapid cytogenetic test to aid the parents in decision-making. In many areas, the option of pregnancy termination is available only until about 20 weeks of gestation. The most rapid method of prenatal cytogenetic diagnosis is fetal blood sampling.

The 25-year-old primigravida could be offered maternal serum α-fetoprotein (MSAFP) testing. She should, of course, be made aware of the screening nature of this test and that further testing will be necessary if the results are abnormal.

The healthy 36-year-old woman is at increased risk for chromosome aneuploidy because of her age, and she should be offered either CVS or amniocentesis. However, since she is already 14 weeks pregnant, the amniocentesis is more appropriate.

80–84. The answers are: 80-A *[Ch 11 II C 2 a]*, **81-C** *[Ch 11 II C 2 d]*, **82-A** *[Ch 11 II C 2 b]*, **83-C** *[Ch 11 II D 2]*, **84-C** *[Ch 11 II F 2]*.
Dietary restriction is useful only if a genetic defect causes accumulation of a toxic metabolite of a dietary component. Dietary restriction cannot replace a missing product.

Activation of a defective metabolic pathway serves to correct a defect whether it is anabolic or catabolic.

Diversion of a substance to an alternate biochemical route results in its elimination by another means. This treatment method is useful if the substance is toxic.

Organ transplantation or somatic gene therapy could, in principle, correct either: (1) the accumulation of a toxic product or (2) the deficit of a necessary one. Replacement of an adenosine deaminase (ADA) gene by marrow transplantation or gene therapy in severe combined immunodeficiency disease caused by ADA deficiency (and subsequent accumulation of deoxadenosine) is an example of (1). Transfer of normal globin genes to a patient with β-thalassemia by marrow transplantation or gene therapy is an example of (2).

85–89. The answers are: 85-E, 86-B, 87-A, 88-D, 89-E *[Ch 1 IV C 1 b (1)–(9)]*.
NOR staining is used to identify nucleolar organizing regions, which are the sites of active ribosomal RNA (rRNA) synthesis on the short arms of chromosomes 13, 14, 15, 21, and 22. G banding is the standard method of cytogenetic analysis employed in most laboratories. The pattern of staining produced by Q banding is similar to that in G banding, with brightly fluorescent Q bands equivalent to darkly staining G bands. In situ hybridization is a method by which labeled DNA probes are made to hybridize with complementary sequences on the chromosomes. In situ hybridization can be used to identify individual single-copy genes.

90–94. The answers are: 90-A, 91-A, 92-D, 93-A, 94-A *[Ch 3 I D 3 a; II A 1 a, b, 2, 4 a]*.
The most common mutation underlying cystic fibrosis (CF) is a deletion of a phenylalanine residue at position 508 of the CF gene, which accounts for about 70% of all CF alleles among people of Northern European ancestry. Major deletions and rearrangements have not yet been detected by Southern blot analysis in CF, but numerous other types of mutations, including nonsense and missense mutations, have been defined.

The most common mutation underlying α-thalassemia is a deletion due to crossing over in the region of the α-globin gene cluster, resulting in chromosomes that lack functional α-globin genes.

Point mutations are the most common cause of β-thalassemia. Many of these mutations decrease the abundance of normal β-globin messenger RNA (mRNA) by either causing some defect in the regulatory region or affecting RNA splicing.

Most of the defects underlying Duchenne muscular dystrophy (DMD) are deletions that account for more than 60% of all mutations. The mutations are not random but are clustered in certain regions of the gene.

Growth hormone deficiency is most often caused by unequal crossing over, resulting in gene deletion for this particular hormone.

95–99. The answers are: 95-E, 96-B, 97-D, 98-D, 99-D *[Ch 4 II A–B]*.
The pedigree of the family follows with the genotypes at the marker locus indicated under each symbol.

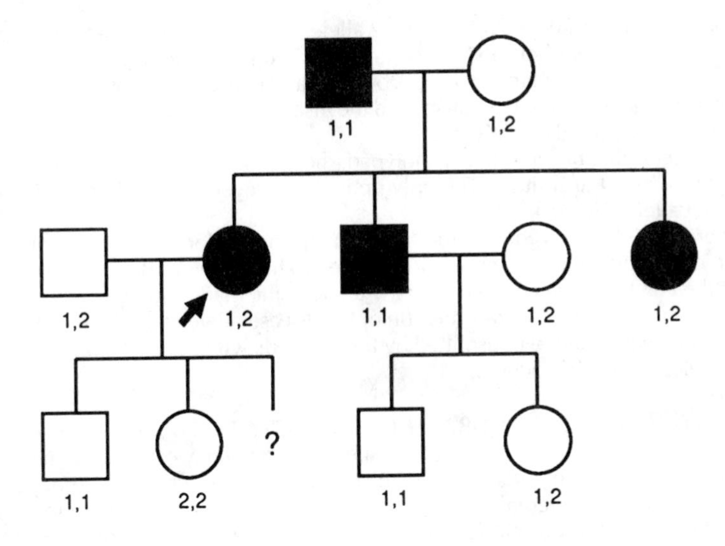

In order to interpret the linkage data, the phase of the linkage must be determined. Each person who carries the mutant gene for Huntington's disease (HD) also carries the *1* allele at the marker locus. The genotypes in the family can be substituted for the symbols, and the pedigree redrawn as shown below. The genotypes at the HD and marker locus on one chromosome are shown above the line for each individual, those on the other chromosome are shown below the line. *H* is used to indicate the abnormal allele at the HD locus; *h* is used to indicate the normal allele.

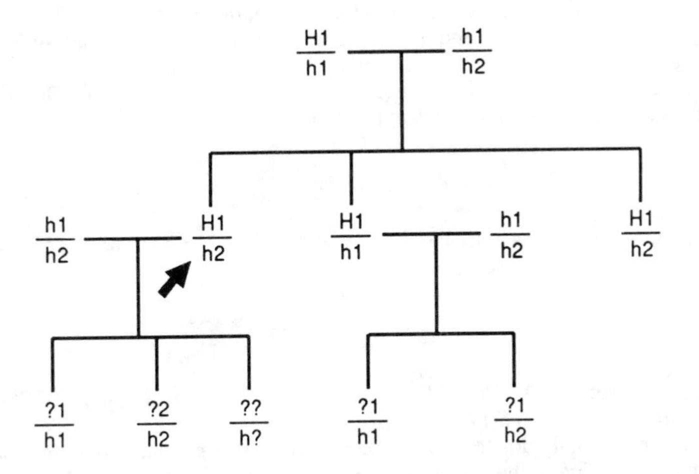

The affected woman's son has inherited a *1* allele at the marker locus and a normal allele at the HD locus from his father. From his mother he has inherited a *1* allele at the marker locus as well. Since the woman carries the disease allele at the HD locus on the same chromosome as the *1* allele at the marker locus, and since the two loci are separated by recombination only 4% of the time [i.e., they are 4 centimorgans (cM) apart], there is a 96% chance that her son inherited the disease allele at the HD locus with the *1* allele at the marker locus. The usual 50% risk of transmitting an abnormal gene does not figure into this calculation because transmission has already occurred. It is known which allele at the marker locus was transmitted; therefore, the likelihood that the disease allele at the HD locus was also transmitted can be calculated from the recombination fraction.

The affected woman's daughter has inherited a *2* allele at the marker locus and a normal allele at the HD locus from her father. From her mother she has inherited a *2* allele at the marker locus. Since the woman carries the disease allele at the HD locus on the same chromosome as the *1* allele at the marker locus, the only way that she could have transmitted the disease allele to her daughter is if recombination had occurred. The chance of this is only 4%.

In the case of the fetus that has not been genotyped, there is no information regarding which allele at the marker locus has been transmitted. The only option is to revert to the chance probability of transmission for the HD allele. This risk is 50%.

The woman's brother is homozygous for the *1* allele at the marker locus. Thus, it cannot be determined from the marker locus which chromosome is transmitted to his children. Both his normal allele and his disease allele at the HD locus are coupled with a *1* allele at the marker locus. Again, the only option is to revert to the chance probability of transmission of the disease allele at the HD locus. This 50% risk is not influenced by the marker allele transmitted by the brother's wife because both of her chromosomes 4 have normal alleles at the HD locus.

100–103. The answers are: 100-D, 101-B, 102-C, 103-A *[Ch 6 I B 6–7].*
X-linked recessive diseases occur in hemizygous males. The frequency of the gene in either males or females is equal to the frequency of the disease in males. Only females can be heterozygous for X-linked recessive diseases because only females have two X chromosomes. According to the Hardy-Weinberg law, the frequency of heterozygous carrier females for an X-linked recessive disease is 2pq, where q is the gene frequency and $p = 1 - q$.

In an X-linked recessive disease with a frequency of 1/10,000 in males, $q = 1/10{,}000$, so 2pq is approximately

$$2 \times \frac{1}{10{,}000} = \frac{1}{5000}$$

In an X-linked recessive disease with a frequency of 1/2500 in males, $q = 1/2500$, so 2pq is approximately

$$2 \times \frac{1}{2500} = \frac{1}{1250}$$

In an autosomal recessive disease, homozygotes for the mutant gene are affected. Thus, in terms of the Hardy-Weinberg law, q^2 is the disease frequency and the square root of this number is the frequency of the mutant gene. The frequency of heterozygous carriers is expressed as 2pq.

In an autosomal recessive disease with a frequency of 1/10,000, $q^2 = 1/10{,}000$ and $q = \sqrt{\frac{1}{10{,}000}} =$ 1/100. The heterozygous carrier frequency is approximately

$$2 \times \frac{1}{100} = \frac{1}{50}$$

In an autosomal recessive disease with a frequency of 1/2500, $q^2 = 1/2500$ and $q = \frac{1}{50}$. The heterozygous carrier frequency is approximately

$$2 \times \frac{1}{50} = \frac{1}{25}$$

104–108. The answers are: 104-B *[Ch 7 II E 3],* **105-C** *[Ch 8 II D 1 b],* **106-E** *[Ch 7 II D],* **107-C** *[Ch 8 II D 1 b],* **108-D** *[Ch 7 II B 1 b; Ch 5 III C 2; Case 5-D].*
Phenylketonuria (PKU) is usually due to absent or low levels of phenylalanine hydroxylase, which converts phenylalanine to tyrosine in the presence of a cofactor, tetrahydrobiopterin. Successful treatment involves reduction, but not elimination, of the amino acid substrate phenylalanine. Children whose PKU is due to abnormal cofactor metabolism do not respond to dietary therapy, because the cofactor disrupts the normal functioning of enzymes other than phenylalanine hydroxylase. Adults with PKU manage on a normal diet because the brain is no longer growing and is less sensitive to harm from substrate byproducts. However, untreated women still have high levels of phenylalanine, which by transplacental transfer can damage a fetus. Although mitochondrial inheritance may be suspected when all children of a mother affected by a disease are also affected, in this case the problem is prenatal exposure. Children who have PKU themselves do not sustain brain damage until after birth, because prenatally the mother maintains the blood phenylalanine at a sufficiently low level through the placenta. Untreated PKU children are often blond because tyrosine is involved in the production of melanin.

109–111. The answers are: 109-C, 110-B, 111-D *[Ch 9 V A 1 b (2) (a), 3, B 3 c, C 3 c; Table 9-6].*
Adoption research in psychiatric disease can be done in different ways, but basically it compares the risk for disease among individuals raised with biologic relatives (who typically are affected) to the risk among individuals raised with adoptive relatives (who usually are not affected). These studies then attempt to separate genetic from environmental contributions to the disease.

A pedigree analysis (family history) is easy to obtain but has drawbacks. If taken by a relatively inexperienced interviewer, an imprecise or inaccurate diagnosis may be recorded. Pedigrees that are taken over the telephone or obtained from only one individual may miss mildly affected relatives, giving lower family morbidity figures.

Family interviews often include unaffected relatives as well as affected relatives and, thus, are more likely to identify mildly affected family members. Testing of family members may be conducted in association with the interviews.

112–116. The answers are: 112-E, 113-A, 114-D, 115-C, 116-B *[Ch 10 III A; IV A; Ch 2 II B 4 a, c; III E-F].*
In a family with an affected child and two similarly affected relatives, X-linked recessive inheritance should be considered. If the affected child is a female, X-linked inheritance would be much less likely. As much detail as possible should be obtained regarding the uncles.

Advanced paternal age is associated only with new dominant mutations, and if an individual has a condition not otherwise classified, this association should be seriously considered.

Recurrent miscarriages may have many causes, only some of which are genetic (e.g., an inherited translocation, aneuploidy in the aborted fetus). Generally, single gene causes have not been identified to cause such miscarriages.

Advanced maternal age is associated with an increased risk of fetuses with trisomies 13, 18, and 21, and it may be associated with an increased risk of fetuses with Klinefelter syndrome. If an infant is born with multiple anomalies to a woman over 35, one of the considerations is a trisomy.

Consanguinity, especially with a relationship of first cousins or closer, is associated with an increased risk of autosomal recessive and multifactorial disorders, as well as an increased risk of stillbirths.

117–121. The answers are: 117-B, 118-A, 119-E, 120-C, 121-D *[Ch 11 V B].*
The patient or guardian should be free to make the decision of whether to enter the trial or whether to discontinue participation at any point (autonomy). All important information regarding the drug, including side effects and complications, should be discussed and the patient's questions answered. If the patient is to be entered in a double-blind trial with placebo, the parents must be made aware that the child may receive a placebo. Such information is all part of veracity. Not only the signed consent form but also the verbal consent are part of the contract with the patient and must be kept by the investigator (fidelity). Beneficence involves decisions on the part of a physician to involve her patients in research. She may believe that there is a good chance for response to the drug. She may believe that the clinical research ultimately benefits all children with muscle disease, and she may believe that undertaking such a trial will not be detrimental to her patient. All of these beliefs concern the patient's best interests. Should the child leave the study, he still has the right to standard care that is equally available to other males with Duchenne muscular dystrophy (DMD) [justice].

122–126. The answers are: 122-E, 123-D, 124-A, 125-C, 126-D *[Ch 1 IV A 1 c, B 2–4].*
Telomeres are repetitive TTAGGG sequences located at the ends of chromosomes; telomeres are important in maintaining chromosome stability. The centromere is a specific DNA–protein complex that becomes associated with microtubules during cell division and is involved in chromosome movement. The centromere is the last part of the chromosome to replicate and the last part to separate at mitosis. Nucleosomes are composed of eight histone polypeptides bound together and wound with 146 base pairs of DNA, an amount that corresponds to about two turns of the helix. Euchromatin contains active regions of the chromosome, while heterochromatin is composed largely of inactive regions. Heterochromatin is generally more condensed than euchromatin.

127–132. The answers are: 127-D, 128-C, 129-E, 130-A, 131-B, 132-F *[Ch 3 I D 1].*
Matings of $Gg \times gY$ and $gg \times gY$ are both rare situations that can result in affected daughters. Since both parents are affected in a $gg \times gY$ mating, all offspring will also be affected. In a mating of $Gg \times GY$, sons and daughters may inherit the mutant allele in equal proportions but the daughters will have normal vision. For a mating of $GG \times GY$, all offspring are normal; in particular, all daughters are homozygous normal since there is no opportunity to inherit a mutant allele from either parent. In a mating of $GG \times gY$, all daughters will inherit the mutant allele from the father, while the sons will inherit the Y chromosome; therefore, all of the children will have normal vision and all of the daughters will be carriers. In

a mating of *gg* × *GY*, all daughters will be heterozygous (*gG*) normal, and all sons will be affected hemizygotes (*gY*).

133–136. The answers are: 133-D, 134-B, 135-C, 136-B *[Ch 7 VI B 1–4].*
The child with short legs has a dysplasia because his structural defect is related to abnormal organization of cells within the cartilage.

The child with only one kidney and the child with chorioretinitis and blindness have disruptions. In both cases, there was destruction of a structure that had already formed, presumably in a normal fashion.

The child with an indentation in his skull has a deformation because his skull has been misshaped by mechanical pressure.

None of these children has a malformation, which is a morphologic defect resulting from an intrinsically abnormal developmental process.

137–141. The answers are: 137-B, 138-E, 139-D, 140-A, 141-C *[Ch 9 IV C 1–5; Figure 9-1].*
The form of congenital adrenal hyperplasia characterized by 21-hydroxylase deficiency may be associated with salt loss, leading to electrolyte abnormalities and severe dehydration. Affected males have normal genitalia, and affected females have normal internal structures. Externally, affected females are masculinized by the excessive androgens and may appear to have a phallic structure with undescended testes. Because of cortisol deficiency, affected children may not survive acute episodes such as that described in the question without replacement therapy.

Complete androgen insensitivity implies that androgen receptors are absent or nonfunctional. Affected males have normal formation of testes with functioning Leydig and Sertoli cells. The testosterone cannot be utilized by the wolffian duct, which regresses. The müllerian duct actively regresses through action of müllerian duct inhibiting substance (MDIS). Therefore, the phenotypic female discussed in the question has no uterus. The external genitalia cannot respond to dihydrotestosterone (DHT) and will be female. An affected infant may present with a hernia that contains the normal testis.

The infant in the question has features typical of Turner syndrome: a cardiac lesion, neck webbing, and possible lymphedema. However, the clitoral enlargement suggests the presence of a Y chromosome in at least some of the cells. Mixed gonadal dysgenesis is a highly variable condition, and if the proportions of the 46,XY cell line are small, the overall phenotype will be female. It is important not to miss the diagnosis, since the gonads should be removed because of the risk for malignancy.

Females with XY karyotypes (those with sex reversal) are a heterogeneous group. In the situation in the question, nonmosaic chromosomes are present with a uterus, which implies a lack of MDIS from the Sertoli cells. Pure gonadal dysgenesis is associated with a normal male karyotype, and, in some cases, deletion of the sex-determining region on the short arm of the Y chromosome (Yp). The gonads may initially start out as ovaries, but the lack of the second X chromosome inhibits continued development. The pelvic mass is worrisome, since it may represent a tumor involving one of the gonads (most likely a gonadoblastoma).

The infant with a 5α-reductase deficiency and an XY karyotype has normal male internal genitalia (i.e., the gonad has developed as a normal testis and produces both testosterone and MDIS). There is suppression of the müllerian ducts, and, therefore, there is no uterus. Because testosterone cannot be converted to DHT, the external structures respond only minimally to hormone stimulation. In this situation, there are normal androgen receptors, but an abnormal testosterone/DHT ratio. This condition is now treatable if recognized early. DHT is available and allows penile growth. The potential for fertility may be reasonable as well.

142–146. The answers are: 142-E, 143-E, 144-C, 145-D, 146-A *[Ch 3 I C 1 a–e, 3 c].*
Cystic fibrosis (CF) is an autosomal recessive disease, and, in most families, the parents of an affected individual are heterozygous. All children of an affected person whose partner is not a carrier are heterozygous. The grandchildren have a 1/2 risk of being carriers. Normal siblings of an affected person have a 2/3 risk of being heterozygous and a 1/3 risk of being homozygous normal. Unrelated, unaffected individuals have a background risk of 1/30.

147–150. The answers are: 147-C, 148-A, 149-B, 150-C *[Ch 7 VI C 1–4].*
Both the child with congenital heart disease and abnormal ears and the child with thrombocytopenia have syndromes, because each has multiple congenital anomalies that have the same etiology and pathogenesis. In the child with congenital heart disease and abnormal ears, the anomalies have resulted from the teratogenic action of isotretinoin. In the child with thrombocytopenia, the anomalies are the consequence of a pair of mutant genes.

The child with spina bifida and club feet exhibits a sequence. The primary error of morphogenesis is the spina bifida. The club feet are a secondary anomaly that developed as a result of the primary defect.

The stillborn infant had a developmental field defect because all of her anomalies resulted from disturbed morphogenesis in a developmental field—the caudal mesoderm of the 3- to 4-week embryo.

None of these children exhibits an association, which is a pattern that is diagnosed only after a syndrome, sequence, and developmental field defect have been excluded. Recognition of an association requires knowledge that the anomalies seen in an individual occur together more often than expected by chance.

151–154. The answers are: 151-C, 152-E, 153-C, 154-B *[Ch 6 II B 1–3, C–E; III B 1–2].*
Eugenics is the improvement of humankind by selective breeding. Eugenic proposals are said to be positive if they seek to increase the propagation of "more desirable" types and negative if they seek to decrease the propagation of "less desirable" types.

New mutations occur when a mutant allele that was not present in either parent is present in a child. New mutations are typically found with a frequency of only 10^{-6} to 10^{-4} per locus per generation. Clinically, they are commonly encountered among people who have autosomal dominant diseases that reduce fertility, because people with such diseases represent a very small subset of the total and all of the new mutations are contained within this subset.

The effect of paternal age is observed among children with diseases due to some new dominant mutations in which the age of the father is substantially older than average.

A gene or gene pair that causes an individual to die or otherwise be incapable of reproducing is said to be a genetic lethal. A lethal equivalent is defined as a gene that, if homozygous, would be lethal to all homozygotes, two genes, which, if homozygous, would each be lethal half of the time, or three genes, which, if homozygous, would each be lethal one-third of the time, and so forth. Each normal person carries three to five lethal equivalents.

Index

Note: Page numbers in italic denote illustrations; those followed by (t) denote tables; those followed by Q denote questions; and those followed by E denote explanations.

E

F

G

H

I

J

K

L

Dr. G Singh

U

V

W

X

Y

Z